AF855802

ECE 2014: Engineering Professionalism

Electrical and Computer Engineering

Virginia Tech

Pearson Custom Publishing

New York Boston San Francisco
London Toronto Sydney Tokyo Singapore Madrid
Mexico City Munich Paris Cape Town Hong Kong Montreal

Pearson Prentice Hall

Editorial Director, ECS: Marcia J. Horton
Senior Editor: Holly Stark
Associate Editor: Dee Bernhard
Senior Marketing Manager: Tim Galligan
Managing Editor: Scott Disanno
Art Director: Greg Dulles
Manufacturing Buyer: Lisa McDowell

Pearson Custom Publishing

Senior Vice President, Editorial and Marketing: Patrick F. Boles
Executive Marketing Manager: Nathan L. Wilbur
Sponsoring Editor: Debbie Coniglio
Operations Manager: Eric M. Kenney
Development Editor: Amy Galvin
Database Product Manager: Jennifer Berry

This special edition published in cooperation with Pearson Custom Publishing.

Printed in the United States of America.

Please visit our web site at *www.prenhall.com/esource*

Attention bookstores: For permission to return any unsold stock, contact us at *pe-uscustomreturns@pearson.com*.

Pearson Custom Publishing is a division of

PEARSON

www.pearsonhighered.com

ISBN 10: 0558273440
ISBN 13: 9780558273446

About ESource

ESource—The Prentice Hall Engineering Source—www.prenhall.com/esource

ESource—The Prentice Hall Engineering Source gives professors the power to harness the full potential of their text and their first-year engineering course. More than just a collection of books, ESource is a unique publishing system revolving around the ESource website—www.prenhall.com/esource. ESource enables you to put your stamp on your book just as you do your course. It lets you:

Control You choose exactly what chapters are in your book and in what order they appear. Of course, you can choose the entire book if you'd like and stay with the authors' original order.

Optimize Get the most from your book and your course. ESource lets you produce the optimal text for your students needs.

Customize You can add your own material anywhere in your text's presentation, and your final product will arrive at your bookstore as a professionally formatted text. Of course, all titles in this series are available as stand-alone texts, or as bundles of two or more books sold at a discount. Contact your PH sales rep for discount information.

ESource ACCESS

Professors who choose to bundle two or more texts from the ESource series for their class, or use an ESource custom book will be providing their students with an on-line library of intro engineering content—ESource Access. We've designed ESource ACCESS to provide students a flexible, searchable, on-line resource. Free access codes come in bundles and custom books are valid for one year after initial log-on. Contact your PH sales rep for more information.

ESource Content

All the content in ESource was written by educators specifically for freshman/first-year students. Authors tried to strike a balanced level of presentation, an approach that was neither formulaic nor trivial, and one that did not focus too heavily on advanced topics that most introductory students do not encounter until later classes. Because many professors do not have extensive time to cover these topics in the classroom, authors prepared each text with the idea that many students would use it for self-instruction and independent study. Students should be able to use this content to learn the software tool or subject on their own.

While authors had the freedom to write texts in a style appropriate to their particular subject, all followed certain guidelines created to promote a consistency that makes students comfortable. Namely, every chapter opens with a clear set of **Objectives**, includes **Practice Boxes** throughout the chapter, and ends with a number of **Problems**, and a list of **Key Terms**. **Applications Boxes** are spread throughout the book with the intent of giving students a real-world perspective of engineering. **Success Boxes** provide the student with advice about college study skills, and help students avoid the common pitfalls of first-year students. In addition, this series contains an entire book titled ***Engineering Success*** by Peter Schiavone of the University of Alberta intended to expose students quickly to what it takes to be an engineering student.

Creating Your Book

Using ESource is simple. You preview the content either on-line or through examination copies of the books you can request on-line, from your PH sales rep, or by calling 1-800-526-0485. Create an on-line outline of the content you want, in the order you want, using ESource's simple interface. Insert your own material into the text flow. If you are not ready to order, ESource will save your work. You can come back at any time and change, re-arrange, or add more material to your creation. Once you're finished you'll automatically receive an ISBN. Give it to your bookstore and your book will arrive on their shelves four to six weeks after they order. Your custom desk copies with their instructor supplements will arrive at your address at the same time.

To learn more about this new system for creating the perfect textbook, go to www.prenhall.com/esource. You can either go through the on-line walkthrough of how to create a book, or experiment yourself.

Supplements

Adopters of ESource receive an instructor's CD that contains professor and student code from the books in the series, as well as other instruction aides provided by authors. The website also holds approximately **350 PowerPoint transparencies** created by Jack Leifer of University of Kentucky–Paducah. Professors can either follow these transparencies as pre-prepared lectures or use them as the basis for their own custom presentations.

Titles in the ESource Series

Design Concepts for Engineers, 2/e
0-13-093430-5
Mark Horenstein

Engineering Success, 2/e
0-13-041827-7
Peter Schiavone

Engineering Design and Problem Solving, 2E
ISBN 0-13-093399-6
Steven K. Howell

Exploring Engineering
ISBN 0-13-093442-9
Joe King

Engineering Ethics
0-13-784224-4
Charles B. Fleddermann

Introduction to Engineering Analysis
0-13-016733-9
Kirk D. Hagen

Introduction to Engineering Experimentation
0-13-032835-9
Ronald W. Larsen, John T. Sears, and Royce Wilkinson

Introduction to Mechanical Engineering
0-13-019640-1
Robert Rizza

Introduction to Electrical and Computer Engineering
0-13-033363-8
Charles B. Fleddermann and Martin Bradshaw

Introduction to MATLAB 6—Update
0-13-140918-2
Delores Etter and David C. Kuncicky, with Douglas W. Hull

MATLAB Programming
0-13-035127-X
David C. Kuncicky

Introduction to MATLAB
0-13-013149-0
Delores Etter with David C. Kuncicky

Introduction to Mathcad 2000
0-13-020007-7
Ronald W. Larsen

Introduction to Mathcad 11
0-13-008177-9
David W. Larsen

Introduction to Maple 8
0-13-032844-8
David I. Schwartz

Mathematics Review
0-13-011501-0
Peter Schiavone

Power Programming with VBA/Excel
0-13-047377-4
Steven C. Chapra

Introduction to Excel 2002
0-13-008175-2
David C. Kuncicky

Engineering with Excel
ISBN 0-13-017696-6
Ronald W. Larsen

Introduction to Word 2002
0-13-008170-1
David C. Kuncicky

Introduction to PowerPoint 2002
0-13-008179-5
Jack Leifer

Graphics Concepts
0-13-030687-8
Richard M. Lueptow

Graphics Concepts with SolidWorks 2/e
0-13-140915-8
Richard M. Lueptow and Michael Minbiole

Graphics Concepts with Pro/ENGINEER
0-13-014154-2
Richard M. Lueptow, Jim Steger, and Michael T. Snyder

Introduction to AutoCAD 2000
0-13-016732-0
Mark Dix and Paul Riley

Introduction to AutoCAD, R. 14
0-13-011001-9
Mark Dix and Paul Riley

Introduction to UNIX
0-13-095135-8
David I. Schwartz

Introduction to the Internet, 3/e
0-13-031355-6
Scott D. James

Introduction to Visual Basic 6.0
0-13-026813-5
David I. Schneider

Introduction to C
0-13-011854-0
Delores Etter

Introduction to C++
0-13-011855-9
Delores Etter

Introduction to FORTRAN 90
0-13-013146-6
Larry Nyhoff and Sanford Leestma

Introduction to Java
0-13-919416-9
Stephen J. Chapman

About the Authors

No project could ever come to pass without a group of authors who have the vision and the courage to turn a stack of blank paper into a book. The authors in this series, who worked diligently to produce their books, provide the building blocks of the series.

Martin D. Bradshaw was born in Pittsburg, KS in 1936, grew up in Kansas and the surrounding states of Arkansas and Missouri, graduating from Newton High School, Newton, KS in 1954. He received the B.S.E.E. and M.S.E.E. degrees from the University of Wichita in 1958 and 1961, respectively. A Ford Foundation fellowship at Carnegie Institute of Technology followed from 1961 to 1963 and he received the Ph.D. degree in electrical engineering in 1964. He spent his entire academic career with the Department of Electrical and Computer Engineering at the University of New Mexico (1961-1963 and 1991-1996). He served as the Assistant Dean for Special Programs with the UNM College of Engineering from 1974 to 1976 and as the Associate Chairman for the EECE Department from 1993 to 1996. During the period 1987-1991 he was a consultant with his own company, EE Problem Solvers. During 1978 he spent a sabbatical year with the State Electricity Commission of Victoria, Melbourne, Australia. From 1979 to 1981 he served an IPA assignment as a Project Officer at the U.S. Air Force Weapons Laboratory, Kirkland AFB, Albuquerque, NM. He has won numerous local, regional, and national teaching awards, including the George Westinghouse Award from the ASEE in 1973. He was awarded the IEEE Centennial Medal in 2000.

Acknowledgments: Dr. Bradshaw would like to acknowledge his late mother, who gave him a great love of reading and learning, and his father, who taught him to persist until the job is finished. The encouragement of his wife, Jo, and his six children is a never-ending inspiration.

Stephen J. Chapman received a B.S. degree in Electrical Engineering from Louisiana State University (1975), the M.S.E. degree in Electrical Engineering from the University of Central Florida (1979), and pursued further graduate studies at Rice University. Mr. Chapman is currently Manager of Technical Systems for British Aerospace Australia, in Melbourne, Australia. In this position, he provides technical direction and design authority for the work of younger engineers within the company. He also continues to teach at local universities on a part-time basis.

Mr. Chapman is a Senior Member of the Institute of Electrical and Electronics Engineers (and several of its component societies). He is also a member of the Association for Computing Machinery and the Institution of Engineers (Australia).

Steven C. Chapra presently holds the Louis Berger Chair for Computing and Engineering in the Civil and Environmental Engineering Department at Tufts University. Dr. Chapra received engineering degrees from Manhattan College and the University of Michigan. Before joining the faculty at Tufts, he taught at Texas A&M University, the University of Colorado, and Imperial College, London. His research interests focus on surface water-quality modeling and advanced computer applications in environmental engineering. He has published over 50 refereed journal articles, 20 software packages and 6 books. He has received a number of awards including the 1987 ASEE Merriam/Wiley Distinguished Author Award, the 1993 Rudolph Hering Medal, and teaching awards from Texas A&M, the University of Colorado, and the Association of Environmental Engineering and Science Professors.

Acknowledgments: To the Berger Family for their many contributions to engineering education. I would also like to thank David Clough for his friendship and insights, John Walkenbach for his wonderful books, and my colleague Lee Minardi and my students Kenny William, Robert Viesca and Jennifer Edelmann for their suggestions.

Mark Dix began working with AutoCAD in 1985 as a programmer for CAD Support Associates, Inc. He helped design a system for creating estimates and bills of material directly from AutoCAD drawing databases for use in the automated conveyor industry. This system became the basis for systems still widely in use today. In 1986 he began collaborating with Paul Riley to create AutoCAD training materials, combining Riley's background in industrial design and training with Dix's background in writing, curriculum development, and programming. Mr. Dix received the M.S. degree in education from the University of Massachusetts. He is currently the Director of Dearborn Academy High School in Arlington, Massachusetts.

Delores M. Etter is a Professor of Electrical and Computer Engineering at the University of Colorado. Dr. Etter was a faculty member at the University of New Mexico and also a Visiting Professor at Stanford University. Dr. Etter was responsible for the Freshman Engineering Program at the University of New Mexico and is active in the Integrated Teaching Laboratory at the University of Colorado. She was elected a Fellow of the Institute of Electrical and Electronics Engineers for her contributions to education and for her technical leadership in digital signal processing.

Charles B. Fleddermann is a professor in the Department of Electrical and Computer Engineering at the University of New Mexico in Albuquerque, New Mexico. All of his degrees are in electrical engineering: his Bachelor's degree from the University of Notre Dame, and the Master's and Ph.D. from the University of Illinois at Urbana-Champaign. Prof. Fleddermann developed an engineering ethics course for his department in response to the ABET requirement to incorporate ethics topics into the undergraduate engineering curriculum. *Engineering Ethics* was written as a vehicle for presenting ethical theory, analysis, and problem solving to engineering undergraduates in a concise and readily accessible way.

Acknowledgments: I would like to thank Profs. Charles Harris and Michael Rabins of Texas A & M University whose NSF sponsored workshops on engineering ethics got me started thinking in this field. Special thanks to my wife Liz, who proofread the manuscript for this book, provided many useful suggestions, and who helped me learn how to teach "soft" topics to engineers.

Kirk D. Hagen is a professor at Weber State University in Ogden, Utah. He has taught introductory-level engineering courses and upper-division thermal science courses at WSU since 1993. He received his B.S. degree in physics from Weber State College and his M.S. degree in mechanical engineering from Utah State University, after which he worked as a thermal designer/analyst in the aerospace and electronics industries. After several years of engineering practice, he resumed his formal education, earning his Ph.D. in mechanical engineering at the University of Utah. Hagen is the author of an undergraduate heat transfer text.

Mark N. Horenstein is a Professor in the Department of Electrical and Computer Engineering at Boston University. He has degrees in Electrical Engineering from M.I.T. and U.C. Berkeley and has been involved in teaching engineering design for the greater part of his academic career. He devised and developed the senior design project class taken by all electrical and computer engineering students at Boston University. In this class, the students work for a virtual engineering company developing products and systems for real-world engineering and social-service clients.

Acknowledgments: I would like to thank Prof. James Bethune, the architect of the Peak Performance event at Boston University, for his permission to highlight the competition in my text. Several of the ideas relating to brainstorming and teamwork were derived from a workshop on engineering design offered by Prof. Charles Lovas of Southern Methodist University. The principles of estimation were derived in part from a freshman engineering problem posed by Prof. Thomas Kincaid of Boston University.

Steven Howell is the Chairman and a Professor of Mechanical Engineering at Lawrence Technological University. Prior to joining LTU in 2001, Dr. Howell led a knowledge-based engineering project for Visteon Automotive Systems and taught computer-aided design classes for Ford Motor Company engineers. Dr. Howell also has a total of 15 years experience as an engineering faculty member at Northern Arizona University, the University of the Pacific, and the University of Zimbabwe. While at Northern Arizona University, he helped develop and implement an award-winning interdisciplinary series of design courses simulating a corporate engineering-design environment.

Douglas W. Hull is a graduate student in the Department of Mechanical Engineering at Carnegie Mellon University in Pittsburgh, Pennsylvania. He is the author of *Mastering Mechanics I Using Matlab 5*, and contributed to *Mechanics of Materials* by Bedford and Liechti. His research in the Sensor Based Planning lab involves motion planning for hyper-redundant manipulators, also known as serpentine robots.

Scott D. James is a staff lecturer at Kettering University (formerly GMI Engineering & Management Institute) in Flint, Michigan. He is currently pursuing a Ph.D. in Systems Engineering with an emphasis on software engineering and computer-integrated manufacturing. He chose teaching as a profession after several years in the computer industry. "I thought that it was really important to know what it was like outside of academia. I wanted to provide students with classes that were up to date and provide the information that is really used and needed."

Acknowledgments: Scott would like to acknowledge his family for the time to work on the text and his students and peers at Kettering who offered helpful critiques of the materials that eventually became the book.

Joe King received the B.S. and M.S. degrees from the University of California at Davis. He is a Professor of Computer Engineering at the University of the Pacific, Stockton, CA, where he teaches courses in digital design, computer design, artificial intelligence, and computer networking. Since joining the UOP faculty, Professor King has spent yearlong sabbaticals teaching in Zimbabwe, Singapore, and Finland. A licensed engineer in the state of California, King's industrial experience includes major design projects with Lawrence Livermore National Laboratory, as well as independent consulting projects. Prof. King has had a number of books published with titles including MATLAB, MathCAD, Exploring Engineering, and Engineering and Society.

David C. Kuncicky is a native Floridian. He earned his Baccalaureate in psychology, Master's in computer science, and Ph.D. in computer science from Florida State University. He has served as a faculty member in the Department of Electrical Engineering at the FAMU–FSU College of Engineering and the Department of Computer Science at Florida State University. He has taught computer science and computer engineering courses for over 15 years. He has published research in the areas of intelligent hybrid systems and neural networks. He is currently the Director of Engineering at Bioreason, Inc. in Sante Fe, New Mexico.

Acknowledgments: Thanks to Steffie and Helen for putting up with my late nights and long weekends at the computer. Finally, thanks to Susan Bassett for having faith in my abilities, and for providing continued tutelage and support.

Ron Larsen is a Professor of Chemical Engineering at Montana State University, and received his Ph.D. from the Pennsylvania State University. He was initially attracted to engineering by the challenges the profession offers, but also appreciates that engineering is a serving profession. Some of the greatest challenges he has faced while teaching have involved non-traditional teaching methods, including evening courses for practicing engineers and teaching through an interpreter at the Mongolian National University. These experiences have provided tremendous opportunities to learn new ways to communicate technical material. Dr. Larsen views modern software as one of the new tools that will radically alter the way engineers work, and his book *Introduction to MathCAD* was written to help young engineers prepare to meet the challenges of an ever-changing workplace.

Acknowledgments: To my students at Montana State University who have endured the rough drafts and typos, and who still allow me to experiment with their classes—my sincere thanks.

Sanford Leestma is a Professor of Mathematics and Computer Science at Calvin College, and received his Ph.D. from New Mexico State University. He has been the long-time co-author of successful textbooks on Fortran, Pascal, and data structures in Pascal. His current research interest are in the areas of algorithms and numerical computation.

Jack Leifer is an Assistant Professor in the Department of Mechanical Engineering at the University of Kentucky Extended Campus Program in Paducah, and was previously with the Department of Mathematical Sciences and Engineering at the University of South Carolina–Aiken. He received his Ph.D. in Mechanical Engineering from the University of Texas at Austin in December 1995. His current research interests include the analysis of ultra-light and inflatable (Gossamer) space structures.

Acknowledgments: I'd like to thank my colleagues at USC–Aiken, especially Professors Mike May and Laurene Fausett, for their encouragement and feedback; and my parents, Felice and Morton Leifer, for being there and providing support (as always) as I completed this book.

Richard M. Lueptow is the Charles Deering McCormick Professor of Teaching Excellence and Associate Professor of Mechanical Engineering at Northwestern University. He is a native of Wisconsin and received his doctorate from the Massachusetts Institute of Technology in 1986. He teaches design, fluid mechanics, an spectral analysis techniques. Rich has an active research program on rotating filtration, Taylor Couette flow, granular flow, fire suppression, and acoustics. He has five patents and over 40 refereed journal and proceedings papers along with many other articles, abstracts, and presentations.

Acknowledgments: Thanks to my talented and hard-working co-authors as well as the many colleagues and students who took the tutorial for a "test drive." Special thanks to Mike Minbiole for his major contributions to Graphics Concepts with SolidWorks. Thanks also to Northwestern University for the time to work on a book. Most of all, thanks to my loving wife, Maiya, and my children, Hannah and Kyle, for supporting me in this endeavor. (Photo courtesy of Evanston Photographic Studios, Inc.)

Larry Nyhoff is a Professor of Mathematics and Computer Science at Calvin College. After doing bachelor's work at Calvin, and Master's work at Michigan, he received a Ph.D. from Michigan State and also did graduate work in computer science at Western Michigan. Dr. Nyhoff has taught at Calvin for the past 34 years—mathematics at first and computer science for the past several years.

Acknowledgments: We thank our families—Shar, Jeff, Dawn, Rebecca, Megan, Sara, Greg, Julie, Joshua, Derek, Tom, Joan; Marge, Michelle, Sandy, Lory, Michael—for being patient and understanding. We thank God for allowing us to write this text.

Paul Riley is an author, instructor, and designer specializing in graphics and design for multimedia. He is a founding partner of CAD Support Associates, a contract service and professional training organization for computer-aided design. His 15 years of business experience and 20 years of teaching experience are supported by degrees in education and computer science. Paul has taught AutoCAD at the University of Massachusetts at Lowell and is presently teaching AutoCAD at Mt. Ida College in Newton, Massachusetts. He has developed a program, Computer-aided Design for Professionals that is highly regarded by corporate clients and has been an ongoing success since 1982.

Robert Rizza is an Assistant Professor of Mechanical Engineering at North Dakota State University, where he teaches courses in mechanics and computer-aided design. A native of Chicago, he received the Ph.D. degree from the Illinois Institute of Technology. He is also the author of *Getting Started with Pro/ENGINEER*. Dr. Rizza has worked on a diverse range of engineering projects including projects from the railroad, bioengineering, and aerospace industries. His current research interests include the fracture of composite materials, repair of cracked aircraft components, and loosening of prostheses.

Peter Schiavone is a professor and student advisor in the Department of Mechanical Engineering at the University of Alberta, Canada. He received his Ph.D. from the University of Strathclyde, U.K. in 1988. He has authored several books in the area of student academic success as well as numerous papers in international scientific research journals. Dr. Schiavone has worked in private industry in several different areas of engineering including aerospace and systems engineering. He founded the first Mathematics Resource Center at the University of Alberta, a unit designed specifically to teach new students the necessary survival skills in mathematics and the physical sciences required for success in first-year engineering. This led to the Students' Union Gold Key Award for outstanding contributions to the university. Dr. Schiavone lectures regularly to freshman engineering students and to new engineering professors on engineering success, in particular about maximizing students' academic performance.

Acknowledgements: Thanks to Richard Felder for being such an inspiration; to my wife Linda for sharing my dreams and believing in me; and to Francesca and Antonio for putting up with Dad when working on the text.

David I. Schneider holds an A.B. degree from Oberlin College and a Ph.D. degree in Mathematics from MIT. He has taught for 34 years, primarily at the University of Maryland. Dr. Schneider has authored 28 books, with one-half of them computer programming books. He has developed three customized software packages that are supplied as supplements to over 55 mathematics textbooks. His involvement with computers dates back to 1962, when he programmed a special purpose computer at MIT's Lincoln Laboratory to correct errors in a communications system.

David I. Schwartz is an Assistant Professor in the Computer Science Department at Cornell University and earned his B.S., M.S., and Ph.D. degrees in Civil Engineering from State University of New York at Buffalo. Throughout his graduate studies, Schwartz combined principles of computer science to applications of civil engineering. He became interested in helping students learn how to apply software tools for solving a variety of engineering problems. He teaches his students to learn incrementally and practice frequently to gain the maturity to tackle other subjects. In his spare time, Schwartz plays drums in a variety of bands.

Acknowledgments: I dedicate my books to my family, friends, and students who all helped in so many ways.

Many thanks go to the schools of Civil Engineering and Engineering & Applied Science at State University of New York at Buffalo where I originally developed and tested my UNIX and Maple books. I greatly appreciate the opportunity to explore my goals and all the help from everyone at the Computer Science Department at Cornell.

John T. Sears received the Ph.D. degree from Princeton University. Currently, he is a Professor and the head of the Department of Chemical Engineering at Montana State University. After leaving Princeton he worked in research at Brookhaven National Laboratory and Esso Research and Engineering, until he took a position at West Virginia University. He came to MSU in 1982, where he has served as the Director of the College of Engineering Minority Program and Interim Director for BioFilm Engineering. Prof. Sears has written a book on air pollution and economic development, and over 45 articles in engineering and engineering education.

Michael T. Snyder is President of Internet startup Appointments123.com. He is a native of Chicago, and he received his Bachelor of Science degree in Mechanical Engineering from the University of Notre Dame. Mike also graduated with honors from Northwestern University's Kellogg Graduate School of Management in 1999 with his Masters of Management degree. Before Appointments123.com, Mike was a mechanical engineer in new product development for Motorola Cellular and Acco Office Products. He has received four patents for his mechanical design work. "Pro/ENGINEER was an invaluable design tool for me, and I am glad to help students learn the basics of Pro/ENGINEER."

Acknowledgments: Thanks to Rich Lueptow and Jim Steger for inviting me to be a part of this great project. Of course, thanks to my wife Gretchen for her support in my various projects.

Jim Steger is currently Chief Technical Officer and cofounder of an Internet applications company. He graduated with a Bachelor of Science degree in Mechanical Engineering from Northwestern University. His prior work included mechanical engineering assignments at Motorola and Acco Brands. At Motorola, Jim worked on part design for two-way radios and was one of the lead mechanical engineers on a cellular phone product line. At Acco Brands, Jim was the sole engineer on numerous office product designs. His Worx stapler has won design awards in the United States and in Europe. Jim has been a Pro/ENGINEER user for over six years.

Acknowledgments: Many thanks to my co-authors, especially Rich Lueptow for his leadership on this project. I would also like to thank my family for their continuous support.

Royce Wilkinson received his undergraduate degree in chemistry from Rose-Hulman Institute of Technology in 1991 and the Ph.D. degree in chemistry from Montana State University in 1998 with research in natural product isolation from fungi. He currently resides in Bozeman, MT and is involved in HIV drug research. His research interests center on biological molecules and their interactions in the search for pharmaceutical advances.

Reviewers

We would like to thank everyone who has reviewed texts in this series.

ESource Reviewers

Christopher Rowe, *Vanderbilt University*
Steve Yurgartis, *Clarkson University*
Heidi A. Diefes-Dux, *Purdue University*
Howard Silver, *Fairleigh Dickenson University*
Jean C. Malzahn Kampe, *Virginia Polytechnic Institute and State University*
Malcolm Heimer, *Florida International University*
Stanley Reeves, *Auburn University*
John Demel, *Ohio State University*
Shahnam Navee, *Georgia Southern University*
Heshem Shaalem, *Georgia Southern University*
Terry L. Kohutek, *Texas A & M University*
Liz Rozell, *Bakersfield College*
Mary C. Lynch, *University of Florida*
Ted Pawlicki, *University of Rochester*
James N. Jensen, *SUNY at Buffalo*
Tom Horton, *University of Virginia*
Eileen Young, *Bristol Community College*
James D. Nelson, *Louisiana Tech University*
Jerry Dunn, *Texas Tech University*
Howard M. Fulmer, *Villanova UniversityBerkeley*
Naeem Abdurrahman *University of Texas, Austin*
Stephen Allan *Utah State University*
Anil Bajaj *Purdue University*
Grant Baker *University of Alaska–Anchorage*
William Beckwith *Clemson University*
Haym Benaroya *Rutgers University*
John Biddle *California State Polytechnic University*
Tom Bledsaw *ITT Technical Institute*
Fred Boadu *Duke University*
Tom Bryson *University of Missouri, Rolla*
Ramzi Bualuan *University of Notre Dame*
Dan Budny *Purdue University*
Betty Burr *University of Houston*
Dale Calkins *University of Washington*
Harish Cherukuri *University of North Carolina –Charlotte*
Arthur Clausing *University of Illinois*
Barry Crittendon *Virginia Polytechnic and State University*
James Devine *University of South Florida*
Ron Eaglin *University of Central Florida*
Dale Elifrits *University of Missouri, Rolla*
Patrick Fitzhorn *Colorado State University*
Susan Freeman *Northeastern University*
Frank Gerlitz *Washtenaw College*
Frank Gerlitz *Washtenaw Community College*
John Glover *University of Houston*
John Graham *University of North Carolina–Charlotte*
Ashish Gupta *SUNY at Buffalo*
Otto Gygax *Oregon State University*
Malcom Heimer *Florida International University*
Donald Herling *Oregon State University*
Thomas Hill *SUNY at Buffalo*
A.S. Hodel *Auburn University*
James N. Jensen *SUNY at Buffalo*
Vern Johnson *University of Arizona*
Autar Kaw *University of South Florida*
Kathleen Kitto *Western Washington University*
Kenneth Klika *University of Akron*
Terry L. Kohutek *Texas A&M University*
Melvin J. Maron *University of Louisville*
Robert Montgomery *Purdue University*
Mark Nagurka *Marquette University*
Romarathnam Narasimhan *University of Miami*
Soronadi Nnaji *Florida A&M University*
Sheila O'Connor *Wichita State University*
Michael Peshkin *Northwestern University*
Dr. John Ray *University of Memphis*
Larry Richards *University of Virginia*
Marc H. Richman *Brown University*
Randy Shih *Oregon Institute of Technology*
Avi Singhal *Arizona State University*
Tim Sykes *Houston Community College*
Neil R. Thompson *University of Waterloo*
Dr. Raman Menon Unnikrishnan *Rochester Institute of Technology*
Michael S. Wells *Tennessee Tech University*
Joseph Wujek *University of California, Berkeley*
Edward Young *University of South Carolina*
Garry Young *Oklahoma State University*
Mandochehr Zoghi *University of Dayton*

Contents

1 Introduction: What Are Electrical and Computer Engineering? 1
Charles B. Fleddermann, Martin D. Bradshaw

2 Electrical and Computer Engineering Specializations 13
Charles B. Fleddermann, Martin D. Bradshaw

3 Engineering Your Career 26
Charles B. Fleddermann, Martin D. Bradshaw

4 Academic Survival Skills 34
Charles B. Fleddermann, Martin D. Bradshaw

5 Engineering Design 42
Charles B. Fleddermann, Martin D. Bradshaw

6 Engineering and Communication 59
Hillary Hart

7 Discovering Ideas and Facts: You Are Leaving Home 69
Hillary Hart

8 Organizing Ideas and Facts: It's the Journey, Not the Destination 86
Hillary Hart

9 Writing: It's about Control 94
Hillary Hart

10 Displaying Data in Written Documents: Are Pictures Always Worth a Thousand Words? 116
Hillary Hart

11 Revising: When Will I Ever Be Finished? 139
Hillary Hart

12 Speaking: Do I Really Have to Stand Up and Talk in Front of All Those People? 156
Hillary Hart

13 Producing Engineering Documents: The Rubber Meets the Road 179
Hillary Hart

14 Project Management 240
Mark N. Horenstein

15 Introduction to Engineering Ethics 277
Charles B. Fleddermann

16 Professionalism and Codes of Ethics 290
Charles B. Fleddermann

17 Understanding Ethical Problems 307
Charles B. Fleddermann

18 Ethical Problem-Solving Techniques 321
Charles B. Fleddermann

19 Appendix: Codes of Ethics of Professional Engineering Societies 338
Charles B. Fleddermann

20 Appendix: Bibliography 351
Charles B. Fleddermann

Index 353

1

Introduction: What Are Electrical and Computer Engineering?

1 Introduction

The rapid developments in technology that transformed society in the late 20th century and continue into the 21st century were fueled by the work of electrical and computer engineers. The Internet, personal computers, and wireless communication are only some of the technologies for which electrical and computer engineers have been responsible. Engineering is both a difficult profession and a very rewarding one. The difficulty arises from the very complex and sophisticated tools that must be mastered in order to be an effective engineer. Learning to use these tools requires years of study and experience. The rewards of engineering come from the accomplishment an engineer feels as his or her ideas are implemented and improving people's lives. Today's electrical and computer engineers have this sense of accomplishment as they see the benefits that society derives from the technological revolution they helped create.

You are embarking upon a great adventure as indented become an engineer and prepare to help society through your work. At this point in your career, it will be helpful to understand what engineering is about and what it takes to be a successful engineer.

Sections

1 Introduction
2 What Is Engineering?
3 What Is Electrical Engineering?
4 What Is Computer Engineering?
5 Other Engineering Fields
6 Where Do Electrical and Computer Engineers Work?
7 The Engineering Profession

Objectives

After reading this chapter, you will

- Learn about engineering as a career
- Understand who engineers work with
- Be aware of the different fields of engineering
- Know about the need for professional registration
- Be able to identify the IEEE as the professional organization for electrical and computer engineers

In this book, we will explore electrical and computer engineering, see what engineers do in their everyday work lives, and find out what you must do to become an engineer. In this first chapter we learn what engineering is and see what types of engineering there are. We also try to get a feel for what a profession is and see how engineering fits in with other professions. This will identify the professional duties and responsibilities of engineers. Finally, we will discuss entrepreneurship and see how engineers often participate in this activity.

2 WHAT IS ENGINEERING?

In order to understand what engineering is and what engineers do, we start with a definition of engineering. Engineering is the application of science and mathematics to produce useful devices and systems. For example, electrical engineers use their knowledge of physics and mathematics to design the electronic components and circuits that allow nearly instantaneous communication anywhere around the world. Chemical engineers use their knowledge of basic chemistry to design efficient processes for converting a barrel of crude oil into the gasoline that powers automobiles. To do these tasks, engineers must have a very solid grounding in the basic sciences and mathematics.

What is the basic knowledge that supports electrical and computer engineering? Most important to electrical and computer engineers are sophisticated mathematical skills. Physical and chemical systems often can only be understood through relatively complex mathematical models. A mathematical model is a representation of a physical system using equations. These equations aid in the understanding of the physical process, allow engineers to predict the behavior of a system in new applications, and help in designing new devices. Engineering would still be in the Dark Ages without the invention of calculus and other mathematical tools.

Physics is one of the basic sciences in which electrical and computer engineers must be well versed. The understanding of how matter is structured and how it reacts to external stimuli is fundamental to understanding complex systems and designing new devices. For example, inside your computer is an array of electronic devices designed to perform certain functions. Many of these devices are fabricated from very pure silicon and are formed into integrated circuits, also known as microchips. Information is stored in a computer in the form of electrons. The presence of electrons in a structure formed in the silicon may be used to signify a "1," and the absence of electrons will signify a "0." Computations are performed by moving these bunches of electrons through various circuits. To design computing devices, engineers must understand the physics of how electrons interact with matter, how they can be stored, and how they can be moved around.

Chemistry also tells us how matter interacts and how complex forms of matter are made. Chemistry has been a branch of science studied by those learning to be electrical engineers.

Computers have become increasingly important in society. This is doubly true in engineering, where calculations and modeling of devices, processes, and structures are increasingly performed by engineers using computers. Many manufacturing processes are controlled by computers, and even consumer devices like the automobile are controlled by onboard computers. Contemporary engineers cannot do their jobs without using computers, and so need a solid grounding in the basics of computer science—how computers work and how to program them to perform useful tasks.

Becoming an engineer involves much more than just studying engineering. Engineering starts with a good fundamental knowledge of the engineering sciences—mathematics, physics, chemistry, and computer science.

So far, we have only talked about the scientific and technical background required to be an engineer. There are also many important nontechnical skills required of engineers. Engineers must be effective communicators, able to convey their ideas through written reports as well as oral presentations. Engineers must also function well as members of a team since engineering projects are rarely performed by an engineer working alone. Rather, projects generally involve teams of engineers from several engineering disciplines who must work together effectively to complete a project. Engineers must also be able to work with nontechnical team members. Engineers must know at least the fundamentals of business principles in their industry since projects that are not economically sound and that do not make good business sense will not go forward.

As we discuss what engineering is, we should also look at others who work with engineers—scientists and technicians. Although engineers use the results of science in their everyday work, most engineers are not scientists. Scientists study nature and try to discover the fundamental laws that govern nature. Their concerns are most often with gathering new knowledge; they are generally less interested in applying this knowledge. Engineers are concerned with how to use scientific knowledge to produce useful devices or processes. In the course of their work, engineers often must design and run experiments to gain the knowledge that they need, so sometimes they do function as scientists. However, the engineer's goal is rarely to gain knowledge alone, but is rather to use that knowledge for practical ends.

Most engineering companies employ a large number of technicians or technologists. These are people charged with building prototypes of new devices, running experiments and recording experimental data, building machines that help in designing new products, and maintaining the machinery needed to support the engineering functions of a company. Technicians generally receive less training than engineers and are not expected to have the sophisticated mathematical skills that engineers require. Of course, there is never a sharp line of demarcation between what engineers do and what technicians do. Depending on the company or the individual engineer, engineers will often do maintenance on machines, run experiments themselves, and build prototypes. Likewise, experienced technicians often perform some of the engineering work required for a new project.

3 WHAT IS ELECTRICAL ENGINEERING?

Electrical engineering is the study of how the interactions of electrons with matter can be exploited for useful applications. For example, the interactions of electrons with solids—their movement through wires or their storage in semiconductor devices—are fundamental to electrical engineering. These interactions can be used to do many useful things. For example, electrons are used to:

- Produce the power for our homes
- Convert radio signals into audible sounds
- Allow physicians to diagnose problems in human bodies
- Convert energy into light in an incandescent lightbulb or in a television screen
- Interact with a gas to emit light; a fluorescent lightbulb is a device for converting the energy of an electron to light through an interaction with a mixture of argon gas and mercury vapor
- Radiate energy, which forms the basis for the transmitting of television or cell phone signals over long distances.

Electrical engineers seek to understand these and other phenomena and to harness them for useful purposes.

4 WHAT IS COMPUTER ENGINEERING?

Computer engineering has emerged as a separate engineering discipline over the last 20 years as electrical engineers interested in computing technology have forged a new field. Since its origins are in electrical engineering, computer engineering shares many similarities with electrical engineering. The focus of computer engineering is on designing and building computers and computer systems. Although at one time this was a fairly narrow field, today it is quite extensive, encompassing not only the design of computer hardware but also the design of networks used to connect computers and the software used to run them.

Sometimes it is hard to see the difference between electrical engineers, computer engineers, and computer scientists. Figure 1 shows the overlap between these areas. Generally, computer scientists are interested in designing the software used by computers. Although they need extensive knowledge about the hardware that makes up the computer, their focus is primarily on constructing algorithms to implement calculations and the efficient design of software to run computers. Computer engineers need to understand the complexities of software—you can't design a computer without knowing the software that will run on it—but focus primarily on the design of the hardware that runs the software.

Electrical engineers generally design and build hardware used in applications other than computers. Of course, many modern electronic systems are computer controlled. Every automobile produced today has a microcontroller, which is essentially a mini-computer, underneath the hood. This is used to monitor and control all of the functions of the engine to ensure that the automobile operates as efficiently as possible. Often an electrical engineer will do the computer design and work on developing the software for these types of applications.

Confused? You should be because many practicing engineers are not able to say where electrical engineering ends and computer engineering begins. Many engineers work in an area that bridges the gap between these two fields. So how do you decide if electrical or computer engineering is the right choice for you? There is no right answer to this question. Many people trained as electrical engineers learn the skills needed to function as a computer engineer on the job. Likewise, many individuals trained in computer engineering make the transition to electrical engineering. A good guide is to

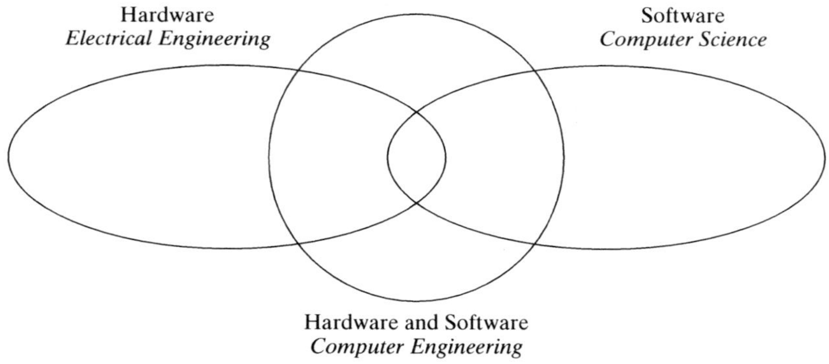

Figure 1 There is significant overlap in the interests and responsibilities of electrical engineers, computer engineers, and computer scientists.

decide how much you like computer programming. Although no engineer can be without good programming skills in the modern engineering marketplace, the computer skills required by computer engineers are often more extensive than those required of electrical engineers. So if you like computers and are interested in how they are put together, computer engineering is probably the right career choice for you.

5 OTHER ENGINEERING FIELDS

Before ending our discussion of the nature of engineering, it will be valuable to mention some of the other fields of engineering. Even if you are sure that you want to be an electrical or computer engineer, you will almost certainly find yourself working on multidisciplinary teams during the course of your engineering career. For example, the design of an automobile requires the skills of mechanical engineers to design the engines and drive trains, electrical engineers to design the electrical systems and controls, computer engineers to work on the computerized monitoring and control systems, manufacturing engineers to design the assembly lines and robots to build the automobile, as well as other engineering and nontechnical professionals. Most engineering projects will require the skills of different types of engineers, so it will be beneficial to at least get a rough idea of what the other types of engineering are so that you will be able to better interact with the other engineers working on your project. The following is a brief description of some of the major types of engineering other than electrical and computer engineering. There are many books that give in-depth discussions of these fields beyond what is presented here.

5.1 Mechanical Engineering

Mechanical engineering focuses on the design and production of mechanical systems. There are many such systems around us in the everyday world. Mechanical engineers design large-scale systems as diverse as automobiles, refrigeration systems, and robots. They also work on very small-scale systems such as computer disk drives and miniaturized medical implants.

5.2 Civil Engineering

Civil engineering is concerned with the design and building of the infrastructure of cities and communities. Civil engineers work on roads, bridges, and sewer systems, and design the structures of large buildings. Civil engineers work closely with architects to ensure that new buildings are structurally sound and can withstand the forces of nature, such as wind and earthquakes, to which they will be subjected. Figure 2 shows an example of some of the work done by civil engineers—a scale model of the design for the reconstruction of the intersection of two interstate highways in Albuquerque, New Mexico.

5.3 Aerospace Engineering

As the name implies, aerospace engineers are concerned with the design of flying machines, including commercial and military aircraft and vehicles launched into space. Aerospace engineers have expertise in the flow of air around airplanes, and design the wings, control surfaces, and the fuselage of airplanes. Figure 3 shows the work of aerospace engineers—a modern airliner during takeoff. Aerospace engineers can also find themselves working in the automotive industry, applying their knowledge of air flow to help design more fuel-efficient automobiles or even faster race cars.

Figure 2 A model of the rebuilt intersection (known locally as "the Big-I") between I-40 and I-25 in Albuquerque, New Mexico. (Photo courtesy of URS Corporation.)

Figure 3 A modern commercial airliner during takeoff. Design of the airframes of commercial aircraft require the work of aeronautical engineers.

5.4 Chemical Engineering

The methods required to make the numerous chemicals used in modern life are designed by chemical engineers. For example, the processes used to refine a barrel of crude oil to turn it into gasoline, plastics, synthetic fabrics, pesticides and herbicides, and other petrochemicals are designed by chemical engineers. Figure 4 shows a large petrochemical plant. In more recent years, chemical engineers have also become involved in working on designs for the complicated processes used to fabricate integrated circuits. Companies such as Intel employ a large number of chemical engineers to design the processes required to make the Pentinum® and Celeron® microprocessor chips. The production systems for medicines are often designed by chemical engineers.

5.5 Nuclear Engineering

Nuclear engineers are trained in the technology used to harness the energy of the atom. They design nuclear reactors for power stations and ships and are helping to design fusion reactors that might one day provide a new source of commercial power. Nuclear engineering also has applications in medicine. Many branches of medicine rely on diagnostic and therapeutic techniques that utilize short-lived nuclear isotopes. Nuclear engineers are involved in the design and running of reactors that produce these valuable materials.

5.6 Manufacturing Engineering

Executing a design is only the first step in getting a new product into the marketplace—the product must then be manufactured. Manufacturing engineers are knowledgeable in manufacturing methods and can design the best way to make a product. They deal with the mechanics of manufacturing such as assembly lines and robots. They also

Figure 4 A typical petrochemical processing plant in the United States.

must consider human factors such as the efficient and safe use of assembly personnel in the design of manufacturing processes. Manufacturing engineering is closely related to industrial engineering and is sometimes considered a subdiscipline of mechanical engineering.

5.7 Biomedical Engineering

An area of engineering that is becoming increasingly important is biomedical engineering. This area incorporates many different activities and crosses boundaries with many of the traditional engineering disciplines. Biomedical engineering studies ways to use engineering principles to design devices that help in the treatment of disease or that can be used to replace biological materials. For example, biomedical engineering meets electrical engineering in the design of diagnostic medical equipment such as ultrasound imagers or patient-monitoring devices. Biomedical engineering meets mechanical engineering in work on prosthetic devices and artificial muscles.

5.8 Other Types of Engineering

There are numerous other types of engineering that tend to be narrowly defined and are sometimes considered to be subdisciplines of those already mentioned. These include mining engineering, maritime engineering, materials engineering, ceramic engineering, and petroleum engineering. Since the names of these fields are fairly self-explanatory, no detailed descriptions of these will be given here.

6 WHERE DO ELECTRICAL AND COMPUTER ENGINEERS WORK?

The answer to the question of where electrical and computer engineers work may seem very obvious. Many of these engineers are employed by the well-known engineering companies such as IBM, Intel, Boeing, or Ford. However, you may be surprised to see the wide variety of other places that employ electrical and computer engineers. Any company that uses electrical- or computer-based control systems as part of their manufacturing equipment will require the services of electrical and computer engineers. For example, clothing manufacturers such as Oshkosh B'Gosh®, brewers such as Anheuser–Busch, and fast-food restaurant corporations such as McDonald's all employ electrical and computer engineers.

Engineers also work for small companies. Many of the innovative new products coming into the market came from very small companies. The federal government recognizes the importance of small companies in fostering innovation in our economy by sponsoring programs specifically designed to aid innovative small businesses. Many engineers prefer the atmosphere in small companies where new ideas can often be brought to market more rapidly than in larger companies.

7 THE ENGINEERING PROFESSION

As you begin your study of engineering, you may not realize that you are preparing yourself to be a member of a profession. What does this mean? In this section we will explore what a profession is and see what rights and obligations professionals have.

7.1 Professions

The terms *profession* and *professional* have many meanings today. Profession is often used synonymously with *occupation* or *job*, and *professional* is used to distinguish someone who is paid for his or her work rather than being an amateur. *Professional* is also used to indicate that someone has a great deal of experience and can be counted on to do a good job. None of these is the sense of the word profession that we examine here. Rather, we will consider a more traditional view of professions.

There are four basic attributes of professions [1]:

1. The work involves sophisticated skills, judgement, and discretion, and it is not routine or capable of mechanization.
2. It requires extensive formal education, not just an apprenticeship or on-the-job training.
3. Special societies or organizations controlled by members of the profession are allowed to play a major role in setting standards for admission to the profession and standards of conduct for members of the profession.
4. The work serves some important aspect of the public good.

As by-products of this, the members of the profession are generally held in high esteem by society and are largely allowed to regulate their own activities. The paradigms, or major examples, of professions in modern society are medicine and law.

Is engineering a profession? Let's examine engineering in light of the definitions given above. Certainly engineering involves sophisticated skills. It requires a very skilled practitioner to design a computer or a wireless telephone network. Judgement is exercised by engineers in their work as they must decide which of several different designs, components, or materials to choose from. Discretion is required as engineers are called upon to keep information on new designs and on their clients strictly confidential. It may seem that engineering can be mechanized. After all, much of the design work done by a modern electrical or computer engineer is done on a computer. However, the computer is merely a tool for engineers. By itself, the computer cannot yet design another computer or even a computer chip. The computer greatly aids engineers in doing these tasks, but engineering is certainly not mechanized.

Does engineering require extensive formal education? Before practicing as an engineer, you must first complete a four-year undergraduate program from an accredited college of engineering. Many engineering jobs require additional classroom work, and most require that skills be continually upgraded through continuing education. This last point is especially important in engineering. Technology changes very rapidly, and as technologies evolve, so must our engineering skills. It is imperative that engineers know how to access information rapidly and learn new knowledge and skills. As a student, one of your goals must be to learn how to teach yourself because many of the skills you will need as an engineer in the future will have to be learned on the job.

All of the engineering disciplines have professional societies that help to set standards for professional conduct by engineers. For electrical and computer engineering this society is the Institute of Electrical and Electronics Engineers (IEEE).

Does engineering serve the public good? This question can be answered by simply looking around you. Many of the indispensable tools of modern life, such as computers, telecommunications equipment, and lighting, were created by electrical and computer engineers. The safety of society has also been enhanced through the work of electrical and computer engineers. For example, automotive safety has improved through anti-lock braking systems designed by engineering teams, including electrical and computer engineers.

Using the definition given earlier, it is clear that engineering is a profession. As you begin your studies of engineering, you should keep in mind that you are embarking on a profession as ancient, as learned, and as important as medicine or law. With the benefits of being a member of a profession—high pay, high status in society, subsidized education, and a degree of self-governance—there are obligations that go along with professional status.

7.2 Codes of Ethics

All professional societies have given thought to the duties and responsibilities of their members. These are generally spelled out in documents called codes of ethics. A code of ethics provides a framework for ethical judgements of the professional. It serves as a means of collective recognition by members of a profession's responsibilities, and it helps create an environment where ethical behavior is the norm. All of the engineering professional societies have codes of ethics, including the IEEE. The IEEE code is included in Appendix B of this book.

It is worth taking a few minutes now to look at the IEEE code of ethics to try to understand some of the obligations of the profession you will be joining. The heart of any engineering code is the responsibility of engineers to ensure that their work will not adversely impact the health or safety of the general public. In other words, in return for our status in society, engineers must promise to hold paramount the safety of the public.

7.3 Licensure

One way in which the engineering profession differs from others is with regard to licensure. Physicians and lawyers all must be licensed by the state in which they reside before being allowed to practice their profession. Most engineers do not need licenses to call themselves engineers and practice their professions. For example, large companies such as IBM employ many electrical and computer engineers who do not have a license. Licensure of engineers, including electrical and computer engineers, is offered by every state in the United States. Generally, though, most electrical and computer engineers don't seek or need a license. Among engineers, only civil engineers are required to be licensed. A civil engineer needs licensure in order to certify that a structure is built according to acceptable standards and will meet state and city building codes. The climate in which engineering is practiced is changing, and there is a trend now towards increased licensure for all engineers.

8 ENTREPRENEURSHIP

In our modern high-technology society, electrical and computer engineers often find themselves involved in starting new companies. In the course of our work we often have the kernel of an idea for a new product. The act of starting a new venture to produce this product or exploit a new idea is called entrepreneurship. Many large corporations seek to facilitate an entrepreneurial environment so that employees will be interested in innovating.

There are three steps to entrepreneurship: invention, innovation, and marketing. As an engineer, you may think that you will only be involved in the first two of these. Engineers clearly have roles in the discovery of new ideas and invention of new products and in working to develop these ideas until they are ready for market. In the modern economy, engineers often participate in commercializing their ideas as well. Many small start-up companies do not have the resources to employ full-time marketing personnel, so this task can fall to engineers. This is especially true in the many start-up companies that only have three or four employees.

The importance of entrepreneurship to modern engineering is manifested in courses on entrepreneurial engineering taught by many schools of engineering in the

U.S. These courses cover topics related to financing new businesses (raising venture capital and finding investors, for example), developing viable business plans, managing small high-tech companies, and marketing ideas in a high-tech economy. In the absence of such courses, many engineering programs encourage students to take some business courses to prepare for the modern marketplace.

APPLICATIONS: BIOGRAPHY

Judith Estrin

In today's fast-changing engineering environment, success often means being both an excellent engineer and a good entrepreneur. Silicon Valley in California is filled with many such people. One of the most successful engineer/entrepreneurs is Judith Estrin, who has held many important technology positions in the computer industry at both large and small Silicon Valley companies. She is perennially listed in *Fortune* magazine's "50 Most Powerful Women in America."

Estrin's education in computing began with bachelor's degrees in computer science and math from UCLA. She later earned a master's degree in electrical engineering from Stanford. While at Stanford, Estrin studied with Vint Cerf, often credited as the "father of the Internet," and she was exposed early on to the possibilities that the Internet held. Her master's thesis from Stanford was on testing of the TCP/IP Internet protocols.

After Stanford, Estrin worked at Zilog, rising to the position of an engineering manager. But her interest in starting her own company led her to a series of high-tech, computer-oriented, start-up companies. Her entrepreneurship began with Bridge Communications, a supplier of Internet routers, bridges, and communication servers. After Bridge was merged with 3Com in 1987, she started Network Computing Devices, a supplier of X terminals and software that allowed PC's to be connected to UNIX-based computers. After NCD, Estrin founded Precept Software. Precept's business was focused on multimedia networking software, and Estrin served as president and chief executive officer. In the spring of 1998, Precept was bought by Cisco Systems, a very successful supplier of Internet networking equipment. When Precept was incorporated into Cisco, Estrin became Cisco's vice-president and chief technology officer, responsible for fostering and managing technical innovation. While at Cisco, Estrin helped Cisco develop products that go beyond routing, their main business, and into other important areas of computer technology, such as computer network security. In addition to being on the board of directors of Cisco, Estrin also serves or has served on the board of directors of Sun Microsystems, Federal Express, Rockwell International, and Disney.

In the spring of 2000, after two years at Cisco, Estrin decided to once again become an entrepreneur. She left Cisco to form another new company, Packet Design, that will focus on Internet infrastructure products.

KEY WORDS

Engineering entrepreneurship

licensure

professions

REFERENCES

[1] Martin, Mike W., and Roland Schinzinger. 1989. *Ethics in engineering*, 2nd ed. New York: McGraw–Hill.

SUMMARY

This chapter has introduced you to the engineering profession. We have seen what it means to be a member of a profession, and have seen where engineering professionals work. We have also learned about other engineering disciplines such as mechanical and civil engineering. Many engineers are also entrepreneurs; we have seen in this chapter what an entrepreneur is and discussed how new high-technology companies are formed.

Problems

1. Interview one of the engineering faculty members in your department at your university. Find out the following information:

   - What is the person's definition of engineering?
   - What is his or her area of specialization?
   - Why did the individual choose engineering as a career?

   Write a brief report on your interview.
2. Repeat Problem 1 by interviewing an engineering faculty member from outside your department at your university.
3. Repeat Problem 1 by interviewing an engineer working in the community outside of the university.
4. Go to the Web site for the professional engineering licensure board for your state. Find out what it takes to become a licensed professional engineer in your state. Write a brief report on what you find.
5. Interview a manager for a local high-tech company in your community. Find out what it takes to be a successful entrepreneur. Write a brief report on your interview.

2

Electrical and Computer Engineering Specializations

1 INTRODUCTION

Engineering is a very diverse profession with numerous specializations such as electrical, mechanical, and chemical engineering. The origins of electrical engineering are in the power industry, developing systems for efficiently generating and distributing electrical power. Since then, electrical engineering has expanded to include many other areas of specialization and has also given birth to computer engineering. The specializations within electrical and computer engineering range from semiconductor-device physics and processing, through electronic circuits and computers, to wireless communication. In this chapter some of these areas of specialization will be described in detail. It is hoped that you will get a feel for the breadth of disciplines within electrical and computer engineering, and that you will also be able to start thinking about which area you want to specialize in during your career.

SECTION

1 Introduction
2 Power
3 Electronics
4 Computers
5 Wireless Communication
6 Microelectronics and Integrated Circuits
7 Control Systems
8 Electromagnetics
9 Signal Processing
10 Image Processing
11 Biomedical
12 Optoelectronics
13 Plasmas
14 Robotics

OBJECTIVES

After reading this chapter, you will

- Be able to list the different specialties within electrical and computer engineering

2 POWER

Perhaps the way in which electrical engineering most touches people's everyday lives is in the area of electrical power. Electrical engineers are involved in many aspects of the design and building of power distribution systems. A typical electrical power transmission line is shown in Figure 1 Many electrical engineers are employed by construction companies to help with the design of generating facilities. In the course of this design work, electrical engineers work closely with mechanical, chemical, and nuclear engineers. Electrical engineers are also involved in helping to supervise the construction of these plants. Once the generating plant is built, there is a need for a power distribution system to deliver the power to businesses and homes. Electrical engineers are also involved in designing the network of transmission lines used to deliver power to customers.

It is important to also mention the aspects of power distribution electrical engineers are not involved in. The actual wiring of buildings and houses is performed by licensed electricians, not engineers, although the codes that establish the standards for how this wiring is done are written by engineers.

Electrical engineers also work on the development and implementation of alternative power sources. There are many potential sources of alternative energy that are being explored including solar photovoltaics, wind turbines, and fuel cells. Much of the research and development in this area is being performed by electrical engineers.

Figure 1. An electrical transmission line. This is part of the network used to transport power from the generating station to homes and businesses.

APPLICATIONS: BIOGRAPHY

Nikola Tesla

Much that we take for granted originated as ideas in the fertile imagination of Nikola Tesla. He was born in 1856, the son of a priest, in the region known now as Serbia. At the early age of five, he built a paddle-wheel device turned by the water of a small stream. He excelled in school partly because of a prodigious memory, memorizing the logarithmic tables while in high school.

Early in his college studies, he complained to his instructor about sparking at the brushes of a DC motor he was using in an experiment. He then told the instructor that it should be possible to build a motor that didn't have this problem by using an alternating current. The instructor told him that such a motor was impossible and that he should not waste his time on such silly ideas. Thus challenged, he put his mind to the problem and invented the concept of a three-phase alternating-current system and a rotating magnetic field. The rotating magnetic field made possible the concept of the three-phase induction motor, which is the backbone of modern industry. This was done while he was in his early twenties.

Tesla immigrated to the United States in 1884, arriving with four cents in his pocket. This was the year that the American Institute of Electrical Engineers (AIEE) was founded. AIEE, along with the Institute of Radio Engineers, founded in 1912, merged in 1960 to form the Institute of Electrical and Electronics Engineers (IEEE). Tesla went to work for Thomas Edison, who was just starting to build DC power distribution systems in New York City. Tesla tried, unsuccessfully, to convince Edison that AC systems would be better. Tesla left employment with Edison over the latter's failure to pay him for improvements he had made in the design of DC motors and generators.

By 1888, Tesla had patented most of the principles that govern the operation of the electrical power systems in place throughout the world today. In fact, it was Tesla who selected 60 Hertz as the operating frequency of the system, as he had calculated that that was the optimum frequency to minimize the losses in the generators and transformers. A short time later, he got the backing of George Westinghouse. At the 1893 World's Fair, the Westinghouse Company demonstrated a practical AC power generation and distribution system. This led to the Westinghouse Company winning the right to build the first electrical generating station powered by the falling water at Niagara Falls. Tesla provided the design work for the entire system, including the first large hydroelectric generators. The success of this endeavor led to the use of Tesla's AC system rather than Edison's DC system.

About this same time, Tesla invented the radio, making his first transmissions in 1893. You may have read that Marconi was the "father of radio." Actually, Marconi had infringed on Tesla's patents in building his system. When Marconi succeeded in spanning the Atlantic Ocean with a a radio signal in 1901, he captured the imagination of the public, and the history books enshrined him, not knowing the actual history. Tesla sued Marconi, and the case lasted for years, finally being one of the few technical cases settled by the Supreme Court of the United States. In 1943, after Tesla had died, the Supreme Court ruled that Marconi had indeed infringed on Tesla's patents.

Tesla invented radio-controlled vehicles and automatic control systems in a radio-controlled submarine, which he demonstrated at the 1898 World's Fair. He also invented fluorescent lighting and digital logic circuits during this time. He conceived of the idea of radar some 40 years before a system was demonstrated in England. In addition, he thought that it should be possible to transmit electrical power through the air using his "Tesla coil," which is the basis of the high-voltage transformer used in modern TV receivers to supply high voltage to the picture tube. He built a laboratory in Colorado Springs, Colorado in 1901 and succeeded in lighting a bank of 200-watt lightbulbs at a distance of 25 miles. Also, he reported the first instance of signals received from sources beyond the earth. He then moved to Long Island and, with the financial backing of J.P. Morgan, began to build a huge structure, called Wardenclyffe, which he planned to use to transmit both radio and power. However, before it was finished, he and Morgan had a disagreement, and the project was never competed after Morgan withdrew his backing.

Tesla is credited with more than 700 patents by the Unites States Patent Office. It is interesting to look at the abstracts of patents from the 1880s and 1890s that were authored by Tesla. In 1960, the International Commission of Electrical Engineers designated that the unit of magnetic field intensity be named the Tesla, abbreviated "T," in his memory.

For additional information about this fascinating genius, the book *Man Out of Time*, by Margaret Cheney, is a very readable biography of Tesla's life. In addition, the web site www.tesla.org leads to a number of interesting sites about Tesla.

3 ELECTRONICS

Another area where electrical engineering directly impacts the lives of consumers is in electronics. Electrical engineers designed the earliest versions of radio and television, and they continue to work on these types of products today. Circuit design has moved beyond these origins to include electronics for a wide variety of devices that impact our lives all day long. For example, both commercial and military airplanes use very sophisticated electronic circuits to help control the aircraft and to help keep the pilot aware of other aircraft in the area. Systems used to control heating and air conditioning in buildings and homes require electronic circuits to operate efficiently. Many large manufacturing companies require the services of electrical engineers to help design one-of-a-kind electronic circuits to efficiently manufacture their product. The functions of modern automobiles are controlled by a variety of electronic devices that help ensure proper engine performance, monitor problems that may occur in the engine, and control safety features such as antilock braking systems.

Electronic design engineers must have detailed knowledge about what devices—transistors, integrated circuits, or resistors, for example—are available and how to combine them into a useful circuit. The results of this type of work are shown in Figure 2, a computer mother board. In performing a design, electrical engineers must ensure that the circuit operates as planned and that it also can be produced economically, will operate efficiently, and will operate correctly for a long time. Electronics engineers must often work closely with manufacturing engineers in designing a product that does its intended job and can be manufactured easily and at a reasonable cost. Electronics engineers often use computer-based design software such as PSPICE® to help in their design work.

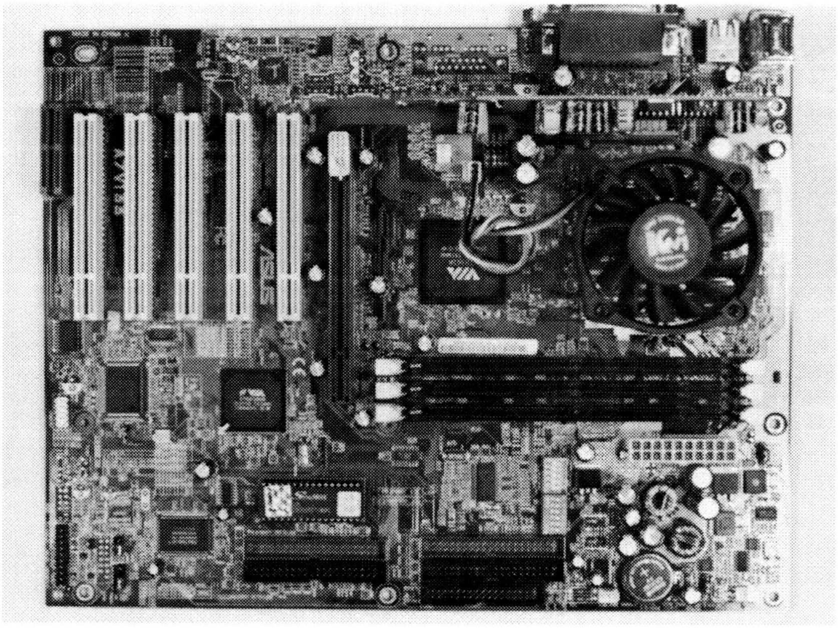

Figure 2. A motherboard for a PC. Electrical and computer engineers contribute to the design of these complex systems. (Photo by M. Bradshaw.)

4 COMPUTERS

Computer engineering has evolved from being one of the specialties within electrical engineering to now being a distinct engineering discipline. Because of its origins and close ties to electrical engineering, computer engineering students are still generally trained in university departments that combine both fields. Computer engineers are concerned primarily with the design of computer hardware, the circuitry that enables a computer to perform its required tasks. Figure 3 shows an array of semiconductor memory chips in a computer. Hardware design involves detailed knowledge of computer architecture, the basic structure of computer systems. It would be impossible to design computers without also having in-depth knowledge of computer software, especially operating systems. Computer engineers are also responsible for designing the networks through which computers communicate with each other.

Computer engineers are employed in many different settings. Obviously, computer companies such as IBM, Gateway, or Dell employ large numbers of computer engineers. However, since nearly all electronic systems today involve some level of computer control, computer engineers find employment in a variety of engineering settings. For example, all modern automobiles have onboard computers that control the operations of the engine. The automobile industry employs many computer engineers to design these systems. Computer engineers work in teams with computer scientists and electrical engineers.

Figure 3. Computer engineers work on the design for computer hardware, such as the circuit boards and integrated circuits shown here.

APPLICATIONS: BIOGRAPHY

Seymour Cray

The world around us would look very different today if not for the development of the supercomputer. These computers have revolutionized both science and industry, allowing very complex and time-consuming calculations to be performed in reasonable amounts of time. Supercomputers, capable of performing billions of calculations per second, are indispensable in weather forecasting and in the design of automobiles and airplanes. The oil industry has used supercomputers to model geologic strata and help predict where oil deposits can be found. The semiconductor industry uses supercomputers to model the performance of circuits and devices before they are made so that design corrections can be carried out inexpensively. The "father of the supercomputer" is Seymour Cray.

Cray was born in Chippewa Falls, Wisconsin, in 1925. He began his engineering career by obtaining a bachelor's degree in electrical engineering from the University of Minnesota in 1950. He also obtained a master's degree from Minnesota in applied mathematics. His professional career began with a Minnesota company called Engineering Research Associates (ERA). ERA had done some work developing computing machines for the Navy, chiefly for cryptography. When Cray started working in the computer industry, computers were still being built with vacuum-tube technology. During the course of his career, he helped design computers during the transition to transistors and finally to integrated circuits. While at ERA, Cray was responsible for the design of the ERA 1103, the first commercially successful scientific computer.

When ERA's interests turned more towards commercial computing, Cray left and joined another fledgling computer company, Control Data Systems in 1957. There, Cray was responsible for the design of many of the firm's most successful computers, including the 1604, 6600, and 7600. Cray rose to become Control Data System's senior vice president.

In 1972, Cray left to form his own company, Cray Research. Here he developed his best known and most widely used products, the Cray-1 and the Cray-2. These machines were the state of the art in scientific computing from the late seventies through the early nineties. While at Cray Research, Cray held many positions in both management and engineering. For example, at various times he was president, chief executive officer, and chairman of the board. In 1981, he stepped aside from his management duties to devote himself full time to work on designing the Cray-2. He left Cray research in 1989 and founded a new company, Cray Computers. Ultimately, this company failed. His last company, SRC Computers, was founded in 1996. Unfortunately, Cray died in an automobile accident in Colorado Springs, Colorado, in the fall of 1996 before SRC had a chance to establish itself.

Although Cray computers are now mostly in museums, Cray's legacy is still visible today. His many innovations in supercomputer technology paved the way for today's generations of scientific computers.

5 WIRELESS COMMUNICATION

Communication systems have long been a major area where electrical engineers have made significant contributions. Electrical engineers have helped design the communications infrastructure of the United States and of the world by designing telephones, switching systems, and telecommunications networks. In more recent times, as many of the functions of communication switching have become computerized and with the rise of the internet, computer engineers have taken a significant role in engineering communication systems. There has been explosive growth in the communications industry in the past decade, fueled by the development of wireless communication systems. Cell phone usage has increased dramatically as more and more people enjoy the convenience of being able to stay in touch wherever they may be. Computers can now be networked without being connected to a telephone line.

Electrical and computer engineers find wide employment in the wireless communication industry. Engineers are needed to design the cell phones, including the important task of designing efficient, yet small antennas. These engineers must work closely with specialists in ergonomics to ensure that the design of the phone does not put undue stress on the user's body. Engineers also design the sending and receiving equipment used to disseminate the cell phone signals. Computer engineers are needed to design computer controls for the sophisticated switching networks required to run a cell phone system.

6 MICROELECTRONICS AND INTEGRATED CIRCUITS

The backbone of the modern electronics, computer, and communication industries are integrated circuits. The invention of the integrated circuit, which contains millions of transistors in a single piece of silicon the size of a thumbnail, revolutionized the way electronic circuits were designed and led to a new field—microelectronics. Microelectronics focuses on the design of extremely small electronic devices and their interconnection to form useful devices. The integrated circuit industry employs numerous electrical engineers. Many of these engineers focus on the special design requirements when circuit elements are shrunk to very small dimensions. Designing a microelectronic circuit involves very different skills than designing the same circuit using discrete components. Other microelectronic engineers work on the processes required to successfully fabricate an integrated circuit. These engineers work at the intersection of electrical engineering and chemical engineering, and need a good working knowledge of the chemistry involved in the fabrication process. Microelectronic engineers work closely with chemical engineers, manufacturing engineers, and production technicians to ensure that the processes for building integrated circuits are working properly.

7 CONTROL SYSTEMS

Control systems are an essential part of many engineered devices. For example, antilock braking systems contain sensors to detect when the driver has applied the brake, other sensors to detect if one of the wheels has locked up and started to skid, and an electronic module to automatically direct the braking function to "pump" the brakes multiple times each second to allow the driver to maintain control of the vehicle. Modern control systems are very sophisticated and can be found on numerous engineered devices from airplanes to washing machines. Electrical engineers design control systems to ensure that complex devices operate efficiently and safely.

Modern control systems often have a computerized control module, so computer engineers also find employment in the controls area. More recently, control systems engineering has been revolutionized by the introduction of "fuzzy" controllers, which rely less on rigid mathematical algorithms to perform the control function and more on qualitative concepts.

APPLICATIONS: BIOGRAPHY

Lofti Zadeh

Lofti Zadeh has spent his entire engineering career in academia, serving on the faculties at Columbia University and at the University of California at Berkeley. He is often called the father of "fuzzy logic," a new way of analyzing complex systems. Professor Zadeh was awarded the IEEE Medal of Honor in 1995 for his contributions to the development of fuzzy logic and its many applications.

During the development of scientific and engineering ideas, occasionally *paradigm shifts* occur as thinking in a particular field undergoes a radical change. The term *paradigm shift* was first coined by Thomas Kuhn, a leading philosopher of science. The change of vision in physics that occurred when quantum theory, based on probability, supplanted the more deterministic Newtonian mechanics is an example of a paradigm shift.

A paradigm shift led by Lofti Zadeh occurred in electrical and computer engineering during the latter half of the 20th century in the area of control system theory. Before Zadeh's work, control theory was based on very precise mathematical techniques. Zadeh, felt that a less precise control strategy based on "fuzzy" concepts would be better able to handle many engineering control problems. Prior to Zadeh's work, complex systems were described and analyzed using numerical variables. In his original paper, published in 1965, Zadeh described a qualitative, linguistic approach to describing these systems. Instead of rigid mathematical modeling, fuzzy logic emphasizes the use of words such as faster and slower, or heavier and lighter. Concepts such as this provide a better way to deal with uncertainty in complicated systems. Interestingly, these concepts also more closely mimic the way the human brain makes decisions than mathematical models do.

Zadeh was born in Azerbaijan, now an independent nation, but then part of the Soviet Union. When he was young, his family moved to Tehran, Iran. He always wanted to be an engineering professor, and he started on that path by studying electrical engineering at the University of Tehran. He then came to the United States, where he studied for his master's degree at MIT. He completed his education at Columbia University, receiving his Ph.D. in 1949. He remained at Columbia, serving on the faculty until 1959, when he accepted a faculty position at the University of California at Berkeley.

Fuzzy theory was not immediately accepted. In fact, it was quite controversial for many years. This is not entirely unusual; it often takes many years for new ideas to be accepted and firmly rooted in the field. Engineers generally like to work only with very precise things, so it is not surprising that a theory based on imprecision would be hard for many engineers to accept. Fuzzy logic received a huge boost when it was incorporated into consumer products by Japanese manufacturers. In 1987, Matsushita Electric applied fuzzy theory to a showerhead that controlled water temperature. Fuzzy logic was then applied to numerous other consumer products from washing machines to cameras to automobiles.

Fuzzy theory has also found acceptance in the academic community, with many books and journal articles written on the subject since it was first introduced. Professor Zadeh is still active in electrical engineering. He is currently a professor emeritus at Berkeley, and he continues to work on fuzzy logic and a new field, "soft computing," that he has also pioneered.

8 ELECTROMAGNETICS

Electrical engineers have harnessed electromagnetic phenomena to build many useful devices. Electromagnetics refers to the propagation of information by time-varying electric and magnetic fields. The design of antennas that send and receive television or cell phone signals, the design of radars that allow aircraft to fly safely, and the development of microwave communication systems are all activities of electrical engineers working in electromagnetics. Given the breadth of devices that utilize electromagnetics, engineers working in this area will work closely with electronics engineers to ensure that the antennas used for radio signals interface correctly with the sending and receiving electronics. They will also work with wireless communication experts to help design the most efficient antennas for cell phone applications.

9 SIGNAL PROCESSING

The signals produced in many applications are often corrupted by the presence of electrical noise. Noise is an unwanted transient voltage or current that is present in all electrical systems. Noise can come from phenomena internal to an electrical device or from external signals that are accidentally picked up. Noise is a problem because it severely limits the ability of a device to detect interesting signals. For example, a radio signal from a distant source cannot be detected since its amplitude is smaller than that of the noise. One of the tasks of signal processing is to try to find ways to pull meaningful signals out of the noise. Signal processing can also involve developing methods for converting analog information into its digital equivalent. An analog signal can be changed to digital by sampling the signal periodically and assigning a digital number to each level of the signal. Signal processing is used to make this conversion. Oftentimes, the information that is generated this way involves a huge amount of information. Signal processing is also concerned with compressing this information—using special algorithms to reduce this data to a more manageable size and to later reconstruct the original data.

Both electrical and computer engineers find employment in signal processing. Efficient processing of signals is vital to computers, but it finds application in many other areas as well. For example, radio astronomers generate large amounts of data during scans of the sky. Electrical and computer engineers help in the development of signal processing tools to improve this process. Signal processing is also important in many medical applications. As medical devices produce data, there is a need for engineers to develop efficient means for processing and utilizing this information.

10 IMAGE PROCESSING

Image processing is similar in many respects to signal processing, but instead focuses mainly on the processing of information in pictorial form. Image processing involves the skills of both computer and electrical engineers. The challenges of image processing involve taking very large amounts of information—a typical image displayed on a computer screen requires megabytes of information—and making that information useful. This might involve developing methods to enhance the contrast in an image to bring out important features or altering the colors in the image to make it easier to detect important information. Storage of information also presents challenges due to the large size of image files on a computer. These files must be compressed, as discussed in the previous section, in order for them to be stored efficiently.

Image processing finds application in numerous areas. For example, engineers with image processing skills are employed by companies making medical diagnostic equipment. Their skills help to make machines that generate high-quality and easily interpreted images that help physicians make accurate diagnoses. Astronomers use image processing to take electronic images of celestial objects and process them to yield useful information. In the automotive industry image processing can be used to understand the combustion processes taking place in an engine. By using specialized optical equipment, engineers can detect the chemical processes taking place within an automobile engine. This information can then be processed to form an image of what is taking place. These types of images help automotive engineers "tune" their engine designs. Electrical and computer engineers are even employed in the entertainment industry, using image processing techniques to create special effects in movies.

11 BIOMEDICAL

The biomedical engineering field employs many different types of engineers, including electrical and computer engineers. Biomedical engineering is concerned with the engineering of new medical diagnostic and treatment tools and new processes for delivering therapy to patients. An MRI system used to help diagnose a variety of diseases is shown in Figure 4. Of course, mechanical and chemical engineers have skills that are directly applicable to medicine as well. Electrical and computer engineers contribute to this field through the design of new electronic equipment, much of it computer based, which allows physicians to do their job better. Specific skills related to biomedical engineering are electronics, signal and image processing, and computer engineering.

12 OPTOELECTRONICS

Optoelectronics is a relatively new field that straddles the line between electronics and optics. Optoelectronics experts have worked to develop optical sensors and optical emitters, including lasers. Optoelectronic components are incorporated into CD and DVD players and laser printers, all of which utilize miniature lasers. Optoelectronics has changed the way communication systems operate. Until recent, all voice and data communications were over wires or microwave links. The development of optical fiber technology has allowed for telephone conversations and data communications such as the

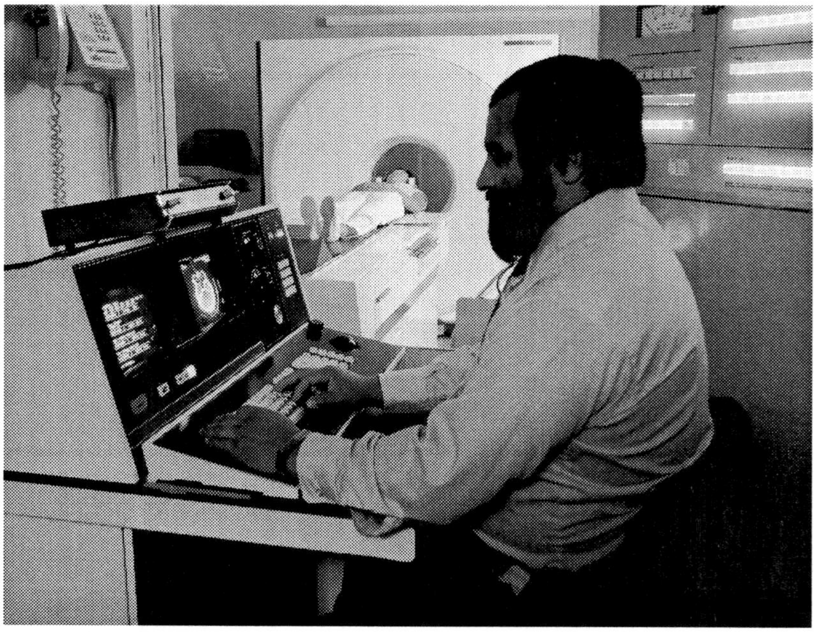

Figure 4. A magnetic resonance imaging (MRI) system used for medical diagnosis. Electrical and computer engineers contribute to the design of such devices.

internet to be carried over optical cables. These optical systems are faster and are able to carry more information than traditional wire technologies.

Optoelectronics experts are also attempting to change the way computers function. Currently, computers operate by directing the flow of electrons in an electronic circuit. There are fundamental limits to how fast this type of computing can be run. In the future, it is possible that computing could be performed by directing the flow of photons, the fundamental unit of light, instead of the flow of electrons. So far, efforts in this direction have only yielded very primitive computing systems, but as advances continue this will change.

13 PLASMAS

Most of you are not familiar with the word *plasma*, or you know of it only as a component of blood. To engineers and physicists, plasmas are sometimes known as the "fourth state of matter," not exactly like a gas, liquid, or solid. For our purposes, a plasma is a collection of electrons, ions, and neutral atoms or molecules, generally in a gaseous form. Even though a plasma is gaseous, the presence of charged particles means that it behaves very differently from a gas. Plasmas are all around us. In fact, most of the matter in space is in the form of a plasma.

Plasmas also have many engineering uses. Many homes and businesses contain plasma-based devices. For example, fluorescent lightbulbs contain a light-emitting plasma of argon and mercury. Neon signs contain plasmas as well. Plasmas are also important in the microelectronics and integrated circuit industry. Plasmas are used to etch transistor patterns into silicon to form integrated circuits. In fact, the high density of devices located on an integrated circuit is only possible when plasma etching is used during the process. Electrical engineers working in plasmas are typically employed by lighting companies such as General Electric and by integrated circuit fabrication companies such as Intel. Plasma engineers also work for companies such as Applied Materials that manufacture the plasma machines required to manufacture integrated circuits.

14 ROBOTICS

Although robots are not utilized in the home (except in the form of toys), robotics is a very important technology in industry. Modern manufacturing processes are highly dependent on robots to perform repetitive tasks quickly and accurately. Robots are found on automobile assembly lines, integrated circuit fabrication facilities, and in food processing plants. A small robotic arm is shown in Figure 5. Both electrical and computer engineers have roles in the design of robotic systems. Robots require very sophisticated control systems to ensure that they perform their tasks correctly. Robots are generally computer controlled, so computer engineers also make major contributions to the design of robotic systems. Many robots are autonomous—they carry their own power supply and are not connected to any external computers. Autonomous robots require communication systems between the operator and the robot, which are based on electromagnetic communication systems. Many specialties of electrical and computer engineering are involved in robotics.

Electrical and computer engineers working in robotics work closely with manufacturing engineers to help design robots for industrial production-line applications. They also work with mechanical engineers who are generally responsible for designing the mechanical aspects of the robot.

Figure 5. A laboratory robotic system. The sophisticated control systems for robotic systems are designed by electrical and computer engineers. (Photo by M. Bradshaw.)

We have seen that there are many different specializations within the field of electrical and computer engineering. We have presented only a partial list of all the different areas of electrical and computer engineering in which you might choose to specialize. Although these specialties sometimes seem very separate, in today's engineering world, engineers work in teams, which incorporate specialists from different engineering fields as well as different specialties within each field. As you progress in your study of electrical and computer engineering, you will have the opportunity to choose a specialty depending on your interests and take a few courses in this area. Keep in mind that many of the specialties listed here did not exist even thirty years ago. And there are many specialties that existed 30 years ago that are not important today. Because electrical and computer engineering change so quickly, it is important to prepare yourself to be able to switch specialties often during your professional career. You can do this by ensuring that you learn how to learn new topics on your own.

KEY WORDS

Engineering specializations

SUMMARY

We have seen that there are many different specializations within the field of electrical and computer engineering. We have only presented a partial list of all the different areas of electrical and computer engineering that you might choose to specialize in. Although these specialties sometimes seem very separate, in today's engineering world engineers work in teams which incorporate specialists from different engineering fields as well as different specialties within each field. As you progress in your study of electrical and computer engineering, you will have the opportunity to choose a specialty depending on your interests and take a few courses in this area. Keep in mind that many of the specialties listed here did not exist even thirty years ago. And there are many specialties that existed 30 years ago that are not important today. Because electrical and computer engineering change so quickly, it is important to prepare yourself to be able to switch specialties often during your professional career. You can do this by ensuring that you learn how to learn new topics on your own.

Problems

1. Talk to a faculty member in your department about new emerging areas of specialization in electrical or computer engineering. What areas does the person see being important in the future? Find out where to get more information on this area, and write a brief report on your findings.
2. Pick one of the areas of specialization listed in this chapter. Go to the library and find out more information on this field. Write a report on the current state of this field. Discuss any new innovations in the field that seem to be important.
3. Write an essay on the specialization within electrical and computer engineering you are interested in. Be sure to answer the question, "Why do I want to be involved in this specialization?"

3

Engineering Your Career

1 INTRODUCTION

So far, we have explored the different types of engineering and the different specialties within electrical and computer engineering. We have also learned a little about the tools and devices that electrical and computer engineers use and methods for solving engineering problems, including design problems. In this chapter, we will look at engineering careers and what it takes to become a successful engineer.

SECTION

1 Introduction
2 Engineering Careers
3 Further Engineering Study
4 Licensure
5 Lifelong Learning
6 Professional Organizations

OBJECTIVES

After reading this chapter, you will

- Be able to describe the types of engineering career choices available to a new engineering graduate
- Explain the advantages of advanced study
- Understand the need for professional registration
- Be able to identify the IEEE as the professional organization for electrical and computer engineers

2 ENGINEERING CAREERS

The opportunities open to newly graduated engineers are as varied as the types of organizations that employ engineers. Electrical and computer engineers find employment in private industry, government agencies, research laboratories, and educational institutions. Figures 1 and 2 show engineers at work looking at a circuit diagram, typical of work within private industry. Within each category, there are careers available at organizations of many different sizes. For example, private employers range from small start-up companies that might employ only a handful of engineers, to large international corporations such as IBM, which employ thousands of engineers. Each type of organization and each size organization has its own advantages and disadvantages. The ultimate decision as to which is right for you depends on where your interests lie and what type of organization you are most comfortable being a part of. Many young engineers find the most satisfaction in starting their own company.

Regardless of the organization that you choose to work for, you will generally start out with very simple projects. Even though you have spent four (or more) years in school learning to be an engineer, you are by no means ready to begin work on complicated projects on your own. The simple projects you will start with will help you gain experience in engineering methods and company practices. If you have been fortunate enough to have had a co-op job or internship during your school years, you will have a head start in knowing how to be an engineer. Most companies will assign new engineers to work with a more experienced engineer who will show you how things are done, check your work, and offer you advice on your career. This type of mentoring relationship is very beneficial to incoming engineers—most people learn best by working with a knowledgeable colleague—and it also benefits the mentor who may be able to learn some state-of-the-art concepts that you learned in school. You will also spend much of your career working with other engineers as part of a team. Figure 3 shows design engineers discussing the development of a new product.

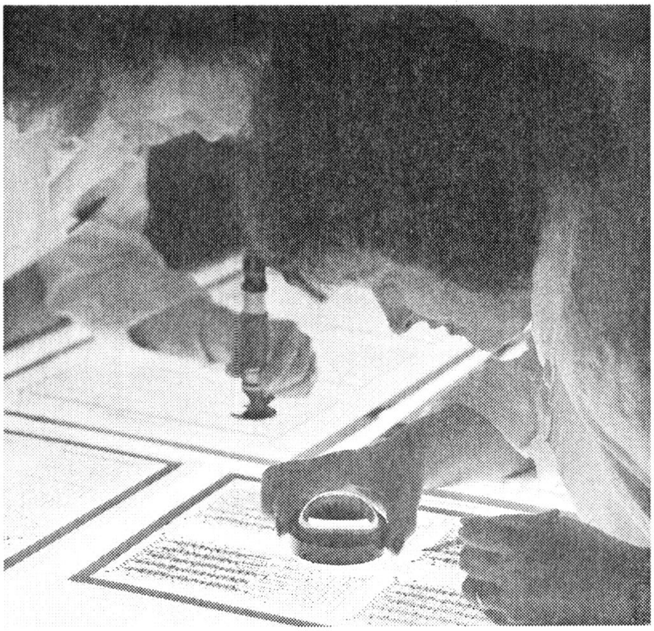

Figure 1. Engineers scrutinizing circuit diagrams.

Figure 2. Examining the layout for an integrated circuit.

Figure 3. Designers discussing product development.

It should be mentioned that not all graduates of B.S. programs in electrical and computer engineering decide to work as engineers. Many are very successful in other fields such as medicine, law, or business. It may surprise you that people trained as engineers would be successful in these other fields. However, the problem-solving skills and work ethic required to successfully earn an engineering degree carry over quite well to these other fields.

3 FURTHER ENGINEERING STUDY

For nearly all of you, a bachelor's degree in electrical and computer engineering will be only the starting point for your engineering education. Many jobs require advanced engineering training. For example, working in a research laboratory or being on the faculty of a college or university generally require a Ph.D. In addition, many companies are interested in seeing their employees upgrade their skills, and they have programs to pay for further education. Generally, engineering education beyond the bachelor's degree takes place in graduate school.

In graduate school, the broad education you received as an undergraduate will be the starting point for further study in a narrower, more specialized field. Many students continue on to graduate school for further training immediately upon completion of their undergraduate program. Others might not feel the need for further education until they have been working in the field for a few years and feel the need to upgrade their skills. Figure 4 shows students working on a computer aided design system designing new electronic circuits. Graduate programs usually lead to the master of science and Ph.D. degrees.

Figure 4. Computer-aided circuit design.

4 LICENSURE

In order to practice most professions, a license, generally issued by the state in which you wish to practice, must be obtained. For example, physicians and lawyers must pass an examination administered by the state in order to be licensed to practice in that state. These licensing requirements exist to help protect the public from unqualified practitioners. Engineering is somewhat unique among the professions in that not all engineers are required to be licensed to practice engineering. Engineers who require licenses are generally those who work on large public projects such as bridges and buildings. Nearly all civil engineers are licensed.

However, most states exempt engineers who work for private companies from the licensure requirements as long as their work does not involve public projects. Most practicing electrical and computer engineers are not licensed since they usually work for private companies. Electrical engineers working on electrical power distribution systems are an exception to this rule.

Even though your job as an electrical or computer engineer might not require that you obtain a license, it is still a very good idea to obtain a license in the state where you reside. *Licensure* is a means for demonstrating a level of competence as an engineer, and thus, it is a nice addition to your resume. How do you obtain an engineering license? The procedure may vary somewhat from state to state, but the general process is the same.

The first step to obtaining your license is to complete your engineering degree from an ABET-accredited school of engineering or from an approved nonaccredited school. (ABET is the Accreditation Board for Engineering and Technology, which accredits engineering programs at schools in the U.S.) Next, you take an examination called the Fundamentals of Engineering (FE) exam. This is an all-day test of your knowledge of the basics of engineering science and the specifics of your own discipline. It is a national test. Most aspiring licensed engineers take this exam during the last year of their undergraduate engineering program. After passing the FE exam, you must practice engineering with supervision from a licensed engineer. This engineer will have to document to the state board that you have had four years of engineering experience and that you are competent in your field. Finally, you must take the state board examination, which is also a nationally administered test. This exam probes your knowledge of your chosen engineering specialty, electrical or computer engineering. Once you have completed this process, you are granted a license and can represent yourself as a "registered professional engineer," denoted by the initials "P.E." after your name.

The requirements for retaining your license vary from state to state as well. In many states, once you become a P.E., you remain one for life unless your license is revoked due to malpractice or incompetence. Although you may be licensed for life, you may be required to renew the license periodically. Some states require that you keep your engineering skills current through some form of continuing education. This may involve attending professional conferences, taking short courses through professional organizations or universities, or even taking regular semester-long courses at a university.

5 LIFELONG LEARNING

Even if you are not a licensed engineer who is required to participate in continuing education in order to maintain your license, staying current in your field will still be important to a successful career. In addition, the rapid pace of change in electrical and computer engineering can render existing jobs obsolete in a very short time, requiring engineers to retrain themselves in new fields continuously throughout their careers.

Even though your course work during your undergraduate career will be directed by your professors, much of the knowledge that you will gain will be obtained on your own. You will need to develop the skills required to research topics and find useful information by yourself. This process will be even more important in your future because frequently you will not have anyone guiding your studies in a new field. What resources are there for obtaining information on new topics and learning new skills by yourself? There are several possibilities.

1. *Magazines and journals*. For any specialty within electrical and computer engineering, there are numerous trade magazines. These publications are generally distributed to engineers free (of course, they are loaded with advertising) and usually contain articles about new developments in the field. These articles are usually not written on a very technical level, but they indicate where more information can be found. Journals contain more technical articles and can be used to study the latest research in the field. Books are also useful in getting up to speed in a new field, although the information in books tends to be older and less up to date compared to the information found in magazines and journals. Books, however, offer the advantage of providing an overview of an entire field in one place.
2. *Technical conferences*. Professional organizations sponsor periodic conferences on nearly any technical topic that you can imagine. At these conferences, experts present their latest results and exchange ideas on new directions that the technology should take. Many conferences also offer short courses that involve intensive class instruction on a particular topic in technology. These courses are generally one week long and offer someone with general engineering training the ability to become knowledgeable in a new field very quickly.
3. *Internet*. The Internet offers extremely easy access to a wealth of information. Embedded among the information on popular culture and all of the advertising, there is also lots of technical information available on the Web. This information can come from educators who post their knowledge on the Web or from manufacturers who provide information on their products, how to use them, and how they work. When accessing information this way, it is important to keep in mind that there is no guarantee that the information you find on the Internet is correct. Information that is found in journals and most books has been reviewed by competent experts in the field to ensure that the statements that are made are valid. No such review process exists for information on the Internet. Before accessing information via the Web, it is essential to evaluate the source of the information and be sure that the source is reputable and can be relied upon to provide accurate information.
4. *University courses*. Sometimes the training that you will require, both technical and nontechnical, can best be obtained in a more formal, traditional setting such as a local college or university. If you want to upgrade your skills by obtaining a graduate degree, these schools are the only means available. A local school of engineering will generally provide in-depth courses and degree programs that you might need in technical areas. For those who are interested in moving into a more managerial role, these schools will also offer business courses and degree programs. Often, companies pay for their employees to obtain new degrees and to upgrade their skills, and they even offer time off from work to obtain additional education that will ultimately benefit

the company. Most universities now offer such courses and degree programs at various industrial sites through distance education technologies. These include delivery of courses via video links where students at remote sites can be part of a class on a distant campus in real time. Many universities also offer courses via the Internet so that students at a remote location can do course work at their own pace in the evenings after work.

6 PROFESSIONAL ORGANIZATIONS

Every branch of engineering has professional societies. For electrical and computer engineers, the IEEE is the professional organization. The IEEE has been in existence for over one hundred years (although with different names over the years) and has over 350,000 members in 150 nations worldwide. The IEEE publishes numerous journals relating to electrical and computer engineering, and it sponsors annual conferences on a variety of topics. The IEEE also serves as an advocate for electrical and computer engineers and the engineering profession. Membership in the IEEE provides you with ready access to the information needed to be successful in your career. It also provides the opportunity to "network" with other electrical and computer engineers employed in organizations other than your own. This networking can be invaluable for obtaining technical information and for finding new job opportunities. The IEEE also helps to establish industry standards for electrical and computer technology, and establishes standards of conduct for those who call themselves electrical and computer engineers through the IEEE code of ethics.

Although most members of the IEEE are practicing engineers, membership is also offered to students at a greatly reduced rate. Nearly every electrical and computer engineering department at universities in the United States sponsor a student branch of the IEEE. It is beneficial to join the IEEE while you are a student to help you in your career, both now and after graduation.

KEY WORDS

advanced study
Career
IEEE
lifelong learning
licensure

SUMMARY

In this chapter we have discussed some of the career paths you might follow after you graduate. In addition, we have mentioned some of the important activities you should pursue including licensure and membership in your professional organizations. Most important, is to remember that your engineering education does not end when you graduate. Rather, you must engage in life-long learning to keep your skills up-to-date and to ensure that you will continue to be an effective member of the engineering profession.

Problems

1. Find out who the president of the student branch of the IEEE is at your university. Find out what it takes to become an IEEE member. Make a brief report to your class on how to become a member.

2. Find out where the career services office on your campus is located. Talk to the person specializing in engineering careers. Find out what engineering companies typically conduct interviews on your campus. Find out when they offer workshops on engineering careers, resume writing, and other job search skills. Write a brief report on your findings.
3. Find out what office on your campus is responsible for coordinating co-op programs. Find out what companies hire co-ops from your school. Find out what the requirements are to participate in the co-op program. Report your findings to your class.
4. Talk to the person responsible for the graduate program in your department. Find what the requirements for admission to the M.S. or Ph.D program are. Find out how many students apply each year and how many are accepted. Report your findings to your class.
5. Interview a faculty member in your department or a working engineer in your community. Find out how that person stays up to date in her or his field. Write a brief report on your findings.

4

Academic Survival Skills

1 INTRODUCTION

There are a number of factors that will influence how well you do in your college or university career. This chapter will discuss some of the more important ones, including study skills, group work, writing and presentations, networking, and searching for information.

SECTION

1 Introduction
2 Study Skills
3 Group Work
4 Writing and Presentations
5 Networking
6 Accessing Information

OBJECTIVES

After reading this chapter, you will

- Be able to describe how to construct a personalized study schedule for your classes this semester
- Be able to develop a set of personalized test-taking strategies
- Be able to explain the concept of networking
- Describe how to find information in your school library and how to use computers to help in this activity

2 STUDY SKILLS

How you study, how much you study, and your strategy for taking exams will determine, to a large extent, how well you do in your studies. For most students, the greatest enemy is time. There never seems to be enough of it. An old rule, but one that has worked for many previous students, is to allow two to three hours outside of class for every hour you spend in class. Note that, in general, laboratory classes are an exception, since there are generally three hours in lab for each hour of credit. We will give an example by showing the required classes for freshman computer engineers at the University of New Mexico [1], typical of most computer engineering programs, in Table 1. While the curriculum at your school may not be exactly the same, the basic ideas will still apply. This freshman schedule assumes that you are a full-time student and that you have the required background to take the first required math course, Calculus I. If you are going to take a lower level math course to get ready for calculus, you will probably be able to take some of your humanities and social science (core) electives, to fill out your schedule until you can take the required math classes. Mathematics *is* the language of electrical and computer engineering, so be certain that you don't get ahead of your math skills in your technical and engineering classes.

Note that this schedule has 16 class hours and 3 lab hours. Using the lower limit of 2 hr of outside work for each hour in class, this would require 32 hr of outside work each week. Using the upper limit of 3 hr of outside work for each hour in class, this would require 48 hours of outside work each week. Add this to the hours in class, and the required lab hours gives a lower limit of 5l hr and an upper limit of 67 hr of total work each week. It should be obvious that going to school is a full-time effort, since you will be spending more time than the customary full-time workweek.

Now, if you have to work while you are going to school, it should be obvious that you *will not* be able to carry a full load of classes. You should keep your total hours, work and school related, down to the 67 hr "upper limit" calculated previously.

Table 2 shows the classes listed in Table 1, put into a time schedule according to the fall 2000 schedule of classes at the University of New Mexico [4]. The core elective chosen for this example was History 101, World History, but any number of other classes could be used, based on your interests. The time available each day was arbitrarily selected as 8:00 A.M. until 10:00 P.M. a 14-hr day, but could be shifted upward or downward depending on whether you are an "early" person or a "late" person.

Now, when the schedule in Table 2 is examined, it is seen that there are "holes" in it during the day. You should find a quiet place where you can study, perhaps at the library, and make up a schedule that shows you studying the same course at the same time each week. Doing the same thing on a regular schedule will program your mind to be ready to work on that subject when it is done on this regular schedule.

TABLE 1 First-Semester Freshman Computer Engineering, University of New Mexico

Course	Hr	Lecture–Lab
Chem 121L, General Chemistry	4	(3–3)
CS 151L, Intro to Computer Science	3	(3–0)
Math 162L, Calculus 1	4	(4–0)
Engl 101, Composition l: Exposition	3	(3–0)
Core elective	3	(3–0)
Totals	17	(16–3)

TABLE 2 Typical Class Schedule for Classes in Table 1

Time	Mon	Tues	Wed	Thur	Fri
800		Hist 101		Hist 101	
900	Math 162		Math 162		Math 162
930	Math 162	Math 162	Math 162		Math 162
1000	Chem 121	Math 162	Chem 121		Chem 121
1100	Engl 101		Engl 101		Engl 101
1200					
1230				Chem 121L	
1300				Chem 121L	
1400				Chem 121L	
1500	CS 151		CS 151	Chem 121L	CS 151
1530	CS 151		CS 151		CS 151
1600					
1700					
1800					
1900					
2000					
2100					

When all of the holes are counted up in the schedule shown in Table 2, it comes out to 50 hrs. Assume that you will need one half hour of time for lunch and another half hour for dinner each day. This reduces the time made available by the "holes" to 43 hrs. This goes a long way toward covering the 48 hours of outside work required by the previous analysis (maximum time) to be done completely on weekdays, leaving the weekends free for rest and relaxation.

Table 3 shows a weekly study schedule, using these assumptions of one hour of time eating lunch and supper each day. Note that the schedule shown in this table leaves the time from 10:00 P.M. on one day until 8:00 A.M. on the next day free. This allows sleeping time, cleanup time, and breakfast each day.

The table shows that the upper limit of 48 hours of outside work required can be accomplished with only 5 hrs needed on weekends. You may want to allocate your time differently, but putting regularity into your study schedule will greatly help in improving your grades in all of your classes.

Most of your classes will require weekly homework or reading assignments. It will be best if you start these on the first available study period after you receive them. This will allow you time to ask questions if you have any problems. Also, make it a practice to visit your professors during their scheduled office hours if you have an extended question that you don't want to ask in class. Your professors want to help you to succeed in their class.

Be sure that you understand the deadline for dropping a class at your school. If you find that you have overloaded yourself, it is much better to drop one or two classes that are requiring too much time. This will allow you to devote more time to those classes in which you are doing well.

If you think that carrying a full load will be too demanding of your time, you may want to cut back on your hours each semester. Doing this means that it will take longer to finish your program. Many students take classes in summer school in order to maintain the required pace. National statistics show that only a very small percentage of students now finish an engineering program in four years.

TABLE 3 Study Schedule for Schedule of Table 2

Time	Mon	Tues	Wed	Thur	Fri	Sat	Sun
800	Engl 101	**CLASS**	Engl 101	**CLASS**	Engl 101		
830	Engl 101	**CLASS**	Engl 101	**CLASS**	Engl 101		
900	**CLASS**	**CLASS**	**CLASS**	**CLASS**	**CLASS**		
930	**CLASS**	**CLASS**	**CLASS**	Hist 101	**CLASS**		
1000	**CLASS**	**CLASS**	**CLASS**	Hist 101	**CLASS**		
1030	**CLASS**	**CLASS**	**CLASS**	Hist 101	**CLASS**		
1100	**CLASS**	Math 162	**CLASS**	Math 162	**CLASS**	Hist 101	Hist 101
1130	**CLASS**	Math 162	**CLASS**	Math 162	**CLASS**	Hist 101	Hist 101
1200	***Lunch***	***Lunch***	***Lunch***	***Lunch***	***Lunch***	***Lunch***	***Lunch***
1230	CS 151	Chem 121	CS 151	**CLASS**	CS 151	Engl 101	Engl 101
1300	CS 151	Chem 121	CS 151	**CLASS**	CS 151	Engl 101	Engl 101
1330	CS 151	Chem 121	CS 151	**CLASS**	CS 151	Engl 101	Engl 101
1400	CS 151	Chem 121	CS 151	**CLASS**	CS 151	Engl 101	Engl 101
1430	CS 151	Chem 121	CS 151	**CLASS**	CS 151	Engl 101	Engl 101
1500	**CLASS**	Chem 121	**CLASS**	**CLASS**	**CLASS**		
1530	**CLASS**	Chem 121	**CLASS**	Chem 121	**CLASS**		
1600	Math 162	Chem 121	Math 162	Chem 121	Math 162		
1630	Math 162	Chem 121	Math 162	Chem 121	Math 162		
1700	Math 162	Chem 121	Math 162	Chem 121	Math 162		
1730	Math 162	Chem 121	Math 162	Chem 121	Math 162		
1800	Math 162	Chem 121	Math 162	Chem 121	Math 162		
1830	Math 162	Chem 121	Math 162	Chem 121	Math 162		
1900	***Supper***	***Supper***	***Supper***	***Supper***	***Supper***	***Supper***	***Supper***
1930	Chem 121	Chem 121	Chem 121	Chem 121	Chem 121		
2000	CS 151	Hist 101	CS 151	Hist 101	CS 151		
2030	CS 151	Hist 101	CS 151	Hist 101	CS 151		
2100	CS 151	Hist 101	CS 151	Hist 101	CS 151		
2130	CS 151	Hist 101	CS 151	Hist 101	CS 151		

You will have to take a large number of exams to finish an engineering program. What is your objective in taking an exam? Simply put, it is to make as many points as possible to improve your grade. In technical courses, where you have to work problems on tests, the following strategy will prove very helpful: First, determine whether all of the problems have to be worked or whether you have a choice of the problems to be worked, say, three out of four. You should have determined beforehand, to help in your preparation, whether the exam will be open book or closed book. Many students have the mistaken impression that open-book tests are "easier" than closed-book tests. They somehow feel that they don't need to be as well prepared, since they can always look up information in the book. This is not a valid approach, because the time spent looking things up would be better spent in actually working the problems.

What you want to do is to make as many points as possible. In order to do this, first read the entire test and find what you consider to be the easiest problem on the exam, and then smash it to pieces! Next, pick the next easiest problem for you, and work it. Leave your hardest problem for last. In addition, while taking an exam, make a time budget. In other words, if there are 100 points on the exam and you have 50 min, you have to put points on the board at a minimum rate of 2 per min. If you have a 25-point problem, you cannot allow yourself to spend more than 12.5 min on that problem. Likewise, a 40-point problem must be completed in less than 20 min. Keep track of your time so that you don't have the unfortunate experience of losing yourself in a problem so completely that you fail even to attempt all of the problems.

Many professors will break a test problem down into smaller parts such that a 25-point problem would have a 5-point part and two 10-point parts. When this is done, be sure to work the problem in the order that it is given, as the answer from the first part will probably be needed to work the second part, and so on. If you manage to finish the exam before time has expired, use the extra time to check your answers.

Should your exam be a multiple-choice exam, determine the scoring method to ensure that you won't be penalized for guessing at answers that you don't know. The best way to attack a multiple-choice question is to answer each question in your mind before reading the answer choices, and then look for that answer. If you don't see it, leave that question and proceed to the next one. Your goal is to get as many correct answers down as fast as possible. Then, go back and work on the questions you have previously skipped. Sometimes a multiple-choice question will have an answer of "none of the above" or "all of the above" as the last choice. Be sure that all of the answers fit these categories before selecting one of these answers.

During class sessions, learn how to outline the material presented by the professor if he or she hasn't already done it. The act of outlining will make you an active participant in the lecture. In addition, it will make your notes more valuable when you study the material at a later time. Also, you will find this skill to be very valuable in taking notes at the many meetings you will attend during your career.

One final hint that will prove greatly useful: There are generally many more problems at the end of each chapter in the textbook than it is possible to put on the homework assignments. Most students only read those problems actually assigned. Many authors use the text problems to expand or extend the material in that chapter. By reading all of the textbook problems rather than just the assigned problems, you will frequently uncover information that will help you to better understand the text material. This simple process will give you an advantage over most of the rest of the class.

3 GROUP WORK

If possible, find friends with whom you can work homework problems and ask each other questions. With personal computers so widely available now, you can even do this via the Internet without having to physically meet. It is best to work at some central location if possible. It should be obvious that you will have to learn the material well enough to do the problems yourself on the tests!

Much of today's engineering work is done by groups, so it is imperative that you learn to work together in such a group. In some of your classes, you may have assigned projects and be assigned to a group to work on this project. In working on group projects, each team member is reasonably expected to handle a proportionate amount of the total work. Group projects often fail because one or more team members don't take their responsibility seriously and let their teammates down. In school, this will probably result in a lower, sometimes failing, grade. In the real world, it will probably result in job loss and the acquisition of a reputation for nonperformance that may make it difficult to obtain further engineering employment.

Regarding employment, a good grade point average (GPA) when you graduate will probably help you get your first job. However, after that, your job performance on your previous jobs will be the criterion on which you will be judged for future jobs. After you have been working, nobody cares what your college GPA was.

4 WRITING AND PRESENTATIONS

As an engineer, you will write hundreds of reports and memos and will be expected to make more presentations. This is the reason that almost all engineering curricula have two or three English classes and perhaps a required speech class. If your curriculum doesn't have a required speech class, you should seriously consider taking one as an elective, if it is permitted in your program.

Look for opportunities to practice your presentation skills. If your school has an IEEE student chapter, join and become active. Most IEEE student chapters sponsor a student paper contest each year, and these offer excellent places to practice your presentation skills in a friendly atmosphere.

While many people feel that engineers are promoted primarily for their technical expertise, most would be surprised to find how important writing and presentation skills are in the engineering workplace. When asked what skills they look for in a newly hired engineer, recruiters from major corporations such as IBM or Hewlett-Packard almost always list communication skills as being the most important. Surprisingly, they say communication is more important than technical skills. It does you little good to have great ideas if you can't communicate them effectively.

The important idea here is that these courses, which are frequently taken very early in your program of studies, are not simply "filler" courses, but are of critical importance to your future success. Give them the attention they deserve now, and you will not be disappointed in the future.

5 NETWORKING

We mentioned earlier in the chapter the concept of group work. Another concept closely allied to this is that of *networking*. The dictionary [3] defines *networking* as "the exchange of information or services among individuals, groups, or institutions." What this means to you is the establishment of an extended group of you and your friends who are interested in the same kinds of problems or activities. Your personal network may consist of friends in your classes or friends at remote locations. By using the e-mail and personal messaging systems available on personal computers, you can share information and ideas with friends who live anywhere in the world. For example, you and your best friend in high school may have both decided to go into CompE after you graduate, but you end up at universities on opposite ends of the country. Since all CompE programs are similar, in order to meet national accreditation standards, you will both be taking pretty much the same classes. You will still be able to "network" with your friend over the Internet and discuss homework problems and class concepts as you encounter them.

After you have finished your program and have gone to work in industry, you and your classmates will have gone in many different directions. However, you can stay in touch as you expand your personal network through new industry contacts.

6 ACCESSING INFORMATION

You will often need to find an elusive piece of information, and the better you know how to use the library at your school, the easier it will be to find what you need. When you encounter a new problem, one of the wisest things you can do is to make a quick literature search to find out if someone has worked on the same problem earlier. If so, you can get a jump on solving your problem. All of the world's technical literature is available in your local library if you know how to find it.

The reference librarian will be of great help to you. The librarian's mission is to try to help people find technical information that is hiding in the most obscure places. Make sure that you get to know this information resource on a first-name basis so that he or she can help you whenever you need it.

Much of the recent technical literature is available on-line. One excellent source is the *Engineering Index*, which has abstracts of technical articles written since 1980. Earlier articles are available in written form in the bound editions that you can find at the library. Your reference librarian friend can give you information on how to access the *Engineering Index* on-line. Many other reference sources are available on-line, including U.S. patents. If you think you have an idea that could be patented, it is a good idea to search the patents to make sure someone hasn't beaten you to it.

Many libraries around the world are also found on the Web. Again, the reference librarian can help you find these sites.

KEYWORDS

finding information, networking, presentations, test taking, time scheduling, writing

REFERENCES

[1] *University of New Mexico 1999–2001 Catalog*. 1999. University of New Mexico, Albuquerque, NM.

[2] *Summer and Fall 2000 Schedule of Classes*. 2000. University of New Mexico, Albuquerque, NM.

[3] *Merriam Webster's Colegiate Dictionary*, 10th ed., 1999. Springfield, MA: Merriam–Webster, Inc.,

SUMMARY

In this chapter, we have discussed a number of topics that will help you to not only survive an EE or CompE program, but also enjoy the process. The person who is most interested in your success is *you*. Apply as many of these suggestions on time management, networking, and test taking as you find useful.

We discussed the importance of being able to write well and being able to present your ideas orally. We showed that it is to your advantage to try to find every opportunity to practice your presentation skills.

Finally, we discussed the importance of becoming acquainted with the reference librarian at your school library. In addition, we pointed out how your personal computer will allow you to find information on-line.

Problems

1. Make up a table, similar to that shown in Table 2, for your own classes this semester.
2. Make up a table, similar to that shown in Table 3, for your personal schedule. Be sure to include any time that you work in the schedule.
3. Taking all prerequisites into account, make up a schedule for your entire program to graduation, showing what courses you plan to take each semester or quarter, including summer sessions. If you don't know what specific technical electives you want to take now, just enter "technical elective."

4. Write a one-paragraph report indicating why you decided to study electrical or computer engineering at your school.
5. Determine whether your school library has a special librarian assigned to science/technical/engineering, etc. If so, meet with that person and write a one-page report telling the details of what you found on this trip to the library. If not, meet at least one of the reference librarians and report on that.
6. Determine whether your school library has access to *The Engineering Index* on-line. If so, go on-line and retrieve three abstracts of articles on a subject of interest to you. If your school doesn't have on-line access to this publication, use the printed version to retrieve the three abstracts. In either case, track down the actual journal for one of your three abstracts and copy the actual article. Turn in your three abstracts and your article.
7. Find a list of all of the IEEE technical transactions that are published, and determine which ones are in your school library. Find the physical location (call number and a simple map) of at least five transactions that are of particular interest to you. Turn in a report indicating what you have found.
8. Write a one-page report describing what you see yourself doing in your electrical or computer engineering career ten years from today.
9. Find out if your school has an IEEE student chapter. If so, find out if they have regular monthly meetings and the names of the officers. Prepare a short report that gives this information.
10. Prepare a list of the instructors of all of your classes, their office addresses, and their office hours. Prepare a short report that gives this information.

5

Engineering Design

1 INTRODUCTION

Design is the process of turning ideas into physical reality. Engineers use their knowledge of engineering science, computer science, physics, chemistry, and mathematics to create new products and to build new devices and systems that meet a human need. The new design may be a radical departure from anything that already exists, or it might be an incremental improvement to an existing design.

Engineering design is an inherently creative process. This is perhaps a surprising statement since most people don't think of engineering as creative at all. It is generally acknowledged that artists, musicians, and writers are engaged in a creative pursuit, but the image of an engineer is more of a technologist than a creative spirit. Nevertheless, engineering requires just as much creativity as art or literature. Developing a new product where none existed before or having the insight to know how to improve an existing design requires a great deal of imagination and creativity.

In this chapter, the process that engineers go through as they create a new design will be described. We will also see some of the things that engineers must take into account while they do their design work.

SECTION

1 Introduction
2 Design Versus Analysis
3 The Engineering Design Process
4 Teamwork
5 Concurrent Engineering
6 Other Design Considerations
7 Engineering Communication
8 Creativity
9 Intellectual Property

OBJECTIVES

After reading this chapter, you will

- Be able to explain the differences between analysis and design problems
- Be able to list the nine steps in the engineering design process and describe what is involved in each step
- Understand the concept of teamwork in the design process
- Be able to explain the importance of good communication skills
- Be able to explain the differences among patents, copyrights, and trademarks.

2 DESIGN VERSUS ANALYSIS

There is a big difference between analytical problems and design problems. Analytical problems are very familiar to students. These are the same types of problems that you probably called "word problems" or "story problems" in elementary school. In engineering, analytical problems are often referred to as *closed-ended problems*. In this type of problem there is generally only one correct answer, often given in the back of the textbook. This is the type of problem that students are generally most familiar with and most comfortable solving. It is nice to know that there is a correct answer, and with the proper diligence, it can be found.

Design problems, in contrast, are *open-ended problems*. There is no unique correct answer that everyone trying to solve the problem will get. The open-ended nature of engineering design is reflected in the diversity of the marketplace. For example, there are several companies that manufacture cell phones. They are all trying to solve the same design problem—to produce a portable, handheld, global-communications device that will be economically accessible to nearly everyone. Although each of these companies is attempting to solve the same problem, the solutions they devise are different from each other, sometimes radically so. No two cell phones is identical in function, cost, capability, or appearance. Each one of these designs solves the basic problem, so all are a solution to the problem. But, are all of these solutions equally good? Of course not. Each solution has its good points and its bad points. In this case, the marketplace ultimately decides which solution is the best. Sometimes the lack of a single right answer worries students, making work on design problems harder and more stressful than it should be. Instead, the lack of a right answer should be viewed as liberating. With a design problem, you are limited only by your creativity, and you are not forced to work within the confines of the "right" answer.

3 THE ENGINEERING DESIGN PROCESS

The engineering design process can be described in many different ways. In this book, a nine-step design sequence will be presented. This basic design sequence is equally applicable to the design of hardware and software. The basic steps used to develop a design are shown in Figure 1. It is important to note that this process is not linear—moving from a starting point directly to an ending point. Rather, design is an iterative process sometimes requiring that the engineer revisit previous steps as the design proceeds. This is indicated in the figure by the arrows that connect the various design steps back to previous steps.

The design process begins with deciding on the needs of the customer. In consultation with the customer, a set of specifications for the final product and a project plan are developed. Up to this point, no real "design" has yet taken place, but rather a great deal of planning and thinking about the nature of the final product has been done. The technical work begins with the development of a block design. This is a functional design, still at a very conceptual level. No actual electrical components or lines of computer code are developed at this step. Also, alternative block designs are developed and evaluated. Next, detailed designs (with alternatives) for each part of the block design are mapped out. After much work and evaluation, the best alternative or alternatives are chosen. The design is then tested to verify that it works properly, and the design is turned over to be manufactured. Finally, it is delivered to the customer. Each of these steps will now be described in more detail.

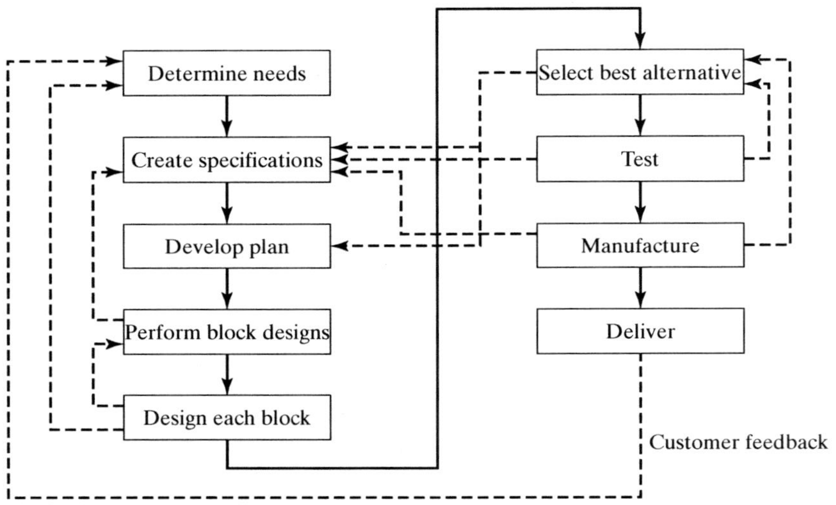

Figure 1. A flowchart showing the steps required to successfully perform an engineering design.

1. Decide on needs
   Having an idea for a new device or product is only the very beginning of the design process. Sometimes this is done by the customer coming to the engineer with a need. Sometimes it is the engineer who has the idea or who sees the need for a new product. At this point, a great deal of thinking and research is done to determine whether the idea is feasible, whether it has already been done, and whether the idea can be sold.

2. Develop product specifications
   In this step of the process, detailed thinking goes on to decide how the device should perform, how much it will cost, what it will look like, and when it will be available. Specifications are often developed in consultation with the customer. Interestingly, the specifications that are laid out at the beginning of the design process are not necessarily the same ones that will exist when the product finally hits the market. Specifications are moving targets. An engineer often finds that one or more of the specifications can't be met and must be changed in order for the project to go forward. Or, the specifications might need to be tightened in order to be competitive in a rapidly changing marketplace. Changing the specifications in consultation with the customer is a normal part of the design process.

3. Develop a project plan
   The project plan is the basic road map used by engineers to efficiently complete a project. The planning stage can often take a great deal of time, but it is nonetheless essential to successful designs. The more complex the project and the more engineers involved in it, the more important good planning is. It sometimes seems like planning is wasted effort and takes time away from actually performing the design. However, careful planning actually makes the design process run more smoothly and efficiently. This is analogous to taking a long trip by car. Most of us wouldn't begin a driving trip from New York to Los Angeles without first consulting a road map. We could

probably complete the trip eventually without planning and without a map. But it's unlikely we would find the fastest route this way. Much time would be wasted in wrong turns and dead ends. Likewise, engineers need to plan how to attack a problem in order to ensure that it will be successfully completed in the least amount of time.

4. Develop a block design
At first, the design is performed at a functional level rather than at a detailed level. This is often done as a block design where functional blocks and their interactions are mapped out. An example of a block design for a radio receiver is shown in Figure 2. At this point the design is still fairly conceptual, with specific details about choices of components made at a later stage in the design process.
In any design it is essential that engineers not limit themselves to just one design concept, but rather develop several ideas simultaneously. The block-design stage is the perfect place to start this. At first, it may seem inefficient and a waste of time to work on more than one design. However, there are several advantages to working on several alternative designs simultaneously. Frequently an idea from one design will be applicable in another design, or ways to combine two designs together to form a better design may become apparent as work on several alternatives continues. Having several fairly well-developed designs can also save time later on. If the design concept that is ultimately chosen turns out to not be feasible, it is smart to have a backup design ready to be implemented in its place.

5. Generate detailed designs of each block
At this stage the engineer is ready to perform detailed designs of each of the blocks mapped out in the previous step. The design at this point is at the component level, with individual transistors, integrated circuits, or lines of software being selected. As with the block-design step, it is important to generate alternative designs at this point, too.

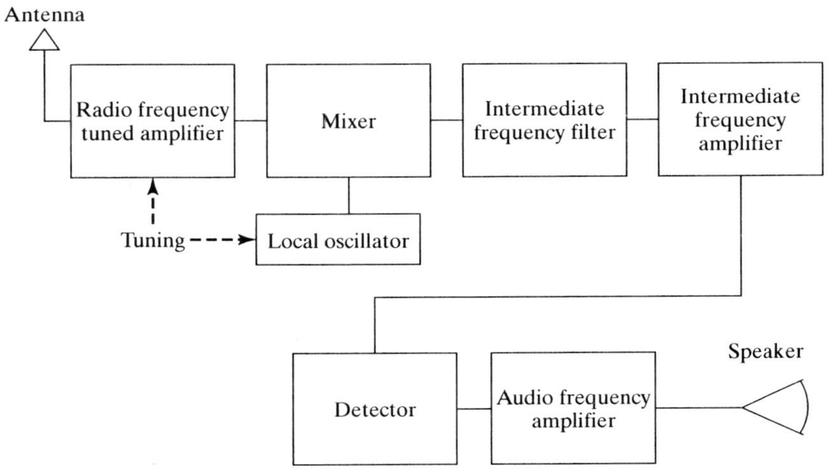

Figure 2. A block diagram for a radio receiver. This is typical of block diagrams used in the engineering design process.

6. Select the best alternative
   After the detailed designs are generated, it is time to select the most promising design from among the alternatives. In order to make an informed choice, engineers will now perform very detailed calculations and analysis to verify that the designs will work as planned. This might also include a circuit simulation using computer software such as PSPICE, which is capable of analyzing complex circuits. Finally, a prototype of the design will be built and extensively tested to verify that it works as planned. A final design will be selected based on this extensive analysis of how the device will perform.
7. Test and verify the design
   After a prototype of the design is built, testing begins. Generally this testing is quite extensive and may include basic tests to see if the design functions as planned and meets the specifications, tests to see how long the design can be expected to last, and also tests to see how it performs in extreme conditions, such as excessive heat or vibration.
8. Manufacture
   An engineer's work doesn't end when he or she has verified that the design is working properly. The design engineer will often work closely with the manufacturing engineers to ensure that the device is fabricated properly and works as planned after the manufacturing process. Also at this stage, design modifications might be required to make the design more easy or less expensive to manufacture.
9. Deliver
   The essential last step is to deliver the finished product or device to the customer.

3.1 A Simple Application of the Engineering Design Process

Without knowing it, you have probably already used the multistep design process outlined in the previous section for making decisions in your life. For example, when you decided where to go to college, you probably went through a version of this process using many of these steps. Some will have used paper and pencil to do a "design," while others may have simply done it in their heads. To help illustrate the engineering design process, let's look at a simple example of how you might use this process in your life—the purchase of a home computer.

The first step is to decide on your needs. You won't necessarily have done this all at once, but you have probably been forming ideas about what you need for a while through seeing what computers other people have, reading about various computers in newspapers or magazines, assessing what you need for your classes, or evaluating the shortcomings of your current computer. Your needs might be for a computer with basic word processing and internet capabilities. You might also want to be able to write your own programs and need a computer capable of running C++. Depending on what you will be doing, you might want to have excellent graphics capabilities.

Once you've decided on your basic needs and made a list of them, you are ready to develop a set of specifications. These might include what type of processor the computer will have and at what speed it will operate. The specifications will include what software will be needed for the tasks you will perform and what kind of graphics card will provide you with the capabilities that you need. And, of course, the overall price that you are willing to pay will be part of your specifications. You probably will want to write down all of these specifications before you begin shopping.

Next you will develop a plan for making your computer purchase. Are you going to shop at electronics stores or a local computer specialist? Are you going to buy the individual components and build the computer system yourself? Are you going to shop for a computer on-line using your current computer? When do you want to have the purchase completed?

It was mentioned before that not every step of the design process is applicable to all situations. In the case of our purchase of a computer system, developing a block diagram and performing a detailed design are probably not applicable. The "design" work was done when you developed your set of specifications. One aspect of these two steps that *is* applicable is identifying alternatives. At this point, you will have decided on one or possibly a few potential systems by different manufacturers that will meet your needs. It is good to have these alternatives in mind in case your first choice is no longer available, or when you "test drive" the system it doesn't perform as well as you thought it would.

The next step is to select the best alternative. This will be based on your research by reading literature produced by the manufacturers, talking with salespersons, or looking on the Web. At this point you might find that it will be necessary to compromise on your specifications. You may not be able to get everything you want in one computer, or you may need to get fewer features in order to get the price that you want. This is part of the engineering design process—assessing trade-offs to arrive at the optimal system.

Testing and verifying your computer "design" is somewhat difficult in this example. You probably can't really see how a new computer performs until you get it home and try out all of its features. Some idea of its performance can be gained from using it at the store where you bought it. Of course, most retailers will let you return the computer up to a few weeks after purchase if it doesn't meet your needs.

The manufacturing and delivery steps in this example consist of unpacking the boxes, connecting the cables, plugging the computer in, and using it!

As you can see, the basic process by which engineering designs are carried out have a great deal of similarity to decision-making processes that we utilize every day. In a sense, you have already been an engineer for many years!

3.2 Example of the Design Process: The Evolution of the Internet

Now it's time to look at the engineering design process as it applies to modern technology. In this example, we will look at how the modern computer network, known popularly as the Internet, was developed. In this example, the design will not be done by an individual or even by people working for the same company. The design of the Internet took many years and the involvement of many different organizations. In describing this, the various steps of the engineering design process will not be spelled out explicitly. Nevertheless, you will be able to see how many of the steps of this process came into play as the Internet evolved.

The Internet started life as a military research project. In the 1960s, while the Cold War raged, there was a concern that during a nuclear war, a few well aimed hits on communication centers could completely disable the military's communications system and restrict the country's ability to retaliate and fight a war. The U.S. government funded studies during the early 1960s on how to solve this problem. The basic idea that came out of these studies was a network with multiple independent nodes. If such a network (or networks) could be built, then when one node in the system went down, the other nodes would still be able to function. Messages could be communicated around bad nodes, rather than be stopped by a nonfunctioning node. In the early thinking about this, each node was to consist of a high-speed computer. This thinking about a multinode network culminated in the funding of a research project by the Defense Department's Advanced Research Projects Agency (ARPA), to determine whether such a network was feasible.

In the fall of 1969, the first node of this network was installed at UCLA, with four other nodes running by the end of the year. The network was called ARPANET. This fledgling network proved its value immediately. Individual computers could be programmed from other sites, and scientists could share their computing resources over the network. The number of nodes in ARPANET grew rapidly with 37 nodes in place by the end of 1972.

One thing became apparent immediately. There was a great deal of enthusiasm among the people with access to the network for using ARPANET as a very fast post-office system. This was the origin of e-mail. Rapidly, most of the traffic on the network became messages, both work related and personal, rather than the shared computer files originally envisioned. In the early seventies, ARPANET grew, aided by its basic structure. The fact that each node was independent of the others and there was no centralized control meant that additional nodes could be added easily. Simultaneously, other computer networks were being developed, most using the same communications protocol as ARPANET.

By 1983 ARPANET had outgrown its military origins. The Pentagon became concerned with the security of ARPANET, especially since it was connected to various other nonmilitary networks. So the military uses of ARPANET were moved onto a different network called MILNET, that was independent of ARPANET. By this time, ARPANET had become only a small part of the overall computer network in the U.S. About this time the network became known as the Internet.

In 1984 the National Science Foundation developed its own network, NSFNET, designed to enhance the communication between scientists around the country. The NSFNET showed continual technical improvement throughout the late 1980s, using higher speed links to connect more powerful computers. Other government agencies got into the act as well, developing their own high-speed networks. Internet development in the 1980s culminated in the disconnecting and retirement of the original ARPANET in 1989. No one even noticed when the "grandfather" of the Internet was removed from the network. The late 1980s also saw commercial involvement in the Internet explode, as many companies and entrepreneurs began to see the vast commercial potential of the Internet.

Growth of the Internet continued throughout the nineties and into the early 21st century. Today the Internet can be accessed from nearly every home in the United States. Children access the Internet in their schools, and most businesses are on the network. There are currently hundreds of thousands of nodes in the current Internet stretching worldwide.

4 TEAMWORK

Most engineering design projects are performed by teams of engineers working together. Often, these teams will involve many different types of engineers. For example, a team formed to develop electronic control systems for automobiles would probably have both electrical and mechanical engineers on it. Having excellent technical skills is essential for any engineer, but to be successful, an engineer must also know how to work successfully as part of a team. Rarely will an engineer be able to work exclusively by herself. In this section, some important aspects of teamwork will be discussed.

Every member of a team brings a different set of skills to a project. These might be different technical abilities or different knowledge. Each team member will also bring to the project different interaction styles and approaches to working with others. For example, some people might be very good at working on detailed technical tasks but not be very good at synthesizing the various parts of a project into a whole. Some people have good leadership skills that help move a team forward, while others are better at working on a piece of the project individually. Some people are good at communicating ideas, while others are better at challenging the thinking of other team members to help develop new ideas. It isn't important that every individual on a team be identical. In fact teams without diversity generally don't function very well. Instead, it is important to have a blend of individual traits that overall will lead to an effective and efficient team. Some companies even administer personality tests, such as the Myers–Briggs type indicator test, to its employees to aid in finding the right blend of individuals when forming teams.

Every team needs people who are competent in the technologies pertinent to the project, but successful teams also need people who are task oriented, who strive for high quality, and who are goal oriented. Teams also need people who are flexible and who can see the "big picture," rather than just the details. All teams need good communicators who can ensure that written and oral reports convey the required information and who can also ensure that everyone on the team understands the project goals. The communicators also facilitate the participation of everyone on the team. It is nice to have someone on the team who is constantly challenging the methods and ideas of the group and the way the project is being handled. All of these traits don't generally show up in any one individual or in all of the team members. On effective teams, there is a blending of personality types that results in all of these traits being present in the team as a whole.

What are some of the characteristics of effective teams? Effective teams are those that have good leadership and a clear purpose. They work informally and have open discussions with an agreement that any disagreements will be civilized and not personal. In good teams everyone participates and everyone listens to other team members. There are very clear roles and work assignments for every team member. And generally, effective teams have a blending of different personal styles.

5 CONCURRENT ENGINEERING

In the previous section it was mentioned that engineering design projects are generally performed by teams of engineers. Often, these teams are multidisciplinary in that there are different types of engineers—electrical, computer, mechanical, or manufacturing—working on the same team. In most corporations today, the engineering design process is performed using a method called concurrent engineering, which is the topic of this section.

In the typical old models of the engineering design process, various tasks were handled sequentially. Engineers were grouped according to specialties, mostly within academic disciplines, such as electrical or mechanical engineering. For example, electrical engineers working for a large corporation might be grouped into departments with names such as digital design, antenna design, or analog design. They would work mostly with other electrical engineers. Of course, the successful completion of a project would require that the engineers interact with other engineers from different departments, but this was typically done at meetings between the two groups.

To illustrate how this system worked, let's look at a part of the design process for a hypothetical computer. The initial block diagram for a new computer would be done by an electrical engineering group. This design would then be passed along to a software design group, which would develop the initial block diagrams for the software. After the software blocks were ready, the electrical engineering group would make the necessary modifications to their block diagrams and begin the detailed design. Simultaneously, the software group would design and develop the required software. Once the detailed electronic design was completed, the plans would be sent to the group who designed the case for the computer. If any problems occurred, there would be meetings between the two groups to reach a solution. Once the plans for the case, the hardware, and the software were completed, the design would be passed to the manufacturing group to ensure that there were no problems with building the system as designed. This process continued with various groups becoming involved as needed until the final design was produced.

This model for the engineering organization had its efficiencies. For example, grouping engineers together by discipline allowed engineers to specialize and become very good at their area. They could also easily help each other out—there was generally another engineer of the same type sitting at the next desk. The disadvantage of this system was that often the time required to complete a design was very long, slowed down by the need to pass the design along from one group to another as new aspects were added. There was not always a group of engineers who had the "big picture" in mind.

The concurrent engineering model changes this completely. In concurrent engineering, the design is performed by multidisciplinary teams. A multidisciplinary team tasked with designing a computer might include digital circuit engineers, power supply engineers, plastics engineers, and software engineers working together on the same team from the beginning of the project. Generally, manufacturing engineers are included on the team as well, to ensure that manufacturing problems are addressed at the beginning of the design process rather than after the design is completed and is ready for manufacture. With concurrent engineering, design teams often include nonengineers as well. Marketing experts or purchasing specialists might also be included on the team to avoid problems in these areas. Sometimes, customers are even invited to participate in the design process as team members to help ensure that the final product will be acceptable.

Concurrent engineering has been shown to greatly reduce the time required to bring a new product to market. This reduction has come about chiefly through addressing issues early in the design cycle that would otherwise not be addressed until the "other" engineers were sent the design. Although the changeover to concurrent engineering has not taken place everywhere yet, most major engineering employers now perform design according to this model. It is important, even at this early stage of your career, that you learn to appreciate what other types of engineers do and learn how to work on multidisciplinary teams.

6 OTHER DESIGN CONSIDERATIONS

There are many aspects to a good design that don't necessarily show up in the product specifications. Even though these might not be specifically mentioned when a project is initiated, a good engineer will keep these in mind as a design project progresses. Many of these fall under the heading of "accepted engineering practice." Engineers are expected to make all of their designs in accordance with this standard. A few of these other considerations will be discussed in this section.

6.1 Safety

Often, issues of safety are not explicitly mentioned in the specifications for a new product. However, it is expected that all products meet minimum standards of safety. It is neither legally nor ethically acceptable for an engineer to participate in a project that is not safe. The code of ethics of the IEEE states explicitly that an engineer is responsible for ensuring the safety of those who will use his or her designs.

Deciding what is safe and what is not is a big challenge for engineers. There are many legal requirements and safety standards that an engineer is required to follow, and these provide a good foundation for producing safe designs. Often, a design is in a new area where little is known about safe design. How does an engineer deal with this? First, an engineer must recognize that nothing is 100% safe. There are ways that any product can be misused, and there are often unanticipated failures in a design. It is the engineer's job to try to anticipate and prevent these things.

How can products be engineered to be safe? Making a design safe should be a natural outcome of the design process. As was mentioned in the basic steps of design, several alternative designs should be generated as part of normal engineering. As each of these designs is evaluated for its ability to meet the specifications, it should also be evaluated for its impact on safety as well. As the decision is being made regarding which alternative to pursue, safety should be given equal weight with all of the other design considerations.

6.2 Environment

As with safety, the protection of the environment is also an important consideration in any design. The IEEE code of ethics requires electrical and computer engineers to produce their designs in accordance with sound environmental principles. This includes ensuring that the generation of hazardous wastes during the production process is minimized. It is also important to ensure that once the product has exceeded its useful life, there are ways to recycle it or to minimize its negative impact on the environment during disposal.

6.3 Design for Testability

Often, the specifications for a product will spell out the test procedure that will be used to verify that the product works as it is supposed to. It is up to the design engineer to ensure that the product is designed in such a way that the testing can be performed easily. It's a good idea to place easily accessible test areas throughout the device, especially on the prototype. This makes it easy to troubleshoot the design or even fix it later. By incorporating testability in the original design, the whole design process can be made more efficient.

6.4 Design for Manufacturability

In today's marketplace, where new products must be brought out in increasingly short times, it is essential to incorporate manufacturability into designs from the start. One of the major costs of producing a product is the cost of assembly. There are many designs that amply meet the specifications, but are not easily assembled. Anything that is complicated to assemble requires specialized and expensive workers, adding cost to the overall design. So design engineers must work closely with the manufacturing specialists to incorporate manufacturability into the design. This often means that at the stage where design alternatives are being evaluated, one criterion that should be used is manufacturability, even though it is not likely to be mentioned in the specifications.

6.5 Esthetics

Esthetics in a design are often felt to be in the realm of artist, not engineers. However, as mentioned before, engineering is an inherently creative enterprise, so engineers should also consider the esthetic aspects of their design. Much of the pleasure of being an engineer comes from seeing your design work well and succeed on the marketplace. Engineers often describe the "warm feeling" they get when they first plug in their new design and see it work. Much satisfaction can also be derived from performing designs that are pleasing to the eye. This can involve laying out components in a pleasing manner or helping to design a unique looking case for a new product. Although esthetics are rarely included in the specifications, it is easy for the engineer to include this in her design and to use it as one of the criteria, although perhaps a lower level one, when choosing among alternative designs.

7 ENGINEERING COMMUNICATION

Having good technical skills is almost worthless without also having the ability to effectively communicate the results of your work. Frequently, engineering students believe that only their technical skills matter, and therefore they don't focus much on courses involving communication, such as writing and composition or speech. This view could not be more wrong. Engineers who wish to advance within their profession must also possess high-quality communication skills. Many of the interactions that engineers have with their manager, and nearly all of the interactions with higher levels of management, occur through memos, reports, or formal oral presentations. Engineers who don't present themselves effectively on paper or in a presentation will be the last to receive promotions and indeed will find it hard to get hired in the first place. Engineers are also responsible for interacting with customers, suppliers, or government agencies. Doing this effectively requires good communication skills. Engineering information is communicated through various types of written documents and drawings. In this section, we will look at various types of engineering communications in depth.

7.1 Engineering Logbooks

One type of communication that engineers do is designed only for the engineer himself. Logbooks, sometimes also called lab books, are a means for recording all the work done on a project in a single place. The engineer uses the logbook to record all calculations that were made, detail all of the design work, record test data, and even to record information about meetings and contacts with suppliers or customers. Generally, the engineer is the only one who will see the logbook, although some companies insist on keeping logbooks when an engineer leaves the company.

At first, it may seem tedious and inefficient to do all of your work in a logbook. However, it is generally very helpful later on to have all of the data pertinent to a project recorded in a single place. This makes it easier to find information about the project later on when the details about it may be hard to remember. Keeping information in a single book is far superior to keeping it on individual sheets of paper that can become lost or destroyed. One important function of a logbook is to satisfy the legal requirements for defending a patent. If there is a dispute about who has invented something first, courts often examine logbooks. If the logbook appears complete, with dated entries, and nothing erased or removed, it will aid in establishing whose work came first.

7.2 Memos

Perhaps the simplest and most often used form of engineering communication is the memo. Although memos tend to be short, the fact that they are used so often makes them a very important document for engineers. Memos are used for a variety of purposes, including setting up meetings, soliciting information from other engineers or suppliers, and briefly communicating the status of a project. Memos might be written to your manager or to other members of your organization. Memos are generally written by the individual engineer.

7.3 Progress Reports

Often, a customer or manager will require that the engineers provide periodic progress reports. These reports detail what has happened during the reporting period and what work is contemplated for the next period. This type of report provides good channels of communication between the engineer and his customer or between the engineer and management, and it allows for timely reporting of any problems that may come up. Periodic reports may be weekly, monthly, yearly, or any other convenient period.

Depending on the type of report and to whom it is sent, an engineer generally prepares this report herself.

7.4 Feasibility Studies

A feasibility study is a much longer document than the previous two. Early in a project, engineers might be required to assess the feasibility of a project and write a feasibility report on their results. A feasibility study requires a lot of background work and preliminary design to answer the important questions about whether the contemplated project can be done, whether it makes economic sense to pursue it, and whether it can be done in a reasonable amount of time. A feasibility study is often sent to a potential customer or upper management of the engineer's company.

7.5 Proposals

In order to begin a project, engineers must often complete a proposal. This is a detailed document laying out the plans and costs to complete a project. Proposals frequently go out to potential customers or to higher levels of management within the company to obtain permission and resources to start a new project. Generally, a customer has received several competing proposals. Although technical considerations are usually most important, the proposal that is most effective in presenting the ideas will garner more consideration. In many companies, especially larger ones, engineers work closely with technical writers to produce high-quality proposals.

7.6 The Engineering Design Report

After a project is completed, an engineering design report is frequently prepared. This is a detailed report on how the design was done and how well it worked. The report might be prepared for the customer who paid for the design, or it might go to the upper management of your company. The engineering design report contains information on the problem being addressed and any background information that was used in working on the design. The report provides information on how the design was performed, how it was tested, and what test results were obtained. The engineering design report might also include recommendations for improvements to the design.

The engineering design report is valuable to the customer, since it gives the information required to modify the design at a future time or to make any needed improvements to the design. It might also be useful to the engineer's own colleagues who might need to do a similar project in the future. It can even be useful a few years later when the engineer is asked to do another similar design; having the old report helps prevent "reinventing the wheel." In larger companies, technical writing specialists will be responsible for writing this report, although the engineers who worked on the project will help with the writing process.

7.7 Technical Manual

Often, a customer will ask for a technical manual, sometimes called an owner's manual. This differs from the engineering design report in that the focus is on how to operate and maintain the device rather than on how the device was designed. A good technical manual will teach the customer how to use the device properly, will show the customer how and when to perform routine maintenance, and will allow the customer to perform simple troubleshooting and repair should there be a problem. Like the engineering design report, the technical manual is generally prepared by a technical writer with significant input from the engineers.

7.8 Engineering Drawings

Frequently, the best way to convey engineering information is through a drawing. The most commonly used type of engineering drawing in electrical and computer engineering

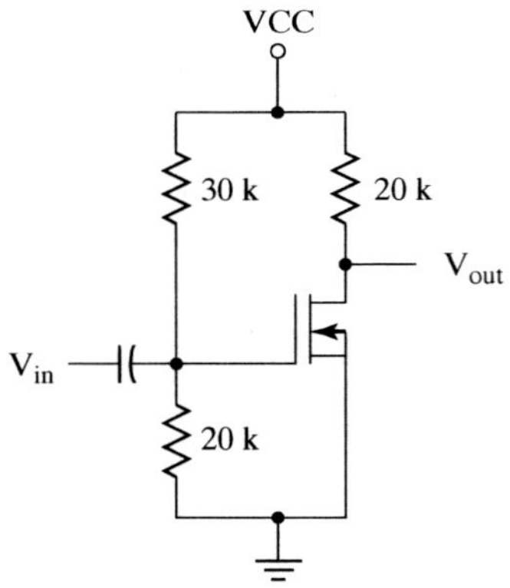

Figure 3. A schematic diagram showing the layout of a circuit. This schematic was generated using PSPICE, a circuit simulation software.

is the schematic diagram, which is a symbolic representation of electrical components and the way they are interconnected. An example of a schematic drawing is shown in Figure 3. Schematics show what components are to be used in a design, how they are interconnected, and what types of voltages or waveforms should appear at various places in the circuit. Schematics allow the designer to visualize his thoughts about the design, and also to allow other engineers to understand a circuit and technicians to properly build the circuit. A schematic drawing is indispensable for anyone trying to repair a circuit that is not working correctly.

7.9 Oral Presentations

Engineering communication can take the form of oral reports. These can be as simple as an informal presentation at a department meeting, or as formal as a talk at a meeting of an engineering professional society. There can also be oral versions of many of the written documents that were described previously. For example, an engineer might be asked to give an oral proposal to a potential customer or to higher management in addition to the written proposal. These oral reports are generally fairly formal. It is important for an engineer to be able to communicate effectively orally as well as in writing.

8 CREATIVITY

It was mentioned earlier that engineering is an inherently creative profession. Bringing a design from concept to finished product requires new solutions and new ideas. Although some people seem to naturally be highly creative, everyone can learn methods to help enhance their creativity. In this section, a few ideas on opening up creativity will be presented.

We will start by discussing the implications of the *split-brain theory* on the creative process. The human brain is divided into two halves, called for convenience the right side and the left side. Scientists who have studied the brain have found that various human activities are housed on different sides of the brain. For example, in right-handed persons the left hemisphere of the brain tends to concentrate verbal and symbolic logical reasoning skills. The right hemisphere concentrates the spatial and holistic reasoning processes. (This is opposite in left-handed people.) Educational processes are dominated by left-brain learning, mostly because these types of activities are easier for a teacher to evaluate and measure. Most of the courses you will take in engineering will emphasize these types of left-brain activities. However, engineering design also requires right-brain processes, especially holistic analysis and synthesis.

Let's look at a few words that can be used to describe the two sides of the brain. Attributes of the left side (in right-handers) are precision, logic, linearity, and order. The right side can be described as experimental, imaginative, creative, and risk taking. Clearly both sides of the brain will be necessary to produce a successful engineering design. Not only are the right-hemisphere skills not generally taught or developed in traditional educational settings, but some would argue that these skills are suppressed. How, then, can an engineer develop some of these skills?

To answer this question, let's first look at some things that can stifle creativity.

1. *Habits*. Often, we are so set in our ways that we don't entertain any creative or new thoughts for very long.
2. *Fear of failure*. Sometimes, we are so scared that we will fail or our ideas will not work, that we suppress these new ideas.
3. *Cultural blocks*. Often, the culture of our organization does not value creativity, but instead emphasizes doing things the same way they have always been done.
4. *Narrow-mindedness*. Many people think very narrowly and can't "think outside of the box" to look at things in new ways.
5. *Negativity*. Many good ideas have been stifled by colleagues who feel threatened by new ideas or who are not themselves capable of creative thinking. You have probably already encountered many of these stiflers of creativity in your life.

The first step to opening up creativity, both individually and for a team, is to recognize the stiflers of creativity that you encounter. Once you understand their nature, many can be overcome or worked around to help open up creativity.

Other solutions range from the personal to the institutional. An easy personal way to open up creativity is to spend some time eating and writing with your nondominant hand. By exercising the other side of your brain this way, you can often stimulate the underused creative side of your brain. Another easy way to open up creativity is to engage in "information gathering." Look at a diverse set of materials to get new ideas and to broaden your thinking. This can be done by looking at different magazines or journals than you normally do, by surfing the Web, or just by asking questions of new people.

Opening up creativity in a team environment is sometimes a little harder, since there are more people involved. The best way to accomplish this is through a process called brainstorming, where the team meets for the sole purpose of coming up with new ideas. The idea is for the team to be freewheeling and to let the new ideas flow. To accomplish this, it is important to not criticize or evaluate the ideas during the brainstorming session, as this might inhibit some team members from contributing. There is plenty of time to evaluate the ideas later. Even wild ideas that would never work may stimulate a useful idea from a different team member. After the ideas have dried up, then the evaluation period begins. Here some ideas are thrown away, while others are combined or modified. Although this may seem like an inefficient use of time, the best new ideas often come out of brainstorming sessions like this.

9 INTELLECTUAL PROPERTY

In the course of design work, engineers often generate intellectual property, which is information that gives the company a competitive advantage over other companies in the same business. Often, when a new company is starting and seeking investors, the first thing potential investors look for is whether the new venture has sufficient intellectual property to be competitive and survive. Intellectual property takes many forms, including information about new designs and processes, financial practices of the company, or even information about a corporation's customers and suppliers. Corporations protect their intellectual property by various legal means, including maintaining trade secrets, patents, copyrights, and trademarks.

Trade secrets are perhaps the most fundamental type of intellectual property. Keeping information within a company prevents any outsiders from benefiting from it. Much of the information that an engineer handles in the course of her work will be labeled "proprietary," "for internal use only," or "secret." When hired by a company, employees frequently are required to sign an agreement specifying that they will not divulge this sort of information to anyone outside of the company. A legendary example of this (and it may be more legend than true!) is the formula for Coca-Cola. It is said that this formula is a closely held trade secret, known to only a handful of people. Similarly, the computer chip manufacturers such as Intel closely guard information on chip yields (the number of good devices on a semiconductor wafer compared with the total number of devices fabricated) so that the competition will not know how well their processes work. If a secret is successfully held within a corporation, this is the best way to protect intellectual property.

A patent is a grant by the United State government (or other governments) giving an inventor the right to exclude others from making, using, or selling the patented invention for a specified period of time. Currently, this protection lasts for 20 years from the date that a patent is first filed. Although a patent gives exclusive rights to the inventor, sometimes obtaining a patent is not the best way to protect intellectual property. The government requires that all patents be published and accessible to everyone. By doing this, you tell the competition what you are doing and allow them to invent their own variations on your patent. Patents are granted to the original inventor, who must be a person, not a corporation. Rights to a patent may be assigned to a corporation, and indeed most companies require employees to assign patents that are generated in the course of their work to the corporation.

There are two types of patents: design patents and utility patents. Despite the name, design patents cover only the appearance of an invention, not its structure or utility. Design patents only run for 14 years. An example of a design patent would be one for the unique appearance of a computer monitor, but the patent would not cover the electronics inside.

Utility patents cover mechanical, chemical, or electrical inventions and can include devices, products, and processes. What can a utility patent be issued for? New and useful processes, machines, manufactures, compositions of matter, or biotechnology can be patented. New and useful improvements on things that have already been patented may also be patented. In order to be patented, things must be useful and operable. Perpetual-motion machines violate the basic laws of thermodynamics, are not "operable," and therefore are not patentable.

What cannot be patented? There are many categories of things that cannot be patented. Devices that have already been invented by someone else (even if they didn't bother to obtain the patent) and devices already for sale or described in a publication more than one year prior to the filing date of the patent cannot be patented. Methods of doing business cannot be patented, nor can the basic laws of science. (Imagine the impact on society if someone were granted a patent for Newton's laws of gravitation!)

Recently, the U.S. patent office began issuing patents for computer software. In the past, software was not patentable, yet, software companies felt the need for some protection for their products. It was never clear whether copyright law was the proper protection for software either. This is an area of law that is still developing and will be unclear for a few more years.

Copyrights exclude others from copying or using your creative work. Copyrights cover books, works of art, music, photographs, and even notes. This type of intellectual property protection is less often used by engineers, although as mentioned before computer software is sometimes protected through copyright. Copyrights have much longer terms than patents, generally for the lifetime of the author plus a fixed number of years thereafter.

A trademark is a mark, word, or symbol that is applied to commercial goods. If the commerce is across state lines, then federal registration of the trademark is possible. Trademarks are unique symbols or product names that a company wishes to retain for its own exclusive use. Trademarks are granted for a period of 10 years, but can be renewed indefinitely as long as the company keeps using it. Trademarks are more often of interest at a corporate and marketing level and rarely are important to an engineer's work.

KEY WORDS

Analysis	communication	design	intellectual property
teamwork	creativity	design process	

SUMMARY

Design is one of the most fundamental activities performed by engineers. In this chapter we have seen how engineers take a design from the initial idea to a finished product. We have discussed some of the important aspects of engineering design including safety, testing, and manufacturability. We have also seen the many ways in which engineers communicate their designs to their customers, managers, and coworkers. Many designs lead to the development of new ideas, concepts, or products that are the intellectual property of the engineer and his employer. We discussed ways to protect this intellectual property.

Problems

1. Apply the steps for the engineering design process to the purchase of an automobile. Explicitly say how each step applies to this purchase.
2. Gather some information (use the Internet and your library) on the development of the personal computer. How did this process fit into the engineering design process described in this chapter?
3. Describe how the development of the Internet fits into the nine-step engineering design process.
4. Create a design for a system to convert your house or apartment building to run totally on solar energy. To do this, analyze the power bills from the last year to determine what your energy needs are. Then, use resources on the Internet or at a library to design a system. Is it economically practical to make this conversion?
5. Study the development of a common consumer electronic device such as the cell phone, VCR, or parts of a computer such as the hard drive or modem. How did this process fit into the engineering design process described in this chapter?

6. Study the current state of the art in electric automobile technology. Compile a list of the current limitations of this technology. In groups, hold a brainstorming session to generate ideas to overcome these limitations.
7. Do a patent search on a modern electronic technology, such as the cell phone or computer hard drives. Patent searches can be performed in your library or on-line at the IBM corporate web site, www.ibm.com. Write a report on the significant developments in this technology over the last year.

6

Engineering and Communication

1 I AM AN ENGINEER, NOT A PROFESSIONAL COMMUNICATOR

Maybe nobody has told you yet, but the truth is this: To be an engineer is to be a technical communicator. Engineering is a problem-solving profession, and clear communication leads to the effective solutions. As an engineer, you will solve problems for people (often, large groups of people). Whether your specialty is civil, mechanical, electrical, chemical, biomedical, computer, aerospace, or any other engineering discipline, you will develop products, services, and built environments needed by some segment of the public. In order to produce those goods and services, you will work collaboratively with dozens of other engineers, experts, and support personnel, designing, analyzing, refining, and building solutions. You, your collaborators, and your clients will have to understand exactly what each of you is doing every step of the way. That is why communication skills are critical to engineering work.

Communication means writing, speaking, and showing visuals to other people, as well as listening to and observing others and reading their work. Of course, most of us think of writing as the most difficult skill to master (although, for some of us, speaking in front of a crowd seems not only difficult, but also terrifying). Yet, writing well is critical for communicating your ideas and solutions. Let us say you have designed a new type of bus shelter for a large city in a rainy part of the United States. Your design is rendered very professionally, either in CAD or as an artist's rendering, and you have submitted the design to the municipal transportation authority. In addition to presenting the

SECTIONS

1 I Am an Engineer, Not a Professional Communicator
2 Writing Demonstrates My Competence as an Engineer
3 Writing and Speaking Can Help Me Discover What I Really Think
4 My Reader or Listener (My Audience) Is Always My Client
5 It Is My Responsibility to Turn Data into Information My Readers Can Use
6 My Audiences Will Change, and so Must My Writing and Speaking

OBJECTIVES

By reading this chapter, you will learn the following:

- why writing and speaking well are critical to your success as an engineer;
- how writing and speaking can help you better understand your work as an engineer;
- for whom you will be writing at different times;
- why it is important to present data in a useful format;
- how to write different parts of a document for different audiences.

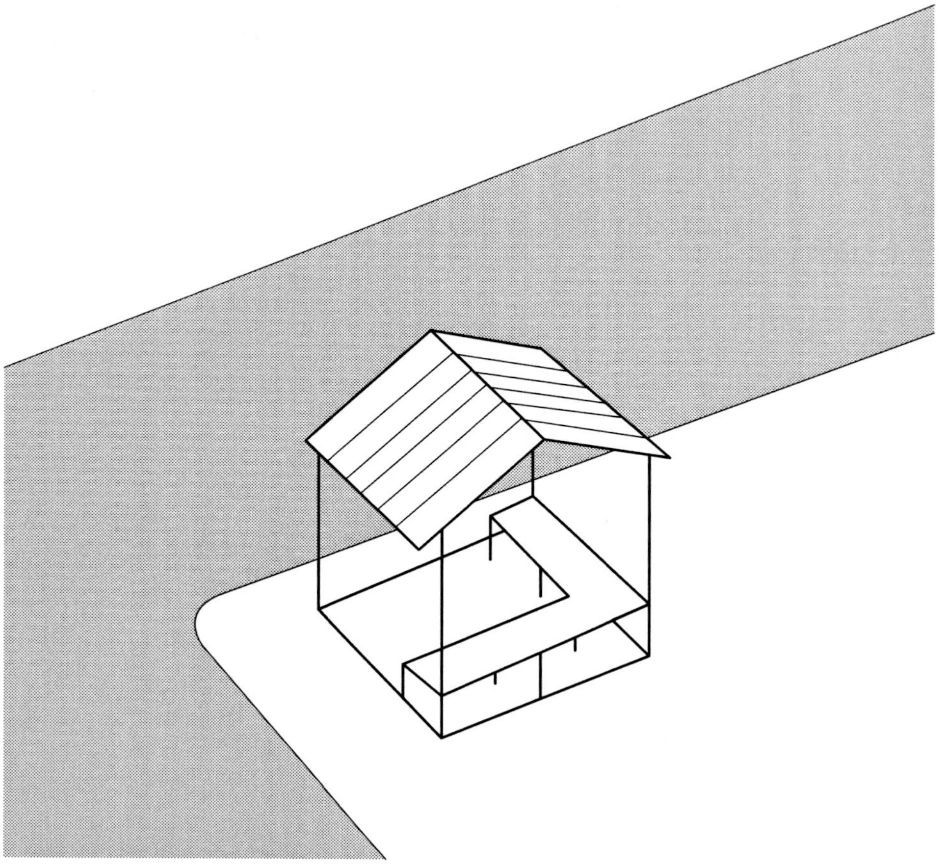

Figure 1. Artist's rendering of proposed bus-shelter design.

design (see Figure 1), you will want to explain in a proposal the main features of the design and the criteria you used to develop it, such as the following:

- Keeps out rain effectively
- Allows enough openness to discourage theft and crime
- Trash does not collect inside

You will use pictures and words to argue that this design should be accepted and built. You may also include a line graph displaying the drop in bus-shelter crime experienced by another city that adopted a similar design. But your design improves on the previous one by being less costly: It can be assembled away from the site and installed on-site within a matter of hours. Sounds good . . . but only as good as you make it sound with your words and make it look with your visuals.

Communication is a part of almost every phase of engineering work. This chapter presents some basic tenets to remember throughout your career.

2 WRITING DEMONSTRATES MY COMPETENCE AS AN ENGINEER

Someone can point to a proposal you write for, as an example, a new type of bus shelter and say, "See what a good solution this is; see this clear explanation of how this design solves the problems." Your competence as an architectural engineer is proven by that proposal. The same concept is true for all the engineering disciplines. Inside your own organization as well, writing demonstrates that you understand the work you do and the jobs of the people with whom you work. Even your informal writing, such as memos and e-mail messages, demonstrates this understanding. When you give people truly useful information, you have shown that you know how they will use that information; otherwise, the information would not be useful. People read information in order to discover what they need to do. Whether your intention is to persuade or simply to inform people, your writing must clearly show them what actions are suggested or required.

RESEARCH SHOWS...

Readers often understand technical writing by forming a "concrete story or event" in their mind. They do this with even the most technical of documents, because they need to *use* the information in some way. They need to know "agents and action." So, the readers in one study actually *restructured* the information to be more active (Flower et al., 1986, p. 45).

3 WRITING AND SPEAKING CAN HELP ME DISCOVER WHAT I REALLY THINK

In the process of producing solutions, you will use many different kinds of oral and written communication. As you move through your career, no matter where you work—whether a large Fortune 500 firm, a small consulting firm, or a public agency—writing and speaking can help you plan, develop, and revise your engineering solutions. Writing can solve problems for you as well as for your reader.

Imagine, for a moment, that you must do some research for your boss on nanocrystalline coatings that protect microchips. There are many of these new coatings on the market, and your company needs information on similarities and differences among coatings before investing research-and-development time and money in adapting a coating for use on its chips. What do you do? How do you start? You probably start by reading as much as you can, collecting information on coatings and trying to organize it. Then you have to look for similarities and differences and make judgments about which coatings fit better with your company as products. How do you capture what you are learning? You probably take notes and make summaries of informative articles and websites. So, you are already writing, even before you have started drafting the report you have to produce. You are also probably talking to people, asking questions and getting their feedback on your findings.

When you do start drafting the report, you probably begin with the lists you have been making of each coating's qualities (strengths and limitations). If you do not have enough information about a particular coating, you will notice, because your list will not be complete. The simple act of putting down words shows you what you need to find out more about. And as you start drafting and making an argument for one kind of coating over the other, you may notice that you never say why, for example, coating 3 will not work. Your argument is thus not complete, because you had assumed early on in your investigation that coating 3 would never work, but forgot to explain why. And then when you start to explain why, you realize that you were mistaken about one of the attributes of coating 3, and it might work after all.

In these and many other ways, writing helps you see your own thoughts in front of you, so you can clarify and finish them. Without writing, you probably cannot know what you understand and what you do not. Just remember that writing happens over time, as a recursive process of outlining, drafting, and revising. You have to give it time. Computers can help us create and revise documents faster, but they probably cannot help us think faster. And your writing is simply the clear arrangement of your thoughts—after you have rearranged them a few dozen times. Remember that when you see someone else's writing, you are seeing a final product; you cannot see the hours of drafting and thinking that went into that written product. So, do not think that most people write faster than you. Almost no one writes very quickly!

EXPERTS SAY...

According to engineer Henry Petroski, there is a clear connection between the practice of engineering and the practice of writing, since both involve creating something new: "Writing no less than engineering consists of ideas to be realized" (p. 10). In his book *Beyond Engineering* (1986), Petroski draws a parallel between an engineer as a bridge builder and as a writer when he describes an engineer as "the designer of a bridge of words" (p. 10). He also explains how writing about bridge building helps the engineer understand the actual process of bridge building better, because writing is "the test of an engineer's understanding of his own theoretical work" (p. 11).

4 MY READER OR LISTENER (MY AUDIENCE) IS ALWAYS MY CLIENT

As an engineering professional, to whom will you be writing and speaking? Those of you who have worked in engineering-related firms, private or public, know that, as an entry-level engineer, you make reports (orally and in writing) to your boss, to colleagues, and sometimes to middle managers from other divisions or offices. For the first few years of your career, you probably will not be writing directly to outside clients. In fact, one survey found that, three years after graduating with a B.S. degree, most engineers were writing more often to people inside their organization than to people outside it (Anderson, 1986). So, in light of this finding, why should you consider your reader a client? Because the other finding from this study was that most of those in-house readers knew less than the writer about the subject matter. So, if you are the writer, your job is to help your readers understand the subject matter and act on that information.

Treat your reader or listener like a valued client—someone with whom you want to maintain a positive professional relationship. And think about how to talk to that client, in writing or by speaking. For example, suppose that you have to present that design for a bus shelter to the city council. How do you explain a design to people who may not look at designs very often? City councils are usually composed of ordinary citizens from many walks of life who happen to care about their city. Do you show them a plan view of your design? A schematic? Will your readers know how to "read" these visuals? Should you refer to the "postmodern use of space"? Will your readers understand that terminology? You would want to examine your communication options ahead of time, including vocabulary, visuals, and even tone of voice.

RESEARCH SHOWS...

In the workplace, chances are good that you know more than your readers—even in-house readers—about the subject matter of your writing (Anderson, 1986). This means that you must explain technical processes, concepts, and vocabulary more than you might think.

5 IT IS MY RESPONSIBILITY TO TURN DATA INTO INFORMATION MY READERS CAN USE

Engineers deal with lots of data; they collect and generate test results, experimental results, specifications, standards, all sorts of numbers, and raw facts. In order to make your communication useful for all your readers and listeners, you need to help them understand the meaning and the relationships of your data. Data are just raw facts. For example, consider the statement, "It is 100°F outside." That piece of data will not be very informative to a child who does not understand what "degrees" means or to a visitor from Indonesia who is not familiar with the Fahrenheit system. Data are bits of information (test results, raw facts) that must be interpreted, organized, and synthesized in order to make sense and to have meaning. As one cognitive scientist put it, "Data plus meaning is information" (Reeves, 1996). Your job is to turn the data you produce into useful information by anticipating the needs of your reader. Sure, many of your readers will be technical folks like you, but they still need to know the relationships among the data you present.

The information, for instance, that air density changes with temperature can be demonstrated only by presenting the correlation between air density and temperature at successive measurements. You would probably use an x–y scatter graph, with a line drawn through the series of points plotted, to convey this information, as in Figure 2. You would *not* give the temperature readings in one paragraph and the corresponding density readings in another paragraph! You could present the data that way, but doing so would make it way too hard for the reader to perceive how rapidly air density changes with temperature (especially by contrast with water density, for instance). And you would be very far from imparting knowledge to your reader that he or she could use later on. The x–y graph provides the best context (a visual context) for the reader to understand the correlations. And then your words can enhance and support the visual context:

"At standard atmospheric pressure, the density of air drops steadily with increasing temperature." Graphical and verbal presentation of data, working side by side—bingo, you have created information, and maybe even knowledge, your reader can use.

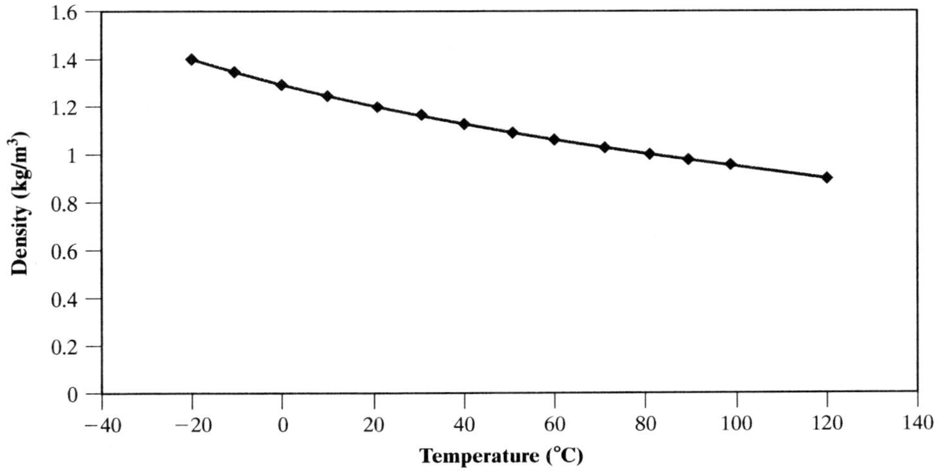

Figure 2. Relationship between density and temperature of air at standard atmospheric pressure. *Source of data:* Crowe, C.T., et al. 2001. *Engineering Fluid Mechanics*, 7th ed.

6 MY AUDIENCES WILL CHANGE, AND SO MUST MY WRITING AND SPEAKING

One of the challenges of learning how to write for engineering work is that, in school, you are not writing for a typical engineer as audience. You professors and teaching assistants are not your boss; they are not paying you to solve problems. Neither are they typical readers of your engineering documents. Unlike your future supervisors, clients, and colleagues, your instructors want to know how much you have learned about what they already know. So it is sometimes difficult for instructors to respond to your writing the way typical readers might, with thoughts such as these:

"What does she mean by 'picocuries'?"
"I thought he was going to give me credit for my part of the design."
"This looks too long to read."
"Doesn't she know we don't use that pollutant-removal process here anymore?"

The reality is that engineers must communicate daily with listeners and readers who have a variety of educational and professional backgrounds. To give you an example, assume for the moment that you are an environmental engineer. For a final report on your study of methods to clean up a particular hazardous-waste site, you may be addressing regulators who are familiar with some of the terminology in your discipline ("*in situ* remediation," for instance), but who do not have a strong background in the science of sampling or of assessing risk. If you then write instructions for the technicians who will implement the cleanup, you will have yet a different audience, one who may not understand the regulatory terms or the science, but who will need clearly written procedures for handling hazardous materials. And sometime later, you may need to help write a description of the remediation process for local news reporters who know very little about engineering. See examples of these three styles of writing in Figures 3 through 5. Each of these audiences has different needs for your documents, needs that must be met if your documents (and presentations) are to communicate meaningfully and get the particular job done.

How many different audiences will you have in your career as an engineer? The answer will depend partly on whether you go into management or not. Chances are that you will become a manager within five years of graduation from college. At the very least, you will be writing for and speaking to people whose role in your organization falls

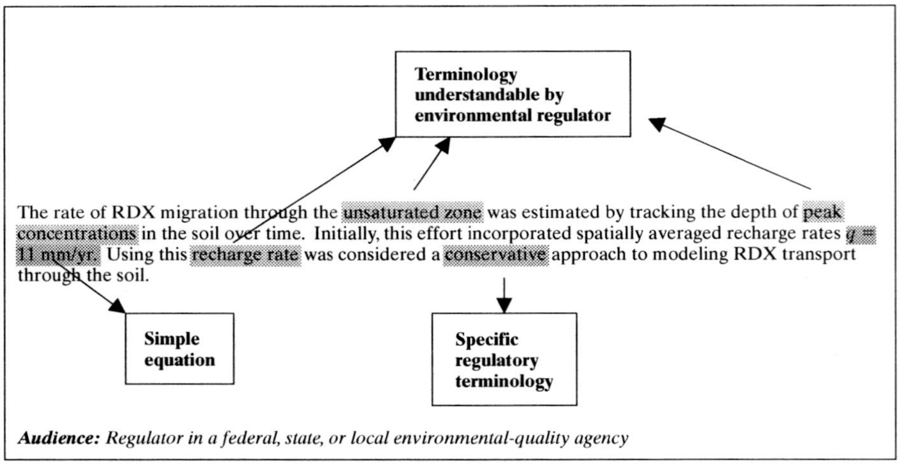

Figure 3. Excerpt from a *technical report* on RDX contamination at a site.

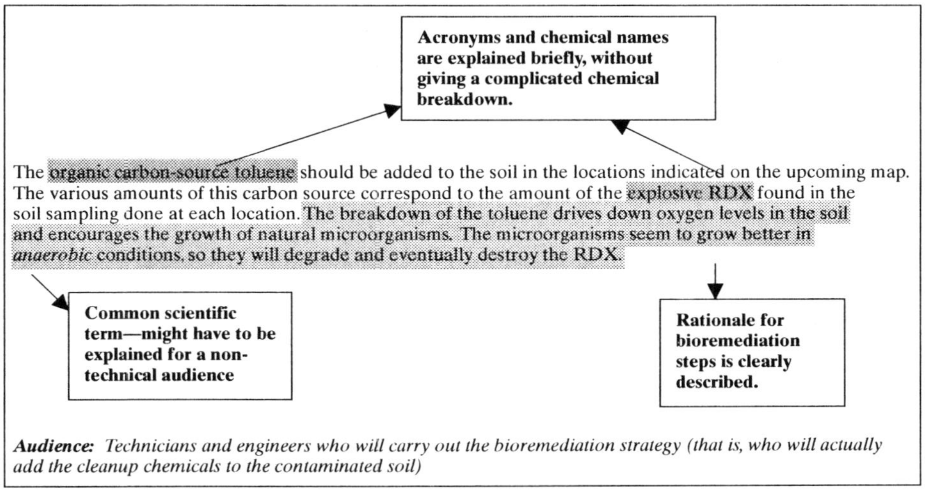

Figure 4. Excerpt from the introduction to a *set of instructions* on implementing a bioremediation system to clean up explosive RDX in contaminated soil.

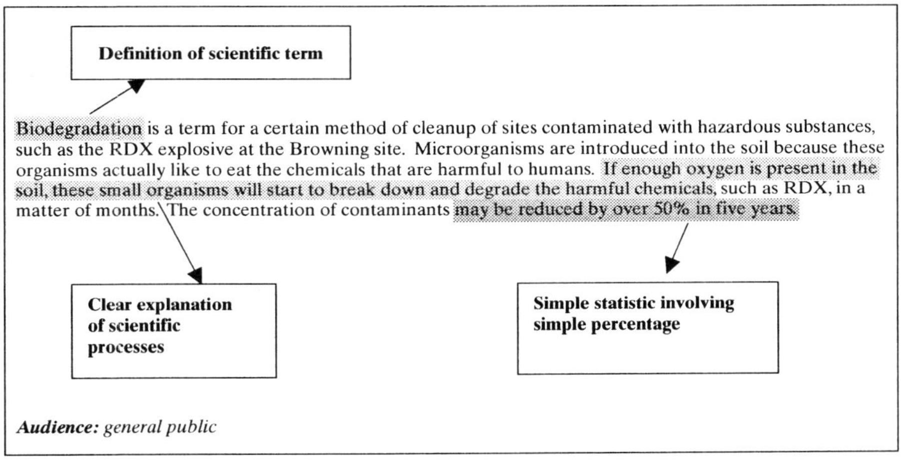

Figure 5. Excerpt from a *newspaper article* on cleaning up RDX contamination.

into one of the three categories defined by Paul Anderson (1986): Decision-makers, advisors, implementers.

Decision makers make things happen in a company or organization. They are ordinarily top management or those managers with some budgetary control. They are "big-picture" people who must allocate resources, keep an eye on the bottom line, and keep the organization moving forward at a healthy, but not out-of-control, pace. They want summaries of information and quick answers to questions such as, "How much will this solution cost?" and "How long will the solution take to implement?" They want to read conclusions and recommendations, not report sections containing lots of details. They want to listen to talks that get to the bottom line quickly.

Advisors are indispensable to decision makers, because advisors *do* care about details and methodology. Their job is to advise the decision makers about the credibility of solutions and of the procedures used to develop those solutions. They want to see descriptions of procedures and discussions of results. They will even check the appendices of reports to make sure that the investigation or proposal follows good science and that the computations are correct.

The final audience type, implementers, usually do not have much decision-making power. Instead, they have to do things on the basis of your writing: recalibrate a machine, build a new component to filter out "bad" air from a manufacturing building, etc. They are often technicians. These readers need clear instructions and rationale for the changes they will be making. They do not need theoretical details, financial justifications, or lengthy discussions of procedures used to arrive at the changes.

There are other audiences inside your own organization who may not quite fit into those categories: colleagues (engineers at the same level as you), support staff, and sales and marketing staff. The first audience is easy to communicate with; the others need clear explanations of concepts and terminology with which they are probably not familiar. And then there are the audiences outside your organization: customers, regulatory agencies, financial institutions, vendors, and perhaps the news media. As a manager, you will be called on to communicate with at least the first four of these audiences and possibly with the last one as well.

PRACTICE QUESTION: DIFFERING AUDIENCES

As an audience for your writing or speaking, how does a workplace engineer (manager or nonmanager) differ from an academic engineer (such as your engineering professors or instructors)?

SUGGESTED ANSWER

Workplace engineers are extremely busy, typically involved in a number of projects, and usually responsible for keeping costs as low as possible for the company and the client. Managers often do not do any engineering work, but rather manage people and other resources. Academic engineers are usually most interested in underlying causes of problems that need solving, and they want to pass their expertise on to younger engineers. Your instructors want to know how much you know and how well you know it; workplace engineers want to know what you know that can help them solve a problem.

SUMMARY

- For engineers, writing is not an extra set of skills that it would be nice to possess, but rather an integral part of doing engineering work every day.
- Writing helps you solve problems by helping you discover what you really think about the connections among the data and facts you have collected.

- *Data* are raw facts and numbers; *information* shows relationships among data and thereby becomes useful for solving problems. Your job as an engineer is to turn data into information.
- You will have many different audiences for your technical writing and speaking, and you must learn to adjust your language and graphics to fit the understanding of the particular audience.

Problems

1. Interview an engineer in the workplace. Choose any sort of engineer; he or she may or may not be working in an engineering firm. Ask him or her to describe a typical day in their workweek: what tasks does he or she perform and how long does each take? Ask especially about how much time your interviewee spends writing and reading, including e-mail and other forms of digital communication. Write up a short summary of the interview and compare with summaries written by other students. How much time (expressed as a range) do engineers spend doing communication tasks?
2. Talk to a family member or a friend who is not an engineer and describe a project you hope to work on, either at work or in college. Or describe the latest concept you just learned in your engineering studies. Be as specific as possible. Then, answer the following questions:

   Does your listener ask questions as you describe the engineering project? If so, write down those questions.
   Can you characterize the questions? Are they primarily requests for definitions, for more description?
3. Based on the questions asked by the non-technical person you interviewed, write a short analysis of the types of information needed by a non-technical audience.
4. Make a sketch of a particular piece of lab equipment in an engineering laboratory. If you don't have access to a lab, choose another available piece of hardware equipment (such as an electric drill or a coffee maker). Show the sketch to a non-technical person and ask him or her to describe how the piece of equipment works, based on your sketch.
5. If the non-technical person had trouble describing how the piece of lab or other equipment works, write a brief analysis of why. How might you redo the sketch, knowing what you know now? If the person had no trouble, write a brief analysis of the strengths of your sketch in communicating practical facts about the piece of equipment.
6. Chances are, your engineering instructor is which kind of audience for you?

   Decision maker
   Advisor
   Implementer

7. Does the following sentence contain "data" or "information"?

   Subjects in the behavioral experiment had a 90% likelihood of pushing the blue rather than the red buzzer.

8. Does the following sentence contain "data" or "information"?

   The findings from this experiment reveal that the number of total fish in the Hudson River has decreased by 23% since the 1950s.

FURTHER READING

Anderson, P.V. 1986. "What Survey Research Tells Us about Writing at Work," in *Writing in Nonacademic Settings*, Odell, L., and Goswami, D., eds. Guilford Press: New York.

This book is a groundbreaking work that looked seriously for the first time at writing in places other than schools. Paul Anderson pioneered research that looked at writing from the point of view of the reader rather than the writer. His essay in this volume is a clear and readable presentation of his study of alumni of Miami University (Ohio) and their experiences with writing at work.

Crowe, C.T., Elger, D.F., and Robertson, J.A. 2001. *Engineering Fluid Mechanics*, 7th ed. John Wiley & Sons: New York.

This is the textbook from which I took the data to construct the graph in Figure 1.2. You can learn a lot about good technical writing by looking at a good textbook for undergraduate students (who have, after all, a variety of educational backgrounds). Notice the way information is broken up into many sections with headings and visual items such as shaded boxes and illustrative figures.

Flower, L, Hayes, J.R., and Swarts, H. 1986. "Revising Functional Documents: The Scenario Principle," in *Writing in Nonacademic Settings*, Odell, L., and Goswami, D., eds. Guilford Press: New York.

Everyone knows that scenarios are good to construct when you are making a video or film, but most technical writers still balk at the idea of telling a story. This article demonstrates persuasively that even knowledgeable readers will reconstruct a passage in their mind to resemble a story with "agents and actions" (p. 45). So we may as well write that way, at least in functional documents such as instructions and procedures.

Petroski, Henry, 1986. *Beyond Engineering: Essays and Other Attempts to Figure without Equations*. St. Martin's Press: New York.

This small book is much beloved by both engineers and communication professionals who are interested in convincing other people that engineering and writing are not mutually exclusive activities; they actually support each other. Written by an engineer with much experience in both industry and academia, the text is down to earth and conversational.

Reeves, W.W. 1996. *Cognition and Complexity*. The Scarecrow Press: Lanham, MD, and London.

This volume is a wonderful little book that explains very readably what cognitive science has to say about how people process information and how we are all coping with the anxiety of information overload. Chapter 2 explains succinctly the similarities and differences among "data," "information," and "knowledge" and why it is important to use clear presentation to turn data into knowledge.

Spilka, R., ed. 1993. *Writing in the Workplace: New Research Perspectives*. Southern Illinois University Press: Carbondale, IL.

This collection of articles extends the work of *Writing in Nonacademic Settings* to present more conclusions about the interplay between writing at work and working. In particular, Spilka's article ("Moving between Oral and Writing Discourse to Fulfill Rhetorical and Social Goals") clearly shows how both speaking and writing help individuals and organizations achieve their professional goals.

Zinnser, W.K. 1988. *Writing to Learn*. Harper & Row: New York.

This wonderful, easy-to-read book presents the connection between writing and thinking. Especially useful to engineering students are Chapters 2 ("Writing across the Curriculum"), 9 ("Writing Mathematics"), and 10 ("Writing Physics and Chemistry").

7

Discovering Ideas and Facts: You Are Leaving Home

1 INTRODUCTION

Whatever the subject of your writing project, you will begin by collecting and organizing a lot of information. You need two kinds of information, both of which often overlap:

1. information to help you understand the issues, facts, problems, and solutions
2. information to be presented to your audience

Most of the information of the first type is derived from your research. Information of the second type may include what you already know as well as some of what your research reveals. The second type of information is what ends up in your document or presentation, and your job is to figure out how much of your research needs to be passed on to the reader or listener.

You cannot do productive research unless you know the purpose of the research project and the audience who needs to know about it. Let us say you are doing a report on noise-reduction methods for an existing apartment complex whose tenants have been complaining that their apartments are noisy. No matter to whom the report will be directed, you have to gather information about acoustic sealants, absorptive materials, how sound waves bounce and reverberate, etc. But if the report is for the apartment-complex owner, rather than the testing lab, you may need to gather somewhat different evidence. For example, you might want to incorporate in your report the graphics from a reputable website showing how sound waves act when they hit various surfaces. (You will, of course, credit your sources fully in the report.) The testing lab, on the other hand, would need no such graphics.

SECTIONS

1 Introduction
2 Framing Your Writing Project
3 Researching
4 Managing and Documenting Information

OBJECTIVES

By reading this chapter, you will learn the following:

- how to plan typical technical documents, including various types of reports and proposals;
- how to begin the research process;
- how to evaluate sources of information;
- how to manage large volumes of information;
- how to document the sources of your information as your project proceeds.

College instructors are usually experts in the subject you are writing about for their classes, so it can be difficult to know how much background explanation to include in assignments for them. On the other hand, instructors also want to know how well you understand the concepts and how much work you have done, so they may actually want more explanation and illustration than a workplace audience would. Discuss the issue of appropriate audience with your instructor; be sure to ask him or her if you should have a workplace audience in mind, such as an owner, a government agency, or a consulting client.

2 FRAMING YOUR WRITING PROJECT

This section discusses some typical writing assignments in engineering classes, along with the procedures for accomplishing the assignments. Note that each step involves reading, writing, or speaking, but only the italicized steps indicate what you may regard as "real" writing. Each of the italicized steps will generate at least one section of a report or proposal.

Lab Reports

Most lab reports document what you did in the laboratory as you followed a particular experimental or testing procedure. You will need to record carefully all data you collect and to note at the time any deviations from given procedure. Your lab report will really benefit from careful and timely note taking. Keep a notebook with you at all times in the lab. Be sure to record all that you observe, even if your teammates are also supposedly recording.

Here is a set of steps for writing lab reports. As with all writing procedures, you may have to loop back and repeat some of the steps (e.g., you may want to graph your data before thinking about what they show). The steps are as follows:

1. Define the law or hypothesis you are trying to prove.
2. Record the data as you go through the procedure.
3. Organize the data, and think about what they show.
4. Create figures and tables to display the data.
5. *Write a first draft of the Results section.*
6. Check your findings with your teammates (if applicable).
7. *Revise and rewrite the report.*

These steps may not be entirely linear, but you definitely want to accomplish steps 3 and 4 before attempting step 5. Do not try to start writing the Results section until you have created the tables and figures that display your data in meaningful relationships. Look at Figure 1. In this case, once you have plotted the values for engineering stress and strain and for true stress and strain, you can easily see at what point the values start to differ. Then your verbal description of what happened becomes easy to write: "At low levels of applied load, the values of engineering and true stress and strain were nearly the same. The differences increased, however, as the magnitude of the applied load increased."

Here is an outline of the sections of a typical lab report, whether academic or industrial:

Introduction (sometimes includes a Background section)
Test Procedure
Results and Discussion
Conclusions
Appendices
Measured Data
Sample Calculations
Notation

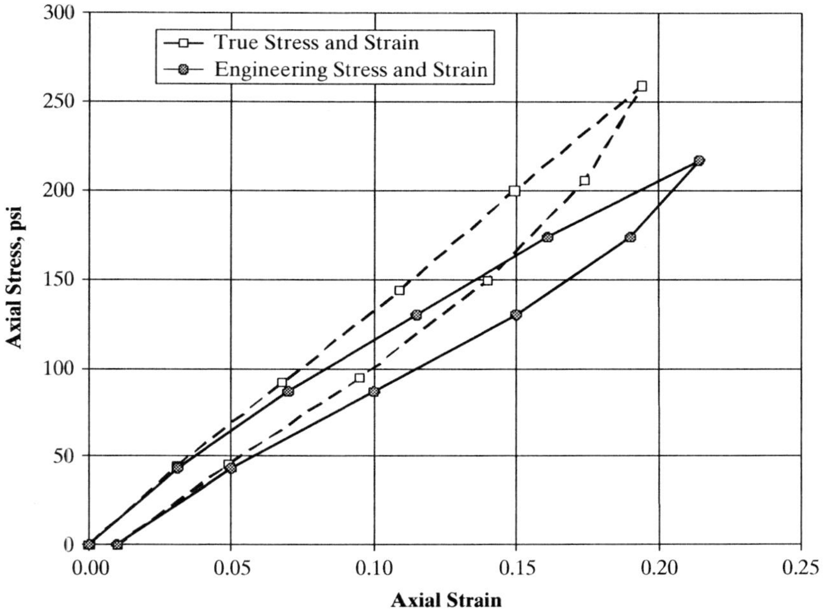

Figure 1. Measured stress–strain curves for rubber specimen.

Some formats for lab reports divide the Results and Discussion section into two sections, and sometimes a Recommendations section is added at the end of the report. But generally, reports on laboratory tests follow an outline close to this one.

In academic lab reports, the reader (your instructor, usually) is most interested in whether you performed the test correctly and learned something from it. Thus, any deviations from established procedure should be carefully noted, and the discussion of the results should show your attempt to organize, see in relationship, and then understand the *significance* of the data you collected. In industry lab reports, the reader is often someone outside your department or organization who needs the information in a hurry in order to make decisions. Thus, for example, the director of engineering for a tire-manufacturing company may need the results of the rubber test in order to assess the strength of the new tire designs being produced by the engineers.

Be sure to think about which engineering problems your findings are likely to help solve; mention those applications in the Conclusions section. If things did not go as planned in the lab, do not worry. Writing up the problems with a particular experiment or test can be very useful to others interested in the same experiment. Failure in engineering can teach us a great deal, and the ideal place for failure is in the lab or under simulated conditions.

Research and Recommendation Reports for Decision Makers

Apparently, there is often a big difference between how students and instructors conceive of a research project. Students feel they are simply supposed to show that they know something, whereas academicians use and do research in order to solve a problem or test a theory (Ford, 1995). Those latter goals require the ability to analyze and synthesize information, not just collect and regurgitate it. And, especially in engineering research, you need to exercise judgment as well as analytical skills in order to develop the best and the safest solution.

RESEARCH SHOWS . . .

Students think that in their research papers, they should show they know something. College and university instructors, on the other hand, believe that the purpose of research is to test a theory or explore a problem (Ford, 1995). This belief means that instructors want to see evidence of synthesis, analysis, and judgment. Clarifying your method of investigation shows that you are *thinking*.

If your research topic is given by your instructor, then you have a head start on the first step in each of the procedures listed next. If not, you will have to come up with a suitable topic for investigation. Presumably, your instructor wants you to demonstrate good research skills as applied to real engineering problems. For sources of good topics to investigate and write about, think about the engineering problems you see all around you—problems with infrastructure, mechanical devices, cars, electronic devices, airplanes, etc. Here is a list of possible topics for a recommendation or research report:

1. Choosing sustainable building methods and materials for a home owner.
2. Evaluating wastewater-treatment methods for a pig farm.
3. Investigating landing-gear systems for a new-model business jet.
4. Evaluating launch vehicles for a communications company's satellite.
5. Comparing robotic-drive platforms over variable terrain for construction automation.
6. Examining the feasibility of tissue engineering for artificial organs.

Notice that in all these topic designations there is an implied client—for example, the owner of the home in #1 or the communications company in #4. In the last two topics, the implied client is the research-and-development division of a company (a construction-automation firm or a biomedical firm, respectively). Engineering work is usually done to solve a particular problem for a particular decision maker. Unless you go into basic research, you will generally not be doing research without a specific audience for the results.

The upcoming box shows a procedure for researching and preparing reports for decision makers. The two writing stages (drafting and revising) are not numbered because you can work on them *at any point in this process.*

HINT

Recommendation Report	**Analysis Report**
1. Define the problem or need.	1. Define the problem or need.
2. Identify the research question.	2. Identify the research question.
3. Define the research methodology.	3. Define the research methodology.
4. Establish selection criteria.	
5. Synthesize and interpret information.	4. Synthesize and interpret information.
6. Create figures and tables to display the data.	5. Create figures and tables to display the data.
Write a first draft.	*Write a first draft.*
7. Reach conclusions.	6. Reach conclusions.
8. Make recommendations.	
Revise and rewrite the report.	*Revise and rewrite the report.*

Adapted from Lay, M., et al. 2000. *Technical Communication.*

A recommendation report is written to help decision makers take action. Most often, the action will be a means of solving a problem, and you, the researcher, will typically be investigating and choosing between courses of action. For an analysis report, the procedure for investigation is the same except that the end result will probably not be an explicit recommendation. In this case, the investigator is probably not choosing between courses of action, so selection criteria are usually not necessary. You might do a report analyzing the various methods for treating water to drinking-water standards, for instance, without choosing the best method for a particular application. So, all the information given here about how to proceed with a recommendation report is valid for an analysis report, minus the steps on developing selection criteria and making explicit recommendations.

In order to recommend a solution or to successfully analyze a problem, you will have to be very precise in your definition of the problem. For a report that analyzes several possible foundations for a planned microchip facility, you might define the potential *problem* for the facility as finding a foundation that does not experience excessive settlement. Too much settlement of the building can damage the delicate machines that produce microchips. You would then need to define "excessive"; in the real world, you would work with the facility's owner to develop the allowable limit. If you defined "excessive" as "settlement greater than 0.25 inches," your project would then investigate only those foundations that reach deep enough into the soil to incur settlement less than this amount.

The *research question* thus becomes, "Which types of foundations are deep enough to cause only that minimum amount of settlement (<0.25 inches) in the clayey soils around the microchip facility?" Answering this question leads you to choose three types of foundation to evaluate, all types of *driven piles* (piles driven deep into the ground): concrete piles, hollow steel piles, and concrete-filled steel piles. So, now you have a refined topic that you can begin researching in earnest. The trick with research reports is to remember that you have to do some up-front research simply to determine your precise topic. Students can quickly get into trouble if their topic is too broad; you cannot, for instance, investigate all types of foundations for all types of buildings. And in the real world, you would hardly ever need to!

How would you collect information about driven piles? Well, you would probably consult an engineering reference book and surf the Web for information on piles. You might also check some engineering journals to see whether any tests have been conducted on the various types of piles. And you would try to talk with engineers who have worked with these piles on other projects. It would be ideal if you could conduct your own tests in a laboratory. You would then be doing what is called "original research." But chances are that you will find the information you need in published sources or by interviewing foundation engineers (experts on the subject matter). You will want to determine early in your project just which research strategies and sources are best for your project and possible for you to use. Here is a list of common research methods used by engineers in academia and in industry:

1. **Conduct tests.**
2. **Observe.**
3. **Solicit expert opinion.**
4. **Collect and synthesize information from published and unpublished sources.**
5. **Make calculations.**
6. **Create preliminary designs.**

Every researcher, whether at a university or in industry, whether a graduate or an undergraduate student, typically uses at least method #4 and often uses method #5. But the other methods may also be possible and useful for you. If, for instance, you are doing a project on possible solutions to the traffic problems at a particular intersection of roads, you do not need to use automatic counting devices to count the cars going through that intersection at peak congestion times, such as rush hours. If you manually count the number of cars within a half-hour interval, for instance, you will have some solid and specific data about the extent of the congestion problem.

HINT

Possible Research Methods for Engineers

Conduct tests.

Observe.

Solicit expert opinion.

Collect and synthesize information from published and unpublished sources.

Make calculations.

Create preliminary designs.

Looking again at the procedure involved in producing a recommendation report, you may wonder why the word "scope" does not appear in any of the listed steps. Much engineering work is defined in terms of scope; the Scope of Work is often a section of a proposal. Actually, accomplishing the first three steps—defining the problem or need, identifying the research question, and defining the research methodology—determines the scope of your investigation, the particular piece of the whole problem for which you will try to find a solution. So, defining your scope of work actually involves at least these three critical steps. It helps to think specifically about where and how you will get information *before* you finalize your scope of work.

Determining the Scope of Your Topic

What information should I read?

Where will I look for it?

on the Web or in print?

To whom will I talk?

What should I observe?

Will I do testing?

Will I administer surveys?

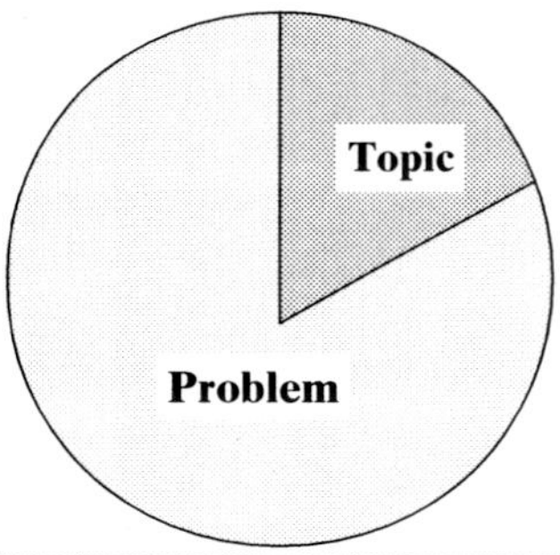

Let us return to the project of investigating alternative foundations for the microchip facility. Because the goal of this project is to choose one type of foundation, you must first understand the constraints put on the possible solutions and then develop a list of *criteria* by which to compare alternative solutions. As stated previously, a big constraint for the foundation investigation is minimal settlement: an allowable limit established by the facility owner's engineers. Since several types of piles can meet that constraint, other qualities come into play in choosing the best foundation—structural strength and cost, for instance. These criteria are what help you select the best alternative for this application.

HINT

Criteria Are Measuring Tools: Measure the Feasibility of Solutions

Standard categories of selection criteria:

- time to implement
- cost
- size,weight, and other physical aspects
- adaptability and compatibility
- efficiency (as a ratio)

HINT

Criteria Need Specific Definition

- **Define your criteria by breaking down the categories into specific solution goals for your project:**
- **What is the ceiling for cost?**
    - **How big is too big?**
    - **What does "efficiency" mean in this case?**
    - **How long is an acceptable implementation time?**
- **Remember that, if any criterion cannot be broken down, it may have to be jettisoned.**

The rest of the report-developing steps involve doing the work you have set out for yourself. Table 1 shows how each of these procedural steps helps you develop particular sections of a standard recommendation report. You can begin writing your first draft at any point during the first six information-gathering and interpreting steps.

TABLE 1 Relationship between Sections of a Typical Report and Particular Cognitive and Writing Activities

Cognitive and Writing Activity	Section of Report
1. Define the problem or need	**Introduction** **1, 2, 3, 4**
2. Identify the research question	
3. Define research methodology	**Methodology** **2,3**
4. Establish selection criteria	**Selection Criteria** **4**
5. Synthesize and interpret information	**Description of Alternative Solutions** **5, 6**
6. Create figures and tables to display data	**Comparison of Alternatives** **5, 6**
Write first draft	
7. Reach conclusions	**Conclusions** **7**
8. Determine recommendations	**Recommendations** **8**
Revise and rewrite	

Here is how the parts of a report evolve from your writing and researching procedure: An explanation of how you defined the problem and identified the research question will probably appear in your *Introduction*. An explanation of the criteria by which you evaluated possible solutions may have its own section, *Selection Criteria*, as might the outline of your research methods in a *Methodology* section. In short reports, the last two sections might be folded into the introduction, but it is wise to limit the Introduction to three pages. Your display and analysis of the data and information you have collected will constitute the body of your report in several sections that capture the essence of your method—*Description of Alternatives* and *Comparison of Alternatives*, for instance. In scientific papers, this analysis is usually presented in a section called *Results and Discussion*. *Conclusions* make up their own section. Sometimes recommendations are included as well—in such cases, the section is called *Conclusions and Recommendations*—and sometimes you will present them in their own section, *Recommendations*.

There is one more, very important, section that usually begins the report: *Executive Summary*. In fact, the executive summary is the most important section of the document; as its title indicates, it may be the *only* section that a busy decision maker reads. Thus, the executive summary should be a stand-alone document; if it includes graphics, for instance, they should not be numbered in sequence with the graphics in the rest of the report. Neither should the executive summary refer specifically to any graphic or page number in the rest of the report. To offer the busy reader the bottom line, the executive summary contains a little bit of all the types of information covered in the report; it presents the problem, the alternatives investigated, and the results and conclusions, but *in far less detail*.

The trick in writing an executive summary is to provide good transitions between sentences, since you are hopping quickly from one topic and type of information to another. Here is the final paragraph from a page-long executive summary for a report comparing deep-foundation types for a proposed microchip manufacturing facility. The boldface words indicate transitions that guide the reader swiftly through this recommendations paragraph:

> Advantage Consultants recommends that EMM, Inc., use hollow steel-pipe piles that are 55 feet long and 14 inches in diameter for their foundation. The final foundation layout costs $405,000, making it more expensive than some of the other solutions. **However**, this solution allows the least amount of settlement while staying within the budget. **Based on the presented research**, Advantage Consultants is confident that EMM, Inc., will achieve the best building foundation with hollow steel-pipe piles.

The executive summary comes first in a document, but it is usually written last, after the rest of the report has been drafted.

For an analysis report, you would probably have fewer sections—no selection-criteria or recommendations sections, for example. For a report on drinking-water treatment methods, you would probably have a *Description of Methods* section. You might also have a *Comparison* section in which you listed the pros and cons of each method, without regard to strict technical, regulatory, or financial criteria.

Understanding how your research and analysis will be written up should encourage you to start drafting before you have quite finished the research. For example, you can write a *draft* of the introduction before you have finished collecting information, because you already know the problem, your goal for a solution, and the approach you are taking to find that solution. You even already know the criteria you will use to evaluate alternative solutions. Remember that this draft will need to be revised later, but drafting the introduction early helps you envision and plot the course of your investigation. Meanwhile, you have written part of your report before you have even collected all the data!

PRACTICE QUESTION: DEFINING TOPIC AND SCOPE

At work, you are usually given the topic and the scope of work for a project, but when you write an unsolicited proposal or when you are in school, you often must define the topic and scope for yourself. How do you go about defining them?

SUGGESTED ANSWER

You begin by collecting as much information as possible about the general topic; then you try to narrow the topic by doing more focused research, often by using search engines. You want to find out whether you can narrow the scope enough to be able to do a thorough job of researching and synthesizing information within a specified amount of time. Before you decide on a particular topic to write about, you should already have done some research.

3 RESEARCHING

You probably think of research as the process of tracking down published information. Actually, you can also uncover much relevant information through unpublished sources, especially through interviews with subject-matter experts (engineers and others who know more than you do about a particular topic or situation). And you can also do your own original research, even if you are an undergraduate student.

Published Information

Fortunately, tracking down published information is relatively easy and fast on the Web. Not all the websites you uncover, however, will have reliable, credible information. You always need to evaluate these sites by noticing the domain (*.com*, *.gov*, etc.), as well as any other publication information you can gather, such as author, date, and sponsoring organization. Basically, you want to discover who published the site and whether the individual or the organization has any bias or any proof of credibility as an expert.

Searches, whether via an electronic database or a Web-based search engine, are an art form in themselves. You can waste a lot of time by using keywords that are too broad. For example, if you type in "pollution," you will get hundreds of thousands of hits. Adding another word like "air" would help somewhat, but you would still have an unmanageable number of sources to check. Your topic should not be that broad anyway. If you are interested in air pollution, you could do a research paper on the contribution of "sewer gases," let us say, to the problem. And you might have to add more keywords to make even that search productive.

As with writing, the process of researching may reveal several items. You may discover what it is you are really interested in (not just air pollution, but radon in particular, let us say, and especially how dangerous it can be in certain amounts). Or you may discover

HINT

Strategies for Researching

- **Be as specific as possible in your search.**
- **Choose the best search engine:**
  - Google loads faster than other engines.
  - Google does not do relevance rating.
- **Consult a librarian.**
- **Set a timetable for research.**
- **People are also sources of information, but they have to be reliable and credible.**

that what you are interested in involves proprietary information that is simply not available. In that case, you would be wise to adjust your topic. Refining your topic early, before you have gone too far with research and writing, is vastly preferable to having to change your topic later.

Here are two of the best sources for engineering information:

Engineering Index (EI Compendex Plus)
This index contains more than 1 million citations, with abstracts from more than 4,500 journals, reports, and conference proceedings covering fields of engineering and technology. It is the best overall source for scholarly engineering information.

Applied Science and Technology Abstracts
Covers 350+ key, international English-language periodicals in the applied sciences and technology. It is a good source for practical (that is, not as research-oriented) information.

Unpublished Information: Interviews

You can learn a lot by talking to people, especially experts in a particular field or experienced managers in any organization. Because experts and managers are usually decision makers and are extremely busy, the trick is to convince them to give you the time. Being a student is an asset, since most professionals want to help students. Interviewing for information is an art. If you can master this art, you can learn a great deal from those folks who have themselves taken years to learn what they know.

HINT

Collecting Information: Interviewing

- Gather necessary information.
- Prepare effective questions:
  - focused
  - few yes-or-no questions
  - one at a time
- Don't interrupt.
- Allow time for response.
- Convey interest:
  - tone of voice and body language
- Write thank-you letters.

Showing interest in the interviewee and enthusiasm for the subject are critical ingredients of a good interview. One way to show interest is to lean slightly forward, toward the "interviewee" or at least to sit upright. Another critical factor is coming in prepared with good, focused questions. If your interviewee talks a lot and wanders off the subject, you may have to carefully and politely push the conversation back on track. But also be prepared to simply listen when he or she is on a roll; listening well usually helps you think of more good questions to ask. Never record a conversation (on the phone or in person) without asking for permission. If you are taking notes, do not lose your focus or slow things up by writing too much down; wait until after the interview, and then *get the information down on paper* as soon as you can.

RESEARCH SHOWS...

How do you know whether someone is an expert? According to Leahey and Harris (1997), aside from possessing "an extensive knowledge base" in their subject area, experts can quickly distinguish relevant from irrelevant information and focus on the former. Remember this caveat when you interview experts: Cut to the chase with your questions. Experts are also flexible and adaptable, "adjusting their strategies to fit a constantly changing situation" (p. 246). In other words, you can learn a lot from experts just by watching them at work. Finally, experts can "simplify, chunk, and organize information efficiently"—just as *you* should.

Unpublished Information: Surveys

There is also an art to designing good surveys. You want to capture *only* the information you want, and you want people to actually complete the survey. So your design should be simple and attractive, and the survey should be as *short* as possible. Never ask people to write out their thoughts for more than one question, and be prepared not to get anything of substance from free-answer questions. The success of your survey should depend on the simple-to-answer questions, not on any essay-type answers you try to elicit. Always consider getting demographic information such as age, sex, and employment, but usually you should not ask for a name, since you want people to feel free to express their real thoughts without worrying that those thoughts can come back later to haunt them. (Remember how many public officials have bemoaned the fact that the press sometimes takes their remarks out of context.) Write an introduction that states the purpose of the survey and the importance of participation. And always include a written thank-you statement.

HINT

Collecting Information: Surveys

- **Use a random, but representative, sample.**
- **Ask short, mainly yes-or-no, simple questions.**
- **Ask very few open-ended questions.**
- **Avoid vague or overly general words.**
- **Be cautious about using jargon and slang.**
- **Use an easy-to-complete, attractive design:**
  – no more than one page.
- **Collect answers yourself whenever possible.**

You can collect some useful information from even a small sample, as long as that sample is random. For instance, for that bus-shelter design project, you might survey riders of the municipal bus system about their preferences in bus shelters, including the features they would like to see. You would probably survey only bus riders (the users of the system), but you would try to vary their age, sex, educational background, and socioeconomic level by administering the survey at several bus stations in very different parts of town.

Evaluating Sources of Information

The integrity of your writing—of the information presented in your writing—depends to a large degree upon the integrity of the sources of that information. If a particular source is an expert whom you have interviewed, you can probably rely on his or her credibility, although you still might want to interview at least one other expert in order to gather corroborating information. If the source is a book or article published by a reputable publisher or a recognized technical journal, and if the work has been favorably cited by other researchers, then you probably have a credible source. But lots of information about technical subjects is available on the Web, and you know that not all the sites you see contain reliable, unbiased information. You must, of course, evaluate these sites by first checking out the domain: *.com* indicates a commercial site; *.org* and *.edu* usually indicate noncommercial sites, and *.gov* indicates a government site. Information on a university or public-agency site will tend to be more reliable than information posted by a private company or a special-interest group. Companies want to sell products or services, and special-interest groups want you to join their cause; on the other hand, universities have a long tradition of encouraging peer review of faculty and student work, and public agencies are accountable to citizens. So, what is published on the latter two types of sites has probably been through an unbiased review process and is therefore usually considered to be reliable information—unless, of course, you are looking on the home page of a faculty member who has not published the information anywhere else. In that case, the site is just as suspect as any site that presents unreviewed opinions. Look for a tilda followed by the person's name (e.g., ~ *smith*) in the directory path of the URL; this designation is a tip-off that the page may be located in a personal directory.

Trace backward in the URL containing information you are scrutinizing in order to discover the home page; somewhere in the site, there should be information about how and by whom the site was developed. Look at the "About" page and check out the site map. Does the organization make sense? Are there some dead links? Internet robots and spiders are helpful aids in the search for good information, but be careful about the URLs they gather, as they look everywhere, even on sites that are no longer maintained or are under construction.

You must also evaluate websites by uncovering any other publication information you can, such as information about the author, date of publication, sponsoring organization, and purpose of publication. Determining some of this publication information can be tricky. If there is no explicitly provided publication date, look on the document-information screen

HINT

Evaluating *All* Sources of Information

1. Does the information answer important questions for your research?
2. Is the information current? Can you determine the date of publication?
3. Are presented facts and figures from reliable sources?
4. Can you detect any biases in the way that the information is presented?
5. Can you determine the author's credentials and qualifications?
6. Is the publisher or sponsor credible? For WWW sources, check the domain.

for a notation of the last date of modification—"Last updated on *mm/dd/yy*." If there is no date anywhere, you must document the date you accessed the site. Otherwise, you will have no proof that that site existed and contained the information you may reference in your writing. So, be sure to keep a dated record of your Web sources, including pathway information, so that you can return to them. You will need this dated record in order to compile a proper list of references and to cite sources responsibly in your documents.

HINT

Evaluating *Internet* Sources of Information

1. Is the site updated regularly?
   – Check dates.
2. Is the site well designed?
3. Is the writing grammatical and clear?
4. Does the source make any biases absolutely clear? Does it make its purpose clear?
   – Check the "About" page.

PRACTICE QUESTION: EVALUATING SOURCES OF INFORMATION

If you were doing a report on the economic and environmental benefits of using acid-gas injection instead of sulfur recovery to dispose of the acid gas produced by natural-gas production, would the following website be relevant to your report?

http://www.epa.gov/air/oaqps/airtrans/acidrain.html

SUGGESTED ANSWER

The EPA article at the website provides some good background information on acid rain, which is produced by pollutants such as those found in natural-gas reservoirs. So, you might use the article in the background section of your report. If the focus of your project is on the benefits of acid-gas injection, however, you will want to move beyond background information quickly and find sources with credible information on this process, how it works, and how much it costs. Don't get hung up on the lengthy literature about acid rain.

4 MANAGING AND DOCUMENTING INFORMATION

Once you have identified and evaluated your sources of information on a topic, you have to keep track of them and determine whether and how the information you have gathered can be used in your document. Let us assume that you are researching foundation alternatives for a planned microchip facility on the outskirts of your town and have determined that three types of driven piles—concrete piles, hollow steel piles, and concrete-filled steel piles—shall be evaluated as possible solutions. We suggest that you follow these steps, because managing information is a **four-step process**:

1. ***Evaluate* the sources and the information itself.** If you have determined that the websites or articles from which you have obtained information on driven piles are reliable sources, you still have to decide whether you are going to use *all* the information you have collected or only the information on two of the three driven piles you started out to examine. By using an initial list of criteria, you may have already determined that concrete-filled steel piles are too expensive and would have to be trucked to the site from

very far away (adding even more to the cost). So you would stop collecting information on concrete-filled piles and use the information already collected to explain in your methodology section why you ended up investigating only two types of piles. You can often determine relevance by checking the information against your outline, but make sure that your outline is keeping up with the changes you are making in your scope of work.

2. **Informally, but thoroughly, *document all sources* of information.** Even if your instructor or boss has not specified that you include a list of references, it is a great idea for you to create one—if only for your own work later on. Chances are that you will be working on a similar topic in the not-too-distant future, and you want to have a record of where you got the most helpful facts and data. So, whether you format a formal list of references or not, always keep a dated record of what information you found where. Use index cards or a word-processing file to keep track of sources. Collect these facts:

author
date
title of work
(title of larger work)
publication information

For electronic sources, add these facts:

electronic address
date of your access

Ask your instructor or boss which referencing style to use. Even within the field of engineering alone, there are many slightly different styles used. For example, some journals want a list of references in the order in which you cited them, with superscript numbers as textual citations, while others want an alphabetized list, with the textual citations containing only the last name of the author(s) and the year of publication. If you are not given a reference book that specifies citation guidelines, consult the website of your field's major professional society and check for something like "A Guide for Authors." The main point here is that when you are in the middle of research, you want to get these facts down and save them in whatever format is most convenient. If you do have to format a formal list of references, try formatting the collected information according to the required referencing style. You may as well alphabetize your documentation right from the start, for instance, if that is how you will organize your reference list in your report or paper.

3. ***Summarize in your own words* the information from each source.** When you immediately write a summary of a piece of writing you have just read, you begin the process of synthesizing the information and extracting out of it what is important for *your* work. Especially when you are less experienced and when you have been reading a lot at one sitting, you quickly lose the ability to distinguish between useful information and unnecessary or bad information. You end up reading blindly, and later you cannot remember whether or why that article seemed useful for your report. Break out of the passive mode; become an active thinker, simply by writing down your own outline of the article's contents and your own thoughts about its quality, findings, and relevance. After summarizing an article, you are one giant step closer to writing your own report than you would be if you just tried to memorize the article's facts or to take notes on what it says.

Try also to link the information in each new article you read with previous information you have collected and summarized. Is this author supporting the claims of others? If not, can you determine why not? Are this article's findings significant, new information, or just careless and superficial thinking? Pay attention to the language of findings. Is the author claiming to have proved something or simply to have noticed a possible correlation between variables? You will need to be equally careful with your own language in making claims and recommendations on the basis of the author's information. Can you say that the study "shows" something conclusively? Are the findings "preliminary" or "conclusive"? If you use inaccurate language in your summary, those inaccuracies will inevitably get translated into your final report. Be careful with your note taking and summarizing. If you do them right, you will save yourself time and greatly improve the credibility of your writing.

4. ***Organize* all that information.** Do not be tempted to start downloading everything you find on your first search. Screen those sources through a further search, using more specific keywords. To manage those piles of information you're collecting, you will want to set up a document-management system, probably on your computer. You have already collected the information you will need to access each source later. Store that documentation in one word-processing file, for both on-line and print sources. You can cut and paste the citations from your library's on-line catalog.

Keep checking the relevance of the information against your outline, and weed out information that does not seem as relevant now as when you started the project. Be careful, though: Don't throw anything away until the project is finished. If you have bothered to download some articles or citations, you might come around again to thinking that they *are* useful after all. So, stash them in a separate file called something like "Outtakes."

SUMMARY

- Any writing project requires intensive planning before you start a draft. You should find examples of the kind of document you need to produce in order to better understand what sorts of information you need to gather and how to frame it.
- There are many sources of information, published and unpublished, and many methods for gathering that information. Become familiar with using search engines, databases, and indexes. Identify the journals and publications cited most often in your field of engineering.
- Experts are good sources of information, but only if you interview them properly. Gather your questions ahead of time, and use body language during the interview to show interest.
- You will need to manage the large volume of information that you will probably collect. Keep track of published sources as you go along, since you will most likely have to document those sources.
- Not all sources of information are credible, especially on the Web. Evaluate information by examining the credibility of the sponsoring institution or group and the author.

Problems

1. Assume you are writing a paper on global warming for an Environmental Engineering conference. On a scale of 1 (most credible) to 5, rate the following

sources in terms of credibility with *your audience*. Look up these URLs and ***justify your ratings***. Use the checklist at the end of section 3.

http://www.greenpeace.org/
(Greenpeace's URL)

http://www.epa.gov/globalwarming/
(The EPA's page on global warming)

http://www.megastories.com/warming/bangla/intro.htm
(An article about the effect of global warming on Bangladesh)

www.skepticism.net/global_warming/
(Links to articles skeptical of environmentalists' claims)

http://www2.msstate.edu/~krreddy/glowar/glowar.html
(Global Warming International Center maintained at Mississippi State University)

2. Interview an engineer, either in the workplace or at your school or college. Choose someone you believe to be an expert in his/her field. Take notes on his or her language and demeanor. Write a description of the words, gestures, tone of voice and/ or facts that tell you this person is an expert.

3. Begin to research a topic in your field by using an internet search engine. Record the exact words you use for each search. What words did you use to find the best source for writing a report on that topic? How many searches did you have to perform to find that source?

4. You work for an auto-industry research company and your job is to find out how many students in your classes drive cars to school and what kind of cars they drive. Design a short survey to capture that information. Use the survey in at least one of your classes.

5. Using the information you gathered in #4, write a short report about the makes and models of cars students are driving and what percentage of students drive rather than use other transportation to school. Can you interpret and explain these findings?

6. Using the information you gathered in #4, conduct a series of interviews with particular students to find out why they drive the cars they do (if they drive at all). Discuss in a short report your findings and your explanation of the interview results.

7. Read an article on a topic in a class you are taking (the article can either be assigned by your instructor or chosen by you). Summarize the article in 200 words or less.

8. Summarize the article you read for #7 in 100 words or less. Did you have to leave out some important information?

9. If you felt you had to leave out important information by doing the summaries in #7 and/or #8, describe the audience(s) you think might really need that information you had to leave out. Would you characterize those people as decision-makers, advisors, or implementers?

FURTHER READING

Chicago Manual of Style, 15th ed., 2003. University of Chicago Press: Chicago.

This is the mothership of style manuals. This big volume contains more than you'll ever want to know about manuscript preparation, documentation, and the publishing process, but it will also have definitive answers for you on all those niggling little questions about style (e.g., Do you write out the number "twelve" or use the Arabic numeral?). In particular, the "Grammar and Usage" and "Punctuation" sections are a terrific reference tool for engineers. Most discipline-specific style manuals in engineering are based on the *Chicago Manual*. See the "Documentation" section for a clear, albeit lengthy, demonstration of the differences between APA and MLA styles. Since you may have been taught the latter in high school, you will want to acquaint yourself with APA style, now that you are in a technical field.

Ford, J.E. 1995. *Teaching the Research Paper: From Theory to Practice, from Research to Writing*. Scarecrow Press: Metuchen, NJ, and London.

This collection of papers is meant for teachers, but it contains some good advice for everyone on how to approach and do productive research. There is a nifty research model on page 141.

Lay, M. Wohlstrom, B; Rude, C; Selzer, J; and Selfe, C. 2000. *Technical Communication*, 2d. ed. Irwin: Chicago.

This is one of the best of the many textbooks on all aspects of technical communication. One of the best chapters for engineering students is Chapter 14, "Reports for Decision-Makers."

Leahey, T.H., and Harris, R.J. 1997. *Learning and Cognition*, 4th ed. Prentice Hall: Upper Saddle River, NJ.

This weighty tome is not meant to be read in its entirety by anyone who is not a cognitive scientist. But it is a surprisingly friendly presentation of everything you might want to know about how humans learn and understand and why we behave the way we do. The most relevant chapter for engineers is the one entitled "Thinking," which offers good strategies for problem solving, reasoning, and decision making.

Ruszkiewicz, J., Walker, J.R., and Pemberton, M.A. 2003. *Bookmarks: A Guide to Research and Writing*, 2d. ed. Longman: New York.

A much-needed guide to the entire process of conceiving and executing a research and writing process, this book takes you from defining initial scope to drafting and finalizing the project. The sections on evaluating sources and on understanding academic responsibility offer advice and information that are hard to find in other places. The book is easy to use, with helpful tabs and nifty illustrations.

Walker, J.R., and Taylor, T. 1998. *The Columbia Guide to Online Style*. Columbia University Press: New York.

This reference book organizes and categorizes the standards for citing electronic documents. It also includes sections on how to create documents electronically for print publication and how to format documents for on-line publication.

8

Organizing Ideas and Facts: It's the Journey, Not the Destination

SECTIONS

1 Introduction
2 Starting to Write
3 Drafting for Myself
4 Using Writing to Think Clearly
5 Using Writing to Solve Problems

OBJECTIVES

By reading this chapter, you will learn the following:

- how to start the writing process;
- how to help yourself in the drafting process;
- to use writing to help yourself write;
- to use writing to clarify your engineering methodology.

1 INTRODUCTION

Just thinking about starting to write a document or put together a talk is scary. No matter how much you know (or think you know) about a subject or a situation, when you try to organize your thoughts and write them down in clear, logical paragraphs . . . well, you quickly realize what a crooked path your thoughts tend to follow. The way we usually think (the way our brains work) is quite different from the way we know we need to write. We think in images, keywords, or clusters of words at most. We do *not* think in sentences and paragraphs; we think *associatively*—one thought reminds us of something else, even if only because we experienced that something else right after the first thought. Thus, the first time you learned what an isosceles triangle is might be forever intertwined with a memory of your first kiss, which happened right after geometry class. So, writing is always an act of translation—translating into a primarily linear form the ideas and concepts that occur to us piecemeal in a recursive, associative, often visual pattern. Therefore, it is important to learn how to become better friends with the process of capturing your thoughts, organizing them, and writing them up.

2 STARTING TO WRITE

So how do we perform this act of translating our thoughts into writing? Not quickly. Writing (or putting together a presentation) is a process that happens over time and goes through several stages. Like most engineering projects, writing any professional document involves these steps (only two of which include what you probably think of as writing):

- Define objectives for writing.
- Define the audience.

- Plan the type and format of writing (e.g., informal letter, formal proposal).
- Draft the document.
- Evaluate the document (with outside help).
- Revise the document.

If you substitute the word "project" for "writing" in this list, the word "plan" for "document," and the word "participants" for "audience," you would be describing any project-management procedure. It may be a while before you become a project manager, but, even as an entry-level engineer, you will work with your team to clarify and find expression for the objectives given by upper management. Whether the desired result of a project is a new design, an improved process, or a list of requirements, the project manager will go through these six steps, although not necessarily in strict sequence. There can be lots of loops back between any two of these steps. The "draft" stage, for instance, results in a test document that you should ask other people to evaluate. If your evaluators cannot understand the purpose of the document (the bottom line, so to speak), you may need to go back to the "plan" stage and brainstorm about how to highlight the purpose better. You may even need to go back to the "define objectives" stage and determine whether you yourself are clear about the ultimate purpose of this project or document. It may sound pretty stupid to talk about a project manager or any experienced professional not being clear about purpose, but confusion and conflicting goals are common to large organizations with dozens of projects being planned simultaneously by hundreds of employees.

The most important point to remember about this list of steps is that *you cannot skip forward in the initial sequence*. For example, you cannot try to write a final draft before capturing your thoughts in a first draft. Even a professional writer cannot do that! You cannot write clearly if you do not know, for instance, to whom you are writing. Take a project to design a city's bus shelters as an example. If you are a consultant responding to a request for proposals published by the city's municipal transit authority, then the first document you write is a proposal to that authority, "selling" the excellence of your conceptual design of the bus shelters. You will probably stress the cost effectiveness of the design, since municipal authorities are responsible to city taxpayers. When you are hired (on the basis of your terrific proposal) to create the final design, you will probably help the transit authority "sell" the design to the public on the basis of how functional and aesthetically pleasing it is. In each case, what you highlight in the document will be somewhat different. So, you need to do some up-front work on defining the audience, purpose, and type of communication *before* you begin a draft of the document.

HINT

Remember: Writing is a *process*, a series of steps.

Sometimes you will repeat steps, but do not try to skip any.

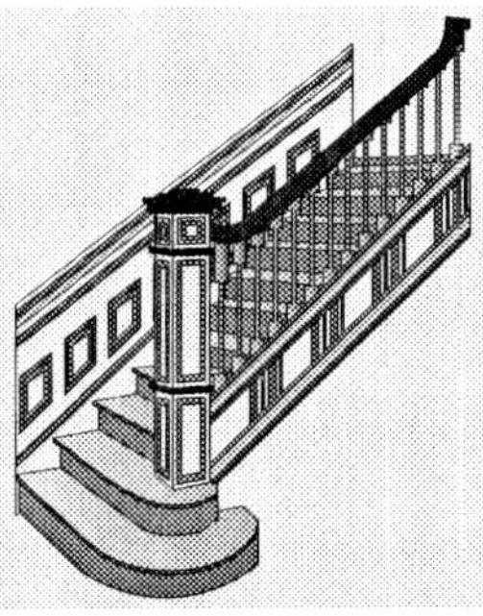

Once you have defined (or worked with others to define) your objectives for writing, the audience you are addressing, and the type of document needed, you are probably ready to take a stab at writing. Whether you feel ready or not, if your research is largely done, you are ready to begin a first draft of your document. Of course, you could convince yourself that your research is not done yet. You can *always* tell yourself that there is more relevant information out there, because there always is. The more you get to know about a subject, the more you realize how much more there is to know. This is the point where having a good definition of your purpose becomes critical for you. If you are writing a design report describing a new prosthetic device for a medical-devices company, you had better know a lot about other, similar devices. But you do not have to know everything about all prostheses ever invented.

So, how do you get started? What strategies can you use to move from researching to writing? In the case of a design proposal, of course, you have the design to start with. And that helps a lot. For the bus-shelter proposal, you can start by describing the design visually and then proceed to list its benefits and compatibility with the urban landscape. You are now well on your way. You can circle back later and write the introduction and supporting sections. So, strategy #1 is this:

1. **Start with the visual.** Create your graphics (designs, figures, and even tables) first. Then write text to describe and connect them together. You will not be starting at the beginning, but who cares? *This is only your first draft.*

Here are some other starting strategies. You can use any or all of them. They all involve using informal writing or speaking to help yourself work up to the first draft. The strategies are as follows:

2. **Start with an outline** that you generate any which way. You can make random lists of ideas, topics, or keywords; you can make doodles; you can talk to someone else and ask him or her to take notes as you speak. An outline is a tool for you to use; it can be as individual as you are. And no one else need ever see it. Some of us learned to outline by using numbers and letters in a linear sequence. Use that format only if it helps you get started writing. Many word processors have an outline feature, but make sure that you are not intimidated by the numbering; keep the outline as free form as possible until you are writing the first draft. Do not make the mistake of thinking you have to begin by writing about the first topic on the outline. You can start with any topic, because *this is only your first draft.*

3. **Start with the research summaries you produced.** This is the point at which you will thank yourself heartily for having taken the time to create paraphrases and summaries of the information you collected. You have already gotten some of *your* words on paper or screen. If you have typed your summaries, you can simply cut and paste them into your first draft. It does not matter which summary or topic you start writing about, because *this is only your first draft.*

4. **Start by freewriting.** This strategy is one of a number of ways to do what is also called *mind mapping* or *mind dumping.* The term *freewriting* was popularized by Peter Elbow in 1973 and 1981 to describe a process he refined for freeing your thoughts from the prison of your mind and getting them down on paper. The technique is simple, if astonishing, for most technical writers. You simply put your hands on the keyboard and type for 10 minutes without stopping. If you cannot think of what to write, you write that down: "I am really stuck here; I cannot think of what to say in this introduction. I know I need to talk about the current solar array and what is wrong with the design. . . ." Sooner or later, you get down some useful information, and you have the beginning of a document to work from. You have now gotten over the hump of beginning your writing. Of course, you might freewrite a methodology section first, rather than starting with the introduction. Why not? *This is only your first draft.*

5. **Start by talking to someone about the subject matter.** Most of us find it easier to talk through our ideas than to write through them. Find a friend or colleague who will listen to your explanation of what you have to write about. When you sit down to write, you will simply be continuing this conversation. You can use this technique later on, too, when you get stuck at a particularly difficult place in your writing. Just walk away from the computer and find a person who will listen.

3 DRAFTING FOR MYSELF

OK, so let us say you have got some kind of outline for your writing project—or maybe just a series of lists and notes loosely tied together. Or you have talked to a good friend or colleague about your project. And you have created some tables and figures to represent data you generated or collected. In other words, you have tried several of the aforementioned strategies—but, nonetheless, you tremble with fear as you open your word processor. It is all very well and good to talk about steps in a process, defining your audience, and using various getting-started strategies, but at some point you just have to sit down and start writing; you have to start typing sentences and paragraphs that will hold together, get the message across, and be grammatically correct (in the end). Almost no one is immune to the terrors of the blank page or the blank screen. When you write, you create something whole (a proposal, say) out of fragments—bits of ideas, concepts, and facts. You are engaged in heavy labor.

EXPERTS SAY...

Peter Elbow says that the main obstacle for most writers is "fear of wrongness" (1981, p. xvii). As long as we have not committed ourselves to words on screen, we can keep the illusion that we really know what we need to know; we just do not want to take the time to write it down. The fact is that you do not know what you know until you write it, even if the draft is rough. As Elbow says, "Clarity is not what we start out with but what we end up with" (1973).

It helps to remember that writing is a process, and the stages cannot be collapsed together or skipped in their initial sequence. This fact means that you have to be realistic in what you demand of yourself. Do not expect to write that report section just right on the first try. Tell yourself that you *will* do several drafts of the document, not because you are a weak writer, but because that is what all sensible writers do. Period. Here are some more suggestions for handling the process of writing:

- **Be good to yourself while you are writing.** Make sure you have water, coffee, tea, or whatever you like close at hand. Take breaks regularly, but not when you are stuck. Find the most comfortable working chair you can, and make sure your keyboard and monitor are at the right heights.
- **Make the process less solitary by getting as much feedback as possible.** Ask friends or colleagues to read drafts of your document (and agree to do the same for them). Or make a deal with a friend or colleague that when you are stuck, you can call him or her and chat about what you are stuck on (e.g., "How do I explain thermal dynamics to an audience who has never studied engineering?").
- **Do not criticize your writing as you write.** Save the criticism for the final revision stage. Especially when you are drafting, stay upbeat and positive

about your own writing abilities: "I can do this; I know I can." Professional writers spend a lot of time talking to each other about their individual strategies for staying cheerful and confident. It is very easy to become sour about your own writing and bog yourself down in editing as you go. It may seem faster to edit as you draft, but in the end this practice will drag you down and make you write slower and slower. And anyway, how can you be a creator (first draft) and a critic (final draft) at the same time? That is like trying to tell a joke when you are angry.

- **If you are collaborating on a document, stay in close touch with your collaborators.** Use e-mail, phone calls, or informal meetings to help you iron out problems and get past sticky spots. Be grateful for collaboration; the hardest thing for most of us about writing is that it is often so solitary. Of course, even when collaborating, you must spend some time writing on your own. But show your work to collaborators as soon as feasible, and get that all-important feedback.

4 USING WRITING TO THINK CLEARLY

Lots of people have said something like, "You do not know what you know until you can write it down clearly." At least some of the time, however, the reason it is difficult to write is that *we are not sure exactly what we want to say*. We may have thought about the concepts and purpose of a document we have to write, but as long as those thoughts stay locked in our head, they remain free-floating, vague, ghostly. Let us take as an example a passage of writing from the introduction of a paper on trying to control ozone in the atmosphere and work backwards into the author's original thoughts on the subject:

> Controlling ground-level ozone is one of the most significant environmental challenges facing the United States. Ozone is a major component of photochemical air pollution, or smog, and causes harm to both human health and the environment. Ozone is formed by atmospheric reactions of volatile organic compounds (VOCs) and nitrogen oxides NO_x) in the presence of sunlight. For more than two decades, emission controls on VOCs and NO_x have been applied, yet ozone concentrations remain at unhealthy levels throughout much of the United States. There are many reasons that the ozone problem remains relatively intractable, but a critical element is the fact that not all VOC and NO_x emission reductions lead to equivalent reductions in ozone formation. *When* and *where* the emissions of VOCs and NO_x occur play a critical role in determining ozone formation rates.

The author has done a lot of research on how ozone forms in earth's close atmosphere and why it is potentially harmful in larger amounts. She also has researched various options for controlling ozone, including controlling emissions of harmful precursors of ozone (VOCs and NO_x), and the ineffectiveness of recent regulations in limiting the amount of ozone produced by these chemicals. So, we can imagine that before she tries to write the introductory paragraph, she lets her thoughts bring to the surface the most important words that keep recurring: "ozone," "controlling," "problem," "emission reductions," and "ozone formation." She jots those words down and starts to construct her introductory paragraph out of them. She also knows that the history of these attempts to control ozone is important, so she inserts a time element: "for more than two decades." And then she realizes that much of her research has involved looking at maps that plot the occurrence of ozone over certain metropolitan areas, and that what is critical to understanding ozone formation is the spatial dimension of how and where harmful emissions are formed. So she needs to use spatial as well as time-oriented words to describe the process: "where" and "when."

Before long, she can construct sentences using those recurring words and string them together using transitions and organizing patterns such as chronology. She is well on her way to writing a clear, readable introductory paragraph, because she has translated her thoughts into straightforward, clear words and phrases. Remember that you can use writing—outlines, jottings, and phrases—to ease the transition from thought to sentence, from mental picture to paragraph. Use informal writing for *yourself*—to help translate your thoughts into writing.

EXPERTS SAY...

Peter Elbow says that freewriting can work to get you started *and* to help your writing flow, because stopping and starting (our usual writing method) creates fragmented writing. Sentences that are freewritten "hang together," even though they still need editing, they have usually captured a whole thought (1998, p. 16).

Engineers can use freewriting to work their way out of hard-to-explain passages and to capture thoughts for the more imaginative types of engineering documents, such as proposals.

5 USING WRITING TO SOLVE PROBLEMS

Writing can actually help you solve problems with your engineering work, especially problems with your methodology and with understanding the full significance of your work. Let us say you are a modeling engineer who models underground reservoirs for oil and gas exploration companies. You use particular models, generated by different software, to solve different issues for clients: size of reservoir, depth, rock and soil formation, action of subsurface water, likely presence or absence of oil and gas over time, etc. You start projects by collecting a lot of field and other data, and then you determine the parameters of the particular model and run it, probably many times. From the moment you start collecting the data that will be inputs to the model, you are involved in analysis that could benefit from being written down. What sort of data are the best as inputs for a particular model? Thinking on paper allows you to sort, categorize, and synthesize multiple types of data on the basis of past experience and of your engineering analysis of the particular situation. In addition, if you write up your thoughts and insights about the data collection, you will already have part of a methodology section written for the final report you have to present to the client. In fact, since these sorts of insights about the data collection are necessary for you to formulate the precise goals of your modeling work for this client, you can now also write a draft of the introduction to the final report. Understanding your overall goals is an excellent way of staying on track with your modeling and number-crunching work. And writing helps you understand (just as understanding helps you write).

Writing can benefit your engineering work throughout the entire project. As often happens, for instance, you may not know whether the model will give you the information you want until you start running it. If the model produces outputs that don't make sense, it could be that you have provided the wrong inputs and need to rethink your methodology. If you have already drafted a data section, it is much easier to pinpoint the data or the calculations that may be faulty or inappropriate. You can then revise your inputs much more easily. Or it may be that your inputs *are* appropriate, but the results (outputs) are very difficult to interpret. Documenting the model's results will help you see connections between them and may lead to the sort of "Aha!" insight in which you suddenly see *why* what happened, happened.

Writing as you perform your engineering work allows you not only to break down the writing process into manageable stages, but also to use what you write to identify mistakes in your approach or in your interpretation. You would not try to keep a string of equations in your head, and neither should you try to keep a string of thoughts in your head. When you interleave *documenting* the work and *doing* the work, you allow each type of activity to improve the other. This approach means writing in increments, using what Herbert Michaelson calls the "Incremental Method" (1990). The trick is to remember that you are using writing to solve problems rather than to impress a reader. Most likely, this method will lead to a draft of the final written product, but that draft will need reorganization, editing, and probably completion. Until that revision stage, however, you are drafting for *yourself* (and perhaps for a trusted content reviewer). You are using writing to help yourself think, analyze, and solve problems.

SUMMARY

- Starting a writing project is quite scary. Everyone should use a start-up strategy that works for him or her. The basic trick is to remember to smother your inner critic at the beginning and bring him or her back into play only when you have finished drafting the document.
- Some writing is done entirely for yourself (lists, initial outlines, etc.); no one needs to see the way you used these forms of writing to get to the stage of actually writing the document. And no one need ever see your drafts, unless you want a review. Writing is an aid for your thinking process, and it can even be combined with that other aid, doodling.
- Writing is a form of problem solving. You can use writing to help plan the direction of your engineering work as well as to frame the document you have to write. Documenting your work *as* you work improves your engineering analysis.

Problems

1. Try one of the start-up strategies described in this chapter. Use the strategy to organize your thoughts and write a first draft of any writing assignment you have currently. Keep a record of how long it takes you to begin writing the first draft (completing at least one paragraph) and then how long it takes to complete the draft. Also record how comfortable you feel with that strategy. Hand in these notes to your instructor.
2. If your chosen strategy in #1 did not work, then, for your own benefit, try another strategy later for another assignment. Determine whether that one is more comfortable for you and prepares you better for writing the draft. Explain why you are or are not better prepared.
3. Think back to a recent writing project. Did you follow the steps listed in Section 1? If you skipped any, do you think your final product was affected negatively? Write up your thoughts on this matter.
4. How might an engineer use freewriting to help himself or herself get started writing a report on how to improve a particular work-flow process?

5. You have been asked to write a report requesting funds for studying solutions to the parking problem at your school, university, or place of work. The report will be written to the school president or the CEO. Brainstorm with another student and produce a list of all the topics you will have to address in such a report.

6. Partner with another student to begin writing the report outlined in #5. List the kinds of information you will have to gather and decide who will gather which information. On what basis did you decide how to divide up the research?

7. For the report requested in #5, talk to a sampling of students about how much time they take to get to work or school in the morning. After the interviews, *first*, plot the results in a graph or chart. *Then*, write a paragraph explaining the visual. Answer these questions:

   How long did it take to write that paragraph? Do you think the writing went faster because you had already created the visual?

8. The first draft of any document is usually written for whom?

   a. The author's boss
   b. The primary audience
   c. The author

FURTHER READING

Elbow, P. 1998. *Writing with Power: Techniques for Mastering the Writing Process*, 2d. ed. Oxford University Press: New York.

—— 1973. *Writing Without Teachers.* Oxford University Press: New York.

These are two of the most influential books on the process of writing ever written. They broke new ground by focusing on the *way* we write rather than just *what* we write. Engineers may not want to try all the exercises in these books, but they will benefit from listening to Elbow's careful articulation of the distinction between the critical and the creative functions of our brain.

Michaelson, H.B. 1990. *How to Write and Publish Engineering Papers and Reports*, 3d. ed. Oryx Press: Phoenix, AZ.

In this handy and brief book, Herbert Michaelson outlines with admirable clarity the organization of engineering papers and reports. The focus is on journal papers, but this is not to say that the book is academic in language or tone. Michaelson's goal is to encourage engineers, wherever they are working, to publish more. Chapter 5, "How to Use the Incremental Method," makes a good case for writing as you proceed with a project; Michaelson points out that publication will be much faster when this technique is used.

9

Writing: It's about Control

Sections

1 Writing for Control Comes *after* Drafting for Content
2 Readable Style
3 Constructing Powerful Paragraphs
4 Grammar and Sentence Structure: Diagnosing and Curing the Problems
5 Words: Picky, Picky
6 Punctuation: Why Should I Care?

Objectives

By reading this chapter, you will learn the following:

- to use a readable style for your technical documents;
- to recognize passive-voice constructions and use them appropriately;
- how to construct logical, organized paragraphs;
- to recognize sentence fragments and avoid them;
- how to choose the correct spelling of commonly misspelled words;
- how to choose the proper verb tense;
- how to control sentences through punctuation.

1 WRITING FOR CONTROL COMES *AFTER* DRAFTING FOR CONTENT

When you draft a document, you are trying to get your thoughts down on paper or screen. You are trying to wrap your arms around the content of what you need to say, the data you need to present, and the arguments you need to make. You should not worry during the drafting phase about grammar, punctuation, and all those other things that the word "writing" may mean to you. If you worry about them too much, you will slow yourself down. You want your draft to be as complete as possible, and *then* you can go back and correct it for writing errors.

When you have completed a good first draft—that is, when you have written something about all the items on your outline or you just believe that you have covered the territory of what you have to say or present—then you are ready to focus on the quality of your writing. If you have already cognitively absorbed the guidelines presented in this chapter, you will of course have a better quality draft than if you know very little about crafting clear sentences and paragraphs. But either way, be easy on yourself and remember that you will still need to get external feedback before you deliver a final document.

2 READABLE STYLE

The style of technical writing has long been thought to be different from the styles of other types of writing, such as creative writing, essay writing, and journalism. And it *is*, although not as different as many people think. As a technical writer, you still have to grab a reader's attention, make logical arguments, and carry the reader along in a flow of words that gets your message across. The difference is that technical writing focuses less on you, the writer, and more on the ideas or facts you are presenting. As with all writing, you are responsible for helping the reader understand the message, but the message in this case often involves complex data and information.

Even in the most technical documents, you will want to be as concrete as possible so that the reader does not have to work too hard to learn the new information you are offering. Research shows that we read at two levels of comprehension as we process information: superficial processing and deeper processing. In the first type, we simply try to retain facts; in the second, we attempt to comprehend relationships among the facts (Der Meer and Hoffmann, 1987, p. 119). In addition, we try to remember new facts by seeing whether they link up with what we already know. If the link is clear (for example, if the author uses transitions and sequences information logically), the reader has good help; if new and important information is highlighted, the reader has even more good help in understanding the content.

RESEARCH SHOWS...

Cognitive psychologists study how people read and understand information. Their studies have shown that these elements enhance the processing of information for the reader:

visual imagery
examples
comparisons
descriptions
questions

(Der Meer and Hoffmann, 1987)

Let us look at an example of clear technical writing. Even if you know nothing about phase-sensitive X-ray imaging, you will probably easily understand this passage from *Physics Today* (July 2000, p. 23):

> Clinical and biological studies stand particularly well poised to benefit from the development of phase-sensitive techniques. Absorption contrast works well in distinguishing between hard and soft tissue: Heavier elements—like calcium in bones and teeth—have a much higher absorption cross section than the lighter elements that constitute soft tissues. However, in many clinical situations, such as mammography, there is a need to distinguish between different kinds of soft tissue—between tumors and normal tissue, for instance. Because the absorption is small to begin with, and differences in density and composition are slight, standard x-ray imaging is not as successful at this task.

Readable style is generally grammatically correct, but it is also a lot more than just that. In fact, grammatical correctness may be less important overall than presenting information logically and leading the reader carefully along the trail of your thought. To lead your readers, you need to prune your writing of deadwood, make the sequence of your thought explicit, avoid ambiguity, and keep the reader focused on the most important points. The keys to good writing, outlined in the upcoming box, are just as important in technical writing as in any other type of writing. They are not so much rules of grammar as strategies for writing clearly and concisely and for helping your reader remember the information.

Passive and Active Voice: What Is All the Fuss?

Perhaps the largest difference between technical and other kinds of writing is the increased use of passive voice in the former. The word "voice" as applied to writing just

PRACTICE QUESTION: CLEAR TECHNICAL WRITING

How many elements listed in the previous "Research Shows..." box can you discover in the passage from *Physics Today*?

SUGGESTED ANSWER

example — comparison

Clinical and biological studies stand particularly well poised to benefit from the development of phase-sensitive techniques. Absorption contrast works well in distinguishing between hard and soft tissue: Heavier elements **—like calcium in bones and teeth**—have a much **higher** absorption cross section **than** the **lighter elements that constitute soft tissues.** However, in many clinical situations, **such as** mammography, there is a need to distinguish between different kinds of soft tissue—**between tumors and normal tissue, for instance.** Because the absorption is small to begin with, and differences in density and composition are slight, standard x-ray imaging is not as successful at this task.

description — specific contrast — example — example

HINT

Keys to Powerful Writing

There are five keys to good writing. The last three contribute to both the big picture (paragraphs and sections) and the smaller picture (individual sentences). The first two keys keep your sentences clear, concise, and communicative. The five keys to powerful writing are

prune your writing
control passive voice
maintain logical linkage
maintain focus
maintain parallelism

means how you "sound" on the page, instead of at the lectern. Your writing has a voice because the reader, in some not-very-well understood way, "hears" the words on your page in his or her head. Voice in writing is different than in talking, because the reader can slow down and "listen" again to your words. So, you have to make sure that your "tone" is exactly what you intend. Both passive voice and active voice can strike the wrong tone when used inappropriately.

What is the difference between passive and active voice, and why should you care? Well, compare these two sentences:

I threw the ball.
The ball was thrown by me.

The first is in active voice, the second in passive. Say these sentences aloud to yourself. The first is the way we would describe the event when talking; the second kind of sentence usually occurs in writing only, not in speaking. The agent (or doer) of the action is clear in the first sentence; the doer may not be so clear in the second version, especially (as often happens with passive sentences) if the doer is left off entirely, as in "The ball was thrown."

When to use which voice is an ongoing issue for technical writers, who often present information in which the doer of the action is not important, as in "The material was heated to a temperature of 100°C." In that sentence, the reader does not care who heated up the water sample, so naming the doer of the action simply wastes the reader's time. We do not need to use the active-voice construction, "The technician heated the material to a temperature of 100°C," because we do not need to know who is the agent of the action of heating (the technician). Teachers of creative and essay writing encourage students to use only active voice, because active-voice sentences are usually clearer and more direct. But technical writers, including you, must decide what they want to emphasize before they can construct a good sentence and use the appropriate voice.

HINT

When to Use Which Voice?

Active Voice	Passive Voice
• to emphasize the agent of the action: Previous researchers established the relationship between chemical qualities and model parameters.	**• to emphasize the object:** The relationship between chemical qualities and model parameters was established by previous researchers.
• to be concise and direct: JMC Consulting recommends that the agency adopt a risk-communication plan.	The water sample was heated to a temperature of 100°C.

Wordiness: Learn to Prune

A big complaint about technical writing is its wordiness. A sentence like the following is typical of much technical writing:

> An important conclusion extracted from the result is that there is a correlation between the mixing length and the volumetric flow.

The sentence is not terribly long, but it wastes words. And those "dead" words get in the way of our quick and easy understanding of the meaning of the sentence. The following is a better way to write it:

> The result demonstrates a correlation between mixing length and volumetric flow.

The revised sentence has 11 words, almost half the length of the original, which has 21 words. The second sentence relates the important conclusion without wasting time announcing that we are about to hear an important conclusion. If the word "important" seems important, there are other ways of declaring importance to a reader, such as italicizing the sentence or making sure that the sentence begins or ends a paragraph.

Our writing models are often not very good, and we pick up habits of using unnecessary words and legalisms. Also, the way we first write something when we are drafting is often the long way around explaining it. *Pruning* is the process of going through your own draft and removing the deadwood—the words that really do not add anything to the meaning you are trying to convey. Just as pruning a tree makes it stronger and eventually more beautiful, pruning your sentences will strengthen what you are trying to convey. See Table 1 for some examples of how to prune common deadwood phrases.

TABLE 1 Words and Phrases that Can often be Pruned

Deadwood	Pruned wood
different aspects	aspects
make a measurement	measure
give consideration to	consider
relative to	about *or* of
it should be noted that . . .	note that . . .
regardless of the fact that	although
due to the fact that	because
subsequent to	after
in the course of	during

PRACTICE QUESTION: PRUNING

How many words can you remove from this sentence without changing the meaning?

The following report outlines the different aspects relating to the current feasibility of constructing a magnetically levitated bullet train in Central Texas.

SUGGESTED ANSWER

This

~~The following~~ report outlines ~~the different aspects relating to~~ the *current* feasibility of constructing a magnetically levitated bullet train in Central Texas.

The word "current" is also a candidate for pruning; it depends on whether you want to emphasize the report's timeliness (perhaps in comparison with another report from last year) or not. Normally, feasibility studies cover only current feasibility.

But do not prune just to make sentences shorter! Shorter sentences are not necessarily better than longer sentences. Consider this string of short sentences:

> This report is a final report. It presents final results for the design of a wastewater treatment plant. The design is for the process sequence, piping, flow distribution, and hydraulics. It includes the pump station and biological-treatment units. It also includes the disinfection and sludge-handling processes for the plant.

Notice how confusing those sentences become as they keep stopping and starting. What does "it" refer to at the beginning of three of the sentences: the report, the design, or some process? "It" is a very small word that does not mean anything by itself. Not only are the sentences choppy, but also we have to keep reading backwards to figure out what "it" means each time. Here, the best solution is to combine some of these sentences, not to shorten them:

> This final report presents the design of a wastewater treatment plant, including the process sequence, piping, flow distribution, and hydraulics. **The design** also includes the pump station, the biological-treatment units, and the disinfection and sludge-handling processes for the plant.

We have lengthened the sentences in this passage, but by combining some sentences, we have actually *shortened the passage*. Productive pruning can be a two-step process in which you actually lengthen a sentence in order to strengthen the passage. Again, pruning is not better for its own sake, but only when it removes deadwood. Here, some of the deadwood is the meaningless word "it" that keeps cropping up, which leads us to the next major problem with much technical writing: empty pronouns.

PRACTICE QUESTION: COMBINING SENTENCES

Consider the following sentences:

> **The population explosion in and around the Austin area has proved to be problematic regarding water-drainage structures. It has particularly affected the Brushy Creek area in southern Williamson County.**

- Does combining these sentences improve their readability?
- What problems does combining the sentences solve?

SUGGESTED ANSWER

> **The population explosion in and around the Austin area, particularly in the Brushy Creek area in southern Williamson County, has proved to be problematic regarding water-drainage structures.**

By taking the geographic details of the second sentence and inserting them into the first sentence, we can keep "population explosion" as the subject of one clear sentence with one predicate ("has proved to be"). Longer sentences may produce a shorter passage! Of course, we can prune this sentence even more—see next Practice Question.

Empty Pronouns

Isn't it annoying to read sloppy sentences that use the word "this" instead of a word with some meaning, such as the following?

> In the United States, the chemical and petroleum industries consume eight million tons of hydrogen annually. **This** is produced from natural gas by a process called steam reforming that extracts hydrogen from hydrocarbons. **This** can occur by one of two methods.

When you use "this" without any noun attached to it, especially at the beginning of a sentence, you are cheating the reader. Guessing what the author means by "this" is a stupid game that we readers would rather not play. In the second sentence of the foregoing example, we would like to be sure that the word "this" refers to hydrogen. (Actually, it refers to the eight million tons of hydrogen consumed by the U.S. chemical and petroleum industries.) Hydrogen, however, is not particularly emphasized in the previous sentence; it is tucked into a prepositional phrase ("of hydrogen"). So why not say, "**This hydrogen** is produced . . ."? And what does "this" refer to in the third sentence? As readers, we prefer the clarity of "**This process** can occur by one of two methods," or better yet, "**Steam reforming** can occur. . . ."

HINT

Simple Rule: Never use "this" by itself at the beginning of a sentence. Always use "this" to modify a noun—for example, "this *process*" or "this *chemical reaction*."

Use of "it" at the beginning of a sentence can also be a problem. Generally, when we begin a sentence with "it," we are delaying getting to the point. For example, consider the sentence, "It is the case that hydrogen occurs naturally." We can prune away the first five words without changing the meaning of that sentence! "It" has its place, but watch out for cases where you are using "it" because you do not know exactly what you want to say or because you are being unthinkingly wordy.

Noun Compression: Let Some Air in

Another barrier for readers of technical writing is long noun phrases, such as *process design revamp project* or *multireservoir water resources systems*. Even if you have seen these phrases before, it is difficult to know exactly what they mean. Here is a sentence that invites you to skip right over it:

> The scope of work provides technical service support to continue to improve the quality of the company's Advanced Science and Technology Plan and **develop a gap-analysis-based, site-specific Integrated Technology Plan** that includes Plant Directed Research and Development Thrusts.

Problems abound with this sentence, including the prevalence of strings of capitalized nouns that mean nothing to a reader unfamiliar with these "plans" and "thrusts." But even the noncapitalized noun phrases are problematic. What does "gap-analysis-based" mean? "Based" is actually an adjective, not a noun, but it does not help us much to discover the meaning. Does the phrase mean an analysis based on identifying the gaps in the current plan? If so, that part of the sentence should read this way:

> develop **a site-specific Integrated Technology Plan that fills in the gaps in the current plan**

If we insert this more understandable phrase into the original sentence, however, we will make the sentence even longer, so perhaps in this case we should split the sentence in two:

> The scope of work provides technical service support to continue to improve the quality of the company's Advanced Science and Technology Plan. **The scope should also** develop a site-specific Integrated Technology Plan that fills in the gaps in the current plan and includes Plant Directed Research and Development Thrusts.

The new passage is longer than the original, but it is much clearer. Of course, we can always work on pruning it. For example, do you think "technical service support" means anything more than "technical support"?

Noun phrases clog up our writing because they try to stuff so much meaning into a small place that there is no room to understand the meaning. Sometimes you have to add words to clarify the relationship of words to each other. Look at this sentence, for instance:

> This report on statewide outdoor burning describes the **activity data collection process**.

Surely some of the nouns in the indicated phrase are being made to act as adjectives, but we are uncertain about what is referring to what. Adding prepositions and pronouns may help to clarify compressed noun phrases. What kind of activity are we talking about? The incidents of outdoor burning? In that case, a better version would look like this:

> This report describes **the process <u>of</u> collecting data <u>on</u> incidents <u>of</u>** outdoor burning statewide.

The solution is to add some prepositions that indicate the relationship of certain words to each other. Here, too, the new sentence is *longer* than the original, but the changes *add* to the meaning rather than detract from it.

HINT

Simple Rule: Shorter is not always better. Clearer *is* always better.

PRACTICE QUESTION: MORE PRUNING

We previously produced the following sentence by combining the original two sentences:

> **The population explosion in and around the Austin area, particularly in the Brushy Creek area in southern Williamson County, has proved to be problematic regarding water-drainage structures.**

Can you make it even more concise (without losing meaning) by pruning it some more?

SUGGESTED ANSWER

Here is the sentence with changes shown and then with the changes incorporated.

The population explosion in and around the Austin area, particularly in the Brushy Creek area in southern Williamson County, has ~~proved to be~~ *created* problem~~atics~~ ~~regarding~~ *for* water-drainage structures.

The population explosion in and around the Austin area, particularly in the Brushy Creek area in southern Williamson County, has created problems for water-drainage structures.

The changes have made this sentence both shorter and clearer.

3 CONSTRUCTING POWERFUL PARAGRAPHS

Here are the three keys to writing clear paragraphs and sections of a document:

- maintain logical linkage;
- maintain focus;
- maintain parallelism.

Maintaining links, focus, and parallelism is a matter of style and emphasis; using these keys requires that you think hard about exactly which bits of information and data are most important.

Maintain Logical Linkage

Good technical writing is not list making. We do not learn anything from lists unless we already know the context for all the items on the list. When you go to the grocery store, you already know that you are buying items for dinner, for instance. But you probably do not write down, "dinner for four people." And yet knowing that information is critical to understanding how much of each item to buy. You cannot really use your grocery list—e.g., chicken, lettuce, tomatoes—without understanding that unwritten context. Good writing helps the reader understand the context by making explicit connections between ideas or pieces of information. These connections are like signposts that say "A is like B" (comparison), "A caused B" (cause and effect), "A is different from B" (contrast), or "A is more important than B" (degree of control). These signposts use conventional words and phrases such as "similarly," "consequently," and "on the other hand" to show the reader where your thought is heading.

PRACTICE QUESTION: MAINTAINING LINKAGE

Here is a paragraph from the Work Remaining section of a progress report. Do you know in what direction the thought is headed in each sentence? If not, can you diagnose what is missing from the paragraph?

> **The evaluation of diamond film based on the final criterion will be completed when the analysis of its qualities is received from the Institute for Advanced Technology (IAT). A request for the resource files on diamond film will be submitted to Bolton Manufacturing. An additional comparison of the results received from IAT will be made with information provided by Bolton. The evaluation of diamond dust will be used for the comparison of the other nanocrystalline coatings.**

SUGGESTED ANSWER

There are no links between these sentences, no words that establish a conceptual connection. Every sentence seems to start the thought process all over again.

Let us take a closer look at the paragraph in the foregoing practice question to understand what has gone wrong:

> The evaluation of diamond film based on the final criterion will be completed when the analysis of its qualities is received from the Institute for Advanced Technology (IAT). A request for the resource files on diamond film will be submitted to Bolton Manufacturing. An additional comparison of the results received from IAT will be made with information provided by Bolton. The evaluation of diamond dust will be used for the comparison of the other nanocrystalline coatings.

The sentences generally describe what will happen, but we do not know the connection between each of these promises. Are they listed in chronological order? Will the most important steps be taken care of first? Do certain steps depend on others and require that the others be accomplished first? What does Bolton Manufacturing have to do before the author can accomplish certain steps? The paragraph is simply a list of unrelated steps and reveals almost nothing about the sequence of this process (except for the word "when" in the first sentence).

Part of the problem with this paragraph is overuse of the passive voice; we are not sure who is doing what when. But even if we retain most of the passive-voice constructions, we can greatly improve the readability of the paragraph by adding links. Lets us look at a revised version of the paragraph. Why is this revision so much easier to understand?

> The evaluation of diamond film based on the final criterion will be completed when the analysis of its qualities is received from the Institute for Advanced Technology (IAT). **In addition**, a request for the resource files on diamond film will be submitted to Bolton Manufacturing, **so that we may compare** the results received from IAT with information provided by Bolton. **Once our evaluation of diamond film is completed**, we can compare diamond film with the other nanocrystalline coatings.

The paragraph now includes some links between sentences and between the concepts of the second, long sentence. We know a lot more about where each sentence is going. What made the difference in this paragraph? Transitional phrases.

HINT

Transitions establish strong links between these elements:
parts of a sentence;
whole sentences;
whole paragraphs.

PRACTICE QUESTION: TRANSITIONS

How many words or phrases can you think of that would fit in the blank at the beginning of the second of the following two sentences?

> ***The project manager will be in Japan for the next two weeks. _____, the project engineers will remain in town at the branch office.***

SUGGESTED ANSWER

Any of these words would make sense as a link for the two sentences:

However
Nevertheless
Therefore
Unfortunately

There are other possible links as well. Each one produces a different meaning. You have to provide the *right* link to the reader to convey the right meaning (the one you intend).

When you consider these two sentences and the relationship between them, you realize that lots of different relationships are possible:

> The project engineer will be in Japan for the next two weeks. _____, the project engineers will remain in town at the branch office.

Each of those relationships would be indicated by a different word or phrase in that blank space at the beginning of the second sentence, such as "On the other hand," "Therefore," or "However." Each of these transitions imparts a different meaning to the sentences. Which is the right meaning? As the writer, you want to establish control and guide your readers to the right meaning. You may want to emphasize contrast ("However") in order to make clear the difference between the situations of managers and nonmanagers. Or you may want to emphasize cause and effect ("Therefore") in order to explain why the engineers cannot travel even locally during the next two weeks. **If you do not establish the correct connection between sentences and phrases, your reader will attempt to establish the connection anyway.** And that might be the wrong connection or meaning. For example, do you want your readers to assume that the missing word in the blank is "Unfortunately"? (This example also shows that sometimes two sentences are better constructed as one: Especially if you use "However," you would want to make one sentence, with a semicolon before and a comma after: "; however,".)

Table 2 lists some useful transitional words and phrases. Note, however, that this list does not begin to cover all the varieties of transitional words and phrases.

TABLE 2 Partial List of Transitions and the Relationships They Indicate

Relationship of One Phrase, Sentence, or Paragraph to Another	Transitional Word or Phrase
Comparison	similarly
Cause and effect	therefore; consequently; as a result
Contrast	but; however
Sequence	first; second; finally
Addition	and; also
Example	for example
Classification	partly; the other half
Spatial relationship	upwind; downwind; to the left; to the right; in back; in front
Relative importance	most importantly; secondarily

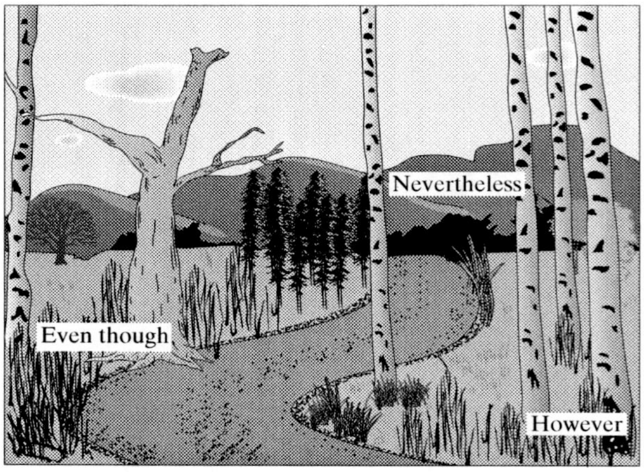

The right transitions guide the reader down the right path.

Transitions also create organization. In a paragraph such as the next example, the transitional phrases establish the organizing pattern—comparison and contrast—that helps us make quick sense of the writing. In this paragraph, the author (in *Physics Today*, August 2002, p. 17) is talking about how a team of researchers is attempting to quantify the amount and distribution of water ice on Mars:

> For the ice's depth distribution, the team considered **two simple cases. In one**, the ice-bearing soil extends right to the surface; **in the other**, a dry layer of variable thickness tops the ice-containing layer. Thermal and epithermal neutrons, thanks to their **different behavior in the two cases**, provide the discrimination.

We can see classification also at work here ("two" cases, "the other").

You will notice that we have already looked at several of these patterns, since making clear the precise relationship among facts, ideas, or events is the main goal of much technical writing. In addition, the patterns often overlap within the same paragraph or section of a document. Which patterns are operating in this paragraph from an article on X-ray imaging (*Physics Today*, July 2000, p. 26)?

> A monochromator throws much of the x-ray flux away, and when used with an x-ray tube source, would make the time needed to record an image unacceptably long for clinical uses. The ability to use the polychromatic output of an x-ray tube is **therefore** important for potential clinical applications. The large beam divergence from such a source has the **additional** advantage of providing magnification, with a corresponding increase in spatial resolution. The spreading beam **also** allows the imaging of large areas in a single step.

Notice that in this paragraph, the transitions are located inside the sentences rather than primarily at the beginning. Nonetheless, they link parts of whole thoughts. By making it easy for the reader to follow the pattern of your thought, you maintain flow and help keep the reader focused on what is important.

Maintain Focus

When it comes to helping the reader focus on critical information, the most powerful writing tool you have is the topic sentence. Think about how you read paragraphs. Usually, you read the first sentence for a clue as to what the paragraph is about. (Every so often, a topic sentence will come second or third, after a sentence or two that link to a previous paragraph.) Then you want to hear mainly about *that* subject; if the paragraph tries to cover too many subjects or if it has no clear topic sentence, you will understandably become confused. The topic sentence is like a welcoming door to a house: If the door is closed and locked, you will feel unwelcome and not go in. If it is slightly open and you peek inside, you will want to know where you are entering. Is this a place of good, clear information?

Consider the following paragraph:

> The State does not appropriate funds to maintain and/or build surface parking lots, parking facilities, or parking garages. All parking on campus, including surface parking and the parking garage, is operated and maintained from the parking system revenues. Revenues include the annual parking permit fee, incomes from citations issued by The University Police, and fees charged to use the University parking garage. The construction of a second parking garage is necessary. To meet the required funding to construct this parking garage, it will be necessary to increase the campus parking revenue by raising the annual parking permit fee. The parking permit fee will increase over a two-year period to accomplish this.

Does the first sentence make you want to "enter" the paragraph? Do you feel that the first sentence is a good indicator of the subject matter covered by the rest of the paragraph? If you answered "no" to both questions, you are not alone. Most people would not read this paragraph all the way through. Knowing the sources of the state's funds does not seem important at the beginning because we are not given a clue as to the real subject matter of this paragraph. Of course, the real subject is bad news: The parking fee will go up. So we can understand why the writer did not want to use the last sentence—*The parking permit fee will increase over a two-year period*—as the topic sentence. But the problem is that hiding bad news may mean that someone simply does not read the entire paragraph and is therefore shocked when receiving a higher bill in the mail. If you really do not want to inform your readers, do not write anything, because badly organized writing will only *misinform* them. If you *do* want to inform them, figure out a way of organizing the paragraph so that you keep the reader focused on the most positive message—in this case, the fact that a garage will be built to help solve the university's parking problems.

PRACTICE QUESTION: TOPIC SENTENCES

Looking at the foregoing paragraph about constructing a new parking garage at a university, can you come up with a topic sentence that emphasizes a positive message? What would be a better second sentence?

SUGGESTED ANSWER

- Here is one suggestion for a better topic sentence:

  To solve the problem of increased traffic on campus, the university is building a second parking garage.

This sentence focuses on the solution to the problem of increased numbers of cars on campus.

- Here is a suggestion for the second sentence:

  To fund this initiative, the parking fee will increase over a two-year period.

Another way to maintain focus in every sentence of a paragraph is to keep the reader's eye on the ball—that is, on the most important pieces of information. Consider this pair of sentences:

> The proposal does not conform to our research goals as outlined in the five-year strategic plan. For instance, the potential for discovering new laser techniques is discussed only briefly and does not form a major part of the proposal.

The topic sentence here seems fine. But the second sentence does not tie in well with the first, in spite of the helpful transitional phrase "For instance." The second sentence is confusing because it does not emphasize at the beginning what is important: the proposal. In the following revision, the noun that is the main subject is repeated in the position of subject:

> The **proposal** does not conform to our research goals as outlined in the five-year strategic plan. For instance, the **proposal** fails to emphasize our potential for discovering new laser techniques; these techniques are discussed only briefly.

Placing "the proposal" in the position of subject of the second sentence greatly increases our comprehension of the content. We know now to keep focusing on the proposal and what it does and does not do.

HINT

Maintaining Focus

- Tie new information to old information.
- Place repeated information in the subject position.

Research has shown clearly that readers want new information to be tied explicitly to information they already know. Study this passage from *Physics Today* (August 2002, p. 21) and notice how seamlessly it weaves new information into the fabric of given information by repeating key words and concepts in the "subject" position:

> After setting a **wave** in motion, the researchers take an **image** of the tray with a video camera. **The wave** is brighter than its unexcited surroundings, and its size and location are easily **measured**. **These measurements** go

> into the feedback step: determining the pointing direction and brightness of the reaction-controlling spotlight. By comparing the current **image** with an image taken two seconds earlier, the researchers can tell whether the **wave** needs a bigger or smaller dose of light to maintain its stability.

This passage also makes good use of parallelism to channel the reader's attention. What is parallelism? Read on . . .

Maintain Parallelism

When you present a series of facts, findings, or steps in a process, you want readers to remember how the items are similar and dissimilar. Mostly, you want them to remember the facts or steps themselves. To make clear the similarity of certain items and to aid reader's memories, you need to use parallel structure (parallelism). A sentence such as this one is unnecessarily wordy and potentially confusing because it does not maintain parallelism:

> The architectural engineer finished the mechanical plan, made five copies, and then it was forwarded to the client.

Of the three actions the engineer took, two are presented in the active voice and one in the passive voice. Maintaining parallelism means being consistent with your grammatical constructions for each item in your list, especially with your verb forms.

To maintain parallelism, you must choose a grammatical form of presentation (based on parts of speech, usually) and stick to it. Here is a set of instructions that is not in parallel form:

Recommended Procedure for Characterizing Emodin:

1. *Obtain emodin from reliable chemical manufacturing company. (See list.)*
2. *Identification should be made with nuclear magnetic resonance spectroscopy.*
3. *Determining purity should be done by elemental analyses.*

In this example, if we agree that the first instruction ("Obtain emodin . . .") is the best form for a set of instructions, then we should stick to that form for the other items. Here, the chosen form sets up each item in the following sequence of parts of speech:

command form of a ***verb/***	***noun*** *as object/*	***prepositional phrase***
obtain/	**emodin/**	**from ... company**

If we rewrite steps 2 and 3 in parallel form with step 1, we come up with something like this revision:

Recommended Procedure for Characterizing Emodin:

1. ***Obtain*** *emodin from reliable chemical manufacturing company. (See list.)*
2. ***Identify*** *emodin by using nuclear magnetic resonance spectroscopy.*
3. ***Determine*** *purity through elemental analyses.*

HINT

Present parallel ideas in parallel form in the following structures:

- lists;
- sentences (series or pairs).

Pay special attention to the first few words of each clause or phrase.
Remember that parallel constructions channel the reader's attention and aid his or her memory.

As we have seen, parallelism clarifies sentences as well as vertical lists. Let us return to our original example:

> The architectural engineer finished the mechanical plan, made five copies, and then **it was forwarded** to the client.

This sentence is easy to fix, right? We simply stick with the past-tense verb form of the first two actions. The final phrase then becomes active instead of passive: ". . . and then **forwarded it** to the client."

Here is an example that includes more than one "and":

> We researched and compared the performance capabilities and FAA records of each aircraft, **as well as conducting** a pilot survey.

The phrase after the comma is not in the same grammatical form as the other items in this sentence list, right? You have to change "conducting" to a past-tense verb: "conducted." Then you can make explicit the fact that the research involved at least two distinct steps—researching and comparing, and conducting a survey—not one big simultaneous step. Readers like processes and information to be explicitly broken down into understandable steps or ideas. Here is one possible revision:

> We researched and compared the performance capabilities and FAA records of each aircraft, and then **we conducted** a pilot survey.

4 GRAMMAR AND SENTENCE STRUCTURE: DIAGNOSING AND CURING THE PROBLEMS

Some problems with sentences are matters of clarity and style, and some problems are matters of grammatical correctness. Focus on the clarity first; then fix the grammar. There are several well-described methods available for pinpointing the weakness and wordiness of so many of our sentences. The best writing guides include Strunk and White's *The Elements of Style* (1979) and Richard Lanham's "Paramedic Method," developed in *Revising Business Prose* (2000). The best methods all work to counteract the vague and bureaucratic writing that Lanham calls the "Official Style." You may think the Official Style is the best way to write, because you see so much business and professional writing that is artificially pumped up with extra words and roundabout explanations. Actually, much of that writing is simply imitating a pompous style to which *those* writers were exposed. The fact is that most technical writers have never been taught how to revise their writing.

Let us look at a sentence on the experimental results of a study of liquid-jet-gas pumps:

> An important conclusion extracted from the result **is** that **there is** a correlation between the mixing length and the volumetric flow.

Thinking about the words "An important conclusion . . . is that there is . . . ," we realize how empty they are. Conclusions should simply be stated, not backed into. If we follow Richard Lanham's advice, we should improve our sentences by always asking the question, "Where is the action?" Most sentences focus on their verbs because we are most interested in what happened (or will happen or could happen). That something simply "is" is rarely enough information about it. The main action of the original sentence might be "extracted." But is the fact that the conclusion was "extracted" really important to the reader? Are we not interested mainly in the conclusion itself, in what the result demonstrates?

Something important has been proven, or at least demonstrated. Can we express that result in a straightforward verb? Consider the following revision:

> The result demonstrates a correlation between mixing length and volumetric flow.

Pruning this particular sentence has also involved changing a passive construction ("conclusion extracted from . . . is") to an active construction ("result demonstrates").

We already know that excessive and inappropriate use of passive voice can be a problem for the reader, but how do you recognize passive voice in your own writing and test whether it is appropriate or not? Let us look at a paragraph from the Work Remaining section of a progress report:

> The evaluation of diamond film based on the final criterion **will be completed** when the analysis of the qualities **is received** from the Institute for Advanced Technology (IAT). A request for the resource files on diamond film **will be submitted** to Bolton Manufacturing. An additional comparison of the results received from IAT **will be made** with information provided by Bolton. The evaluation of diamond dust **will be used** for the comparison of the other nanocrystalline coatings.

The passive-voice constructions are printed in boldface. To test for passive voice, just check each verb and see whether you can easily ask the question, "By whom?" For instance, "will be completed" brings up the question, "By whom?" If you have read the progress report up until this point, you have a pretty good idea that the agent of the action of "completing" is the writer, so you are probably not confused at that point in the paragraph. But you are forced to keep asking yourself, "By whom?" for every single action in this paragraph. There are, however, other potential agents of action (IAT and Bolton), so eventually you get tired of having to answer that darned question at least once in every sentence. Excessive use of passive voice wears us out! Reading is time consuming enough without having to worry about who is the agent of the action all the time!

Of course, not all uses of passive voice are wrong. Especially in technical writing, we may not need to know who is the doer of the action. Sometimes the sentence itself will answer the question, as in this example:

> The report **was disseminated** by the Department of Justice.

In that case, the verb is still passive, but the question gets answered, and so we are not confused. But if many sentences take this form, the document will be longer than it needs to be. As a writer, think about what is more important: the object of the action or the doer of the action.

Of course, putting verbs in active voice means inserting what is missing: the agent of the action. If the agent is the writer, technical style has long dictated that the writer may not appear as "I." In many contemporary journals, however, "we" is now perfectly acceptable. If you are told that you cannot use "I" or "we," then you may want to restructure some of your sentences to include active verbs nonetheless. Consider the following example:

> The evaluation of diamond film based on the final criterion will be completed when the analysis of its qualities is received from the Institute for Advanced Technology (IAT). A request for the resource files on diamond film will be submitted to Bolton Manufacturing. An additional comparison of the results received from IAT will be made with information provided by Bolton. The evaluation of diamond dust will be used for the comparison of the other nanocrystalline coatings.

We may revise this paragraph as follows:

> The evaluation of diamond film based on the final criterion **will be completed** when the analysis of its qualities **is received** from the Institute for Advanced Technology (IAT). Once Bolton Manufacturing **responds** to a request for resource files on diamond film, an additional comparison of the results received from IAT **will be made**. The evaluation of diamond dust **will be used** for comparison with the other nanocrystalline coatings.

Of the five verbs with a subject in the foregoing paragraph, four are still in passive voice, but one is now in active voice—"responds." Turning just one verb form into active voice helps the flow of this paragraph and helps prune away other unnecessary words. Of course in this case—an industry progress report—there is no reason not to use "we." The writer is probably doing work for a client.

Another test for good writing is determining the number of prepositional phrases. Check out the original version of the last two sentences again:

> An additional comparison of the results received from IAT **will be made** with information provided by Bolton. The evaluation of diamond dust **will be used** for the comparison of the other nanocrystalline coatings.

In these two sentences, totaling only 33 words, there are seven almost-consecutive prepositional phrases. Those phrases slow down our reading, because prepositions do not involve action. They do establish relationship; something belongs *with* something else or comes *after* it. But too many in a row pull us away from the main thread of action that we are always trying to follow. As a rule, stick to no more than three prepositional phrases in a row. If we look again at our revised sentences, we can make them even stronger if we get rid of a couple of those phrases:

> Once Bolton Manufacturing sends us its resource files on diamond film, an additional comparison with the results received from IAT will be made. The evaluation of diamond dust will then provide a comparison with the other nanocrystalline coatings.

HINT

Remember: No more than three prepositional phrases in a row!

This report describes ***the process of collecting data on incidents of** outdoor burning statewide.*

That is the limit!

5 WORDS: PICKY, PICKY

Paying attention to every word you use makes for good writing. You want to be sure that you are saying exactly what you mean. Be sure to keep a list handy of the words you know you tend to use improperly or are unsure of. See Table 3 for good examples of words that trip up many people.

Verb Tenses

Many technical writers get confused about which verb tenses to use in which sections of a report. Do we say, "research shows" or "research showed"? Proposals are easier in this regard because they describe a plan for doing something in the future: "We will do x, y, and z." Here is an outline of sensible assumptions about verb tense and when to use which.

TABLE 3 Words That are Often Misspelled or Misused

Diction: Watch out for these pairs of words!	
affect/effect	"affect" is usually the verb "effect" is usually the noun
among/between	"among" for more than two
cite/site	"cite" is the verb form of "citation" "site" is the place
criteria/criterion	"criteria" is plural
data/datum	"data" is plural
fewer/less	"fewer" refers to number: *fewer tons* "less" refers to volume: *less capacity*
imply/infer	"Imply" is your intention "Infer" is your discovery

Past tense is used mainly to describe the work you or others did to come up with the findings you have. Present tense is used in technical communication to indicate that you are speaking of something that is (apparently) eternally true—$e = mc^2$—or that you are speaking of what the reader is reading: "This report analyzes lab-test results for the air samples." Future tense is used solely for statements that project into the future with plans or visions. Figure 1 gives more examples of how to use past, present, and future tense in specific sections of a report or paper.

Past Tense:

- **describes what you did**
- **speaks biographically of another researcher's actions**
- **sets up a historical continuum**

Use mainly in Procedures, Methodology, and Background sections (also in parts of the Introduction).

Example:

"Researchers **have** long **known** that microbes can destroy contaminants in soil." (historical continuum)

Present Tense:

- **states theory or established knowledge**
- **says what your document does**

Use mainly in Introduction and Conclusions.

Example:

"This research shows that microbes **can destroy** contaminants in soil." (conclusion)

Future Tense:

- **outlines recommendations for future work or action**
- **envisions a future state of affairs**

Use mainly in the Recommendations section.

"Implementing this water treatment system will ensure safe drinking water.

Do not use this tense to state what your report does!

"This report ~~will~~ present the results of the study."

"This report presents the results of the study."

Figure 1. Uses of past, present, and future tense in engineering reports and papers.

6 PUNCTUATION: WHY SHOULD I CARE?

Do you think of punctuation as a set of unfathomable rules that mostly do not matter, except that teachers like to make corrections on your papers? If so, you are not alone: Many students get conflicting messages about various marks of punctuation, especially commas. Should the comma always follow an introductory clause, as in the following sentence?

> After the initial tests were run, the material was sealed in an airtight container.

Should the comma always follow the last item in a series, coming before the final "and" or "or," as in the following sentence?

> The three steps of this process are gas separation, sulfur recovery, and excess gas disposal.

Actually, neither of these questions has one right answer; as a rule of grammar, the comma is generally considered to be optional in both cases. Practicing writers, however, will tell you that using the comma in these two places often helps reader comprehension and provides fewer confusing choices for the writer. In particular, the final comma in a series sometimes has to be there, such as if the items themselves in the series contain "and." See the following example, in which the final comma is necessary to avoid confusion:

> Engineers learned to isolate **and** separate the gasses, identify **and** recover the sulfur, **and** dispose of the excess gas.

Using the final comma routinely means that the writer doesn't have to worry about whether it is required or not. At least one revered style guide, however, advises *not* using the comma when the only "and" is the final "and" (Strunk and White, 1979).

The main reason to bother about punctuation is to control and clarify your meaning in any given sentence. Even hyphens can be critical in conveying the correct meaning. For instance, which of these phrases is correct?

fast-sailing ship
fast sailing ship

Well, of course, it depends on whether you are referring to a ship (any kind) that sails fast or a sailing ship (in particular) that is fast. Many engineering disciplines have noun phrases that are commonly used without hyphens; remember to think of your audience before you omit the hyphens from a phrase such as this: *antibiotic resistant infections*. If you want to be sure that no reader will mistakenly assume that you are talking about resistant infections *caused by* antibiotics, then you should write the phrase this way: *antibiotic-resistant infections*. Check with your instructor or boss about the punctuation of commonly used phrases in your field, and read around in your field's technical journals to see how the phrases are handled there.

In general, the best practice is to familiarize yourself with punctuation rules as outlined in some reference guide and then simply place the guide on your desk for easy reference when writing. Here is an annotated list of a few of the best guides:

Strunk, W., and White, E.B. 1979. *The Elements of Style*, 3d. ed. Macmillan: New York.
Though originally written almost 50 years ago, this concise guide to writing style is a bible for many technical professionals. (Actually, the original book was privately printed by William Strunk in 1918!) It covers only those elements and issues of punctuation that are most relevant, according to the authors—issues such as

whether to incorporate the comma in a series before the final "and." But because the book is short and written in the style it recommends—direct, brief, and precise—its "essential" rules of punctuation, grammar, and style are easy to remember and have stood many, many writers in good stead for decades.

***Chicago Manual of Style*, 15th ed., 2003. University of Chicago Press: Chicago.**
This is the mothership of style manuals. This big volume focuses on style and publication issues, but also has an extensive chapter on punctuation (almost 40 pages long) that is a terrific reference tool for engineers.

Harris, M. 2003. *Prentice Hall Reference Guide to Grammar and Usage*, 5th ed., Prentice Hall: Upper Saddle River, NJ.
This volume is more reader friendly than the mammoth *Chicago Manual*; it uses tabs, section indices, and other formatting techniques to make finding information easy. The chapter on punctuation is comprehensive without being overwhelming and includes useful exercises.

Nagle, J.G. 1996. *Handbook for Preparing Engineering Documents: From Concept to Completion*. IEEE Press: New York.
This handbook contains a nice summary, in an easy-to-read chart form, of the major punctuation rules and usage.

SUMMARY

- The keys to effective writing are to prune away the deadwood (vague and useless words), control passive voice, and maintain focus by linking thoughts together clearly.
- Empty pronouns drive readers crazy; "it or "this" at the beginning of a sentence means nothing without a specific noun to modify. Combining sentences can be a good way to avoid empty pronouns.
- If you can write a good paragraph, you have got it made. Effective paragraphs have a topic sentence that acts like a miniature thesis, and all the sentences that follow fit together logically and derive from that topic sentence. The direction of thought *moves*.
- Transitional words and phrases connect ideas and link data together. Making connections is what writing technical information is all about. Otherwise, you are just making lists.
- Use a good style guide to pinpoint and fix grammatical errors and weaknesses of style in your sentences.

Problems

1. In the following paragraph, change the passive constructions (italics) to active verb constructions. Passive constructions are discussed in section 1.

> **The evaluation of diamond film based on the final criterion *will be completed* when the analysis of the qualities *is received* from the Institute for Advanced Technology (IAT). A request for the**

resource files on diamond film *will be submitted* to Bolton Manufacturing. An additional comparison of the results received from IAT *will be made* with information provided by Bolton. The evaluation of diamond dust *will be used* for the comparison of the other nanocrystalline coatings.

2. After changing the passive verb constructions in the paragraph in #1, answer these questions:

   Does this exercise force you to add additional words?
   Are those words mainly transitions, or do they have another function?
   How do these additional words help keep the paragraph moving along?

3. Write a list of all the words or phrases that could fill in the blank in the following pair of sentences. Which transitional word or phrase is best and why?

   The technician heated the material to a temperature of 100° C. ___________, the material changed color slightly.

4. Rewrite the pair of sentences in #3 as one sentence. How did you punctuate it? Is the information clearer when expressed as one sentence or as two sentences?

5. Add correct punctuation to the following sentence:

   Spending limit balances will carry forward for some financial accounts however we will remove expenditure limits against IT accounts on August 31 2004.

6. True or false? In a report, you use past tense to state theory or established knowledge.

7. True or false? The word "criteria" is used properly in the following sentence:

   The design criteria is one reason we decided to shorten the length of the rod.

8. Which word in the following sentence is used improperly?

   Among the two of us researchers, there can be no disagreement about methodology.

FURTHER READING

Ballstaedt, S-P and Mandl, H. 1987. "Influencing the Degree of Reading Comprehension," in *Knowledge Aided Information Processing*, Van Der Meer, E., and Hoffmann, J., eds. Elsevier Science Publishers: Amsterdam
This collection of essays on recent theories of cognitive psychology is far too technical for most readers, but the cited article gives a good idea of the complex interplay of elements—memory, prior knowledge, semantic analysis, and levels of processing, just to name a few—that enable us to understand what we read.

Chicago Manual of Style, 15^{th}ed., 2003. University of Chicago Press: Chicago.
This is the mother ship of style manuals. This big volume contains more than you'll ever want to know about manuscript preparation, documentation, and the publishing process, but it will also have definitive answers for you on all those niggling little questions about style (do you write out the number twelve or use the Arabic numeral?). The sections on "Grammar and Usage" and "Punctuation" are a terrific reference tool for engineers. Most discipline-specific style manuals in engineering are based on the *Chicago Manual*.

Day, C. 2002. "Remotely Sensed Neutrons and Gamma Rays Reveal Ice beneath the Martian Surface" and "Choreographing Wave Propagation in Excitable Media," *Physics Today*, August 2002, vol. 55, no. 8, pp. 16–17 and 21.

Reading science and engineering magazines intended for a mixed audience (technical and semitechnical) can provide models for good technical writing. *Physics Today* is one such magazine; each issue contains thoughtful articles on very technical subjects written for anyone with an interest in physics or in any of the fields deriving from physics (such as engineering). See next entry for another example of good technical writing.

Fitzgerald, R. 2000. "Phase-Sensitive X-Ray Imaging," *Physics Today*, July 2000, vol. 53, no.7, pp. 23–26.

Lanham, R. 2000. *Revising Business Prose*, 4th ed. Allyn & Bacon: Needham Heights, MA.

Richard Lanham's book contains a wonderful discussion of how to tighten your sentences and strengthen your meaning. He calls this procedure the "Paramedic Method" for diagnosing and curing wordy sentences. Read the whole book (it is short), but look especially at the first chapter, "Who's Kicking Who" (if you caught the error in grammar, you are right—Lanham wants to get your attention).

Strunk, W., and White, E.B. 1979. *The Elements of Style*, 3d. ed. Macmillan: New York.

This book is the last word for many technical professionals on style and grammar. It covers only those style guidelines the authors considered most important or most often abused, but the book's frank, direct, and directive style has delighted many people and helped them remember rules such as "Do not join independent clauses with a comma." The effect of this book has been extraordinary for almost a century. It was originally written (by William Strunk) in 1918!

Zinsser, W. K. 1990. *On Writing Well: An Informal Guide to Writing Nonfiction*, 25th anniversary ed. Harper Collins: New York.

William Zinsser is a prolific writer whose books and journalism pieces cover subjects from jazz musicians to the art of Roger Tory Peterson. This book on writing has become a classic for its crisp demonstration of the value of simplicity in writing and for its explanation of writing as a transaction between author and reader. Chapter 15 covers science writing and technical writing.

10

Displaying Data in Written Documents: Are Pictures Always Worth a Thousand Words?

1 INTRODUCTION

A picture is said to be worth a thousand words. And so is a table, a scatter plot, or a bar chart. Figures and tables are often the heart of an engineering document. They display the data, and the exact relationships among data, that will give life and meaning to all interpretation and conclusions. In addition, representing data graphically allows us to show relationships among the data quickly and effectively. For example, it would take far longer to try to explain in words all of the relationships expressed in Figure 1.

If you were to explain all parts of Figure 1, you would have to write something like this:

> *Air-quality modeling integrates emissions data (from multiple sources), meteorological information, and chemical mechanisms to determine pollutant production over time and space. Emissions are produced by natural organisms as well as by man-made sources such as automobiles. The pollutants produced are many, and the current work is concerned with how these pollutants interact to produce ozone in concentrations greater than current standards deemed safe for human health. This work is concerned primarily with ozone concentrations in the Houston, TX, area.*

However, you cannot make the mistake of thinking that a good technical communicator does not need words to describe the significance of these relationships. A picture may be worth a thousand words, but it does not necessarily replace all words. Some people are not used to reading figures and tables, and they need to be oriented to what the figure shows. Other people know how to read figures, but they are reading quickly and could thereby miss the significance of the information revealed by the graphic. Using figures or

SECTIONS

1 Introduction
2 Displaying Your Data
3 Figures and Tables: Which Is Which?
4 Guidelines for Designing Figures and Tables
5 Integrating Text with Figures and Tables

OBJECTIVES

By reading this chapter, you will learn the following:

- the importance of displaying data
- how to design and construct appropriate representations of your data for written documents
- how to integrate graphical representations into your documents
- how to explain in words the significance of each graphic.

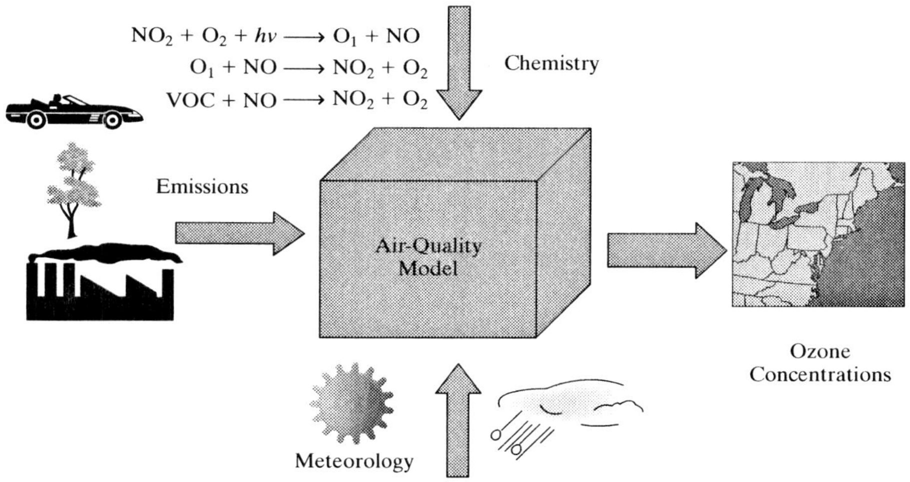

Figure 1. A model for ozone concentrations in the Houston, TX, area.

tables *and* words allows you to communicate very effectively. By combining modes of delivery, you allow readers to use parallel processing in understanding the importance of your information: They can use both their visual and their verbal understanding. Especially when dealing with complex data, you need the double power of words plus graphics to ensure that readers and listeners will get the total picture. Some people are more visual in their orientation toward learning, and some people are more verbal. Most engineers tend to be visual, but your readers may not be engineers.

HINT

Why use graphics in technical documents?

- **to show complex data in a simplified form**
- **to show a lot of data in one place**
- **to emphasize relationships**
- **to help the reader remember**
- **to allow parallel processing of information**

Graphics are also a very important part of presenting information verbally. It is difficult to imagine a technical presentation that does not exhibit well-projected, readable figures. The design rules governing presentation graphics, however, are somewhat different from those for graphics in written documents. This chapter deals with how to design and use graphics in *documents*.

2 DISPLAYING YOUR DATA

As an engineer, you will collect data in various ways. Sometimes you will model a system or process, providing certain inputs and extracting outputs. Sometimes you will use mathematical calculations to test a hypothesis or prove a theory. And very commonly, you will use experiments, in the lab or in the field, to yield information that helps solve problems. Let us posit a situation in which you work for a manufacturing company as an

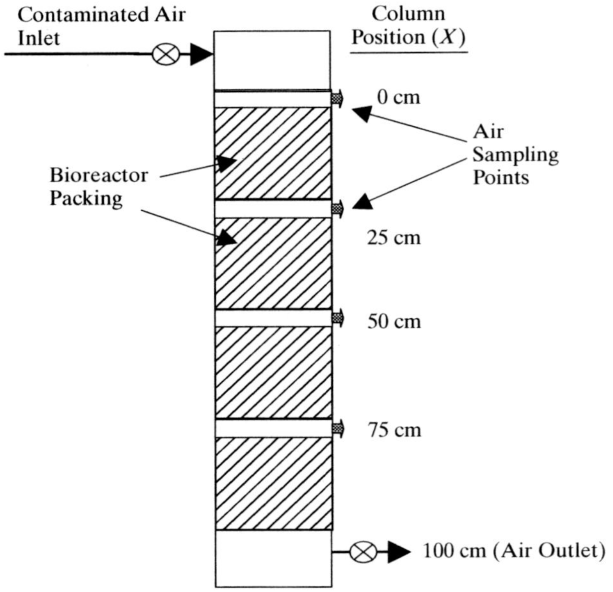

Figure 2. Schematic of the experimental bioreactor.

air-pollution control expert. Your job is to find economical ways of cleaning the gas waste stream produced by the manufacturing process. Biofiltration is a promising technique that uses microorganisms to reduce pollutants in the gas stream, but you need to find out how large a biofiltration system would be needed at the plant and how much it would cost. A typical system uses a packed bed of biological material inside a column through which the gas stream is run for treatment. Figure 2 is a schematic identifying key components of the system.

You set about designing and conducting a series of experiments to show the relationship between the size of the system and the amount of pollutants in the gas stream that is destroyed by the microorganisms. You run the experiments, collect the data by taking samples of the gas at various positions within the column, and record the data. At that point, your communication work begins, because you must present the data to various managers within your company and help them come to a decision about what size system to use. You must first display the data for your own benefit, so you can determine at what position in the column enough pollutants are destroyed that the column has done its work and is no longer necessary; a shorter column makes for a cheaper treatment system. Then you must show these data and this relationship (between position in the column and concentration of pollutants) to others in a comprehensible, visual form. But how?

Well, to understand the data yourself, you will use a spreadsheet program such as Microsoft® Excel. You will enter the sampling data (concentration of pollutants) that you recorded, with three samples for each position in the column. Let us say you recorded data in triplicate at the inlet point, at the outlet point, and at 25, 50, and 75 cm from the inlet of the column. Since you will want to take the average of the three readings for each sample, your data sheet would look like Table 1 (where pollutant concentration is measured in milligrams per cubic meter).

TABLE 1 Excel Worksheet Containing Sample Data for an Experimental Biofiltration System

Position (cm)	0	25	50	75	100
Concentration (mg/m³)					
sample 1	101.0	68.0	45.5	28.8	15.0
sample 2	98.4	66.5	43.0	27.3	13.1
sample 3	97.7	71.2	46.3	22.3	16.8
Average (mg/m³)	**99.0**	**68.6**	**44.9**	**26.1**	**15.0**
Standard Deviation	1.74	2.40	1.72	3.40	1.85

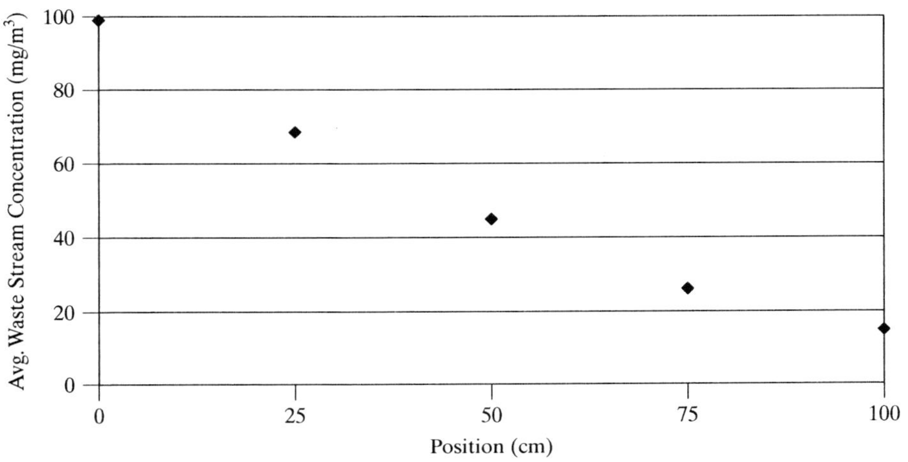

Figure 3. Relationship of position in the biofiltration column and pollutant concentration of the gas waste stream [data points only].

Once you have entered the data in the spreadsheet, you can create many kinds of graphs to show the relationship between pollutant concentration and position in the column. You could show simply the data points, as in Figure 3 (remember that the points on the graph are averages of the three samples taken at each position).

If you want to show more clearly the *trend* in the data, you could connect the data points (as in Figure 4a) or fit the data by using a regression model (as in Figure 4b). Excel contains a tool called "trendline" that provides various regression models. Be sure to indicate the equation of the trendline on the figure.

If you wanted to show more clearly the *scatter* in your data (emphasizing the fact that not every one of the three samples taken at the same position had exactly the same concentration), you could use error bars, which plot the standard deviation of the data at each point, as in Figure 5.

So, now you would be ready to put together a report for a variety of audiences, from the other project engineers to the head of your division, and even the chief financial officer. Of course, your presentation or report should begin with a schematic of the system, showing the column and where the gas stream enters and exits. Remember, we began this discussion about displaying data with a drawing of the treatment column. Be sure to use visuals to orient your reader to as many aspects of your experimental conditions as possible.

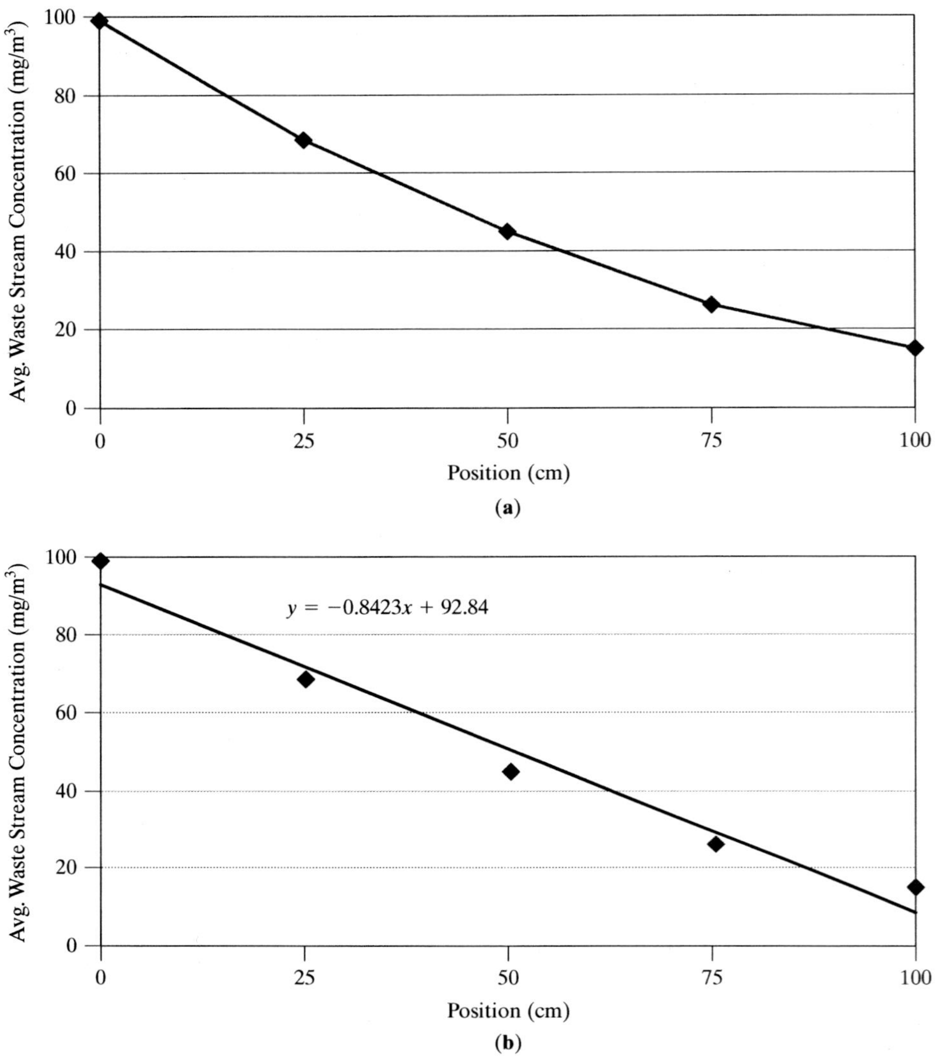

Figure 4. Relationship of position in the biofiltration column and pollutant concentration in the gas waste stream: (a) data connected from point to point; (b) linear regression fit to the data.

3 FIGURES AND TABLES: WHICH IS WHICH?

Although there are many different types of graphical display, in engineering documents these types are usually grouped into one of two categories for purposes of labeling and numbering: figures and tables. So, first of all, let us answer the question, "What is a figure and what is a table, and why?"

Tables

Tables are compilations of reference data, usually displaying numbers and/or keywords. The data are listed in columns (vertical elements) and rows (horizontal), with headers clearly delineated. Table 2 shows a typically designed table.

Notice that the units of measurement for flow rate are given in the column heading, so that they do not have to be repeated in each row. In written documents, all tables are

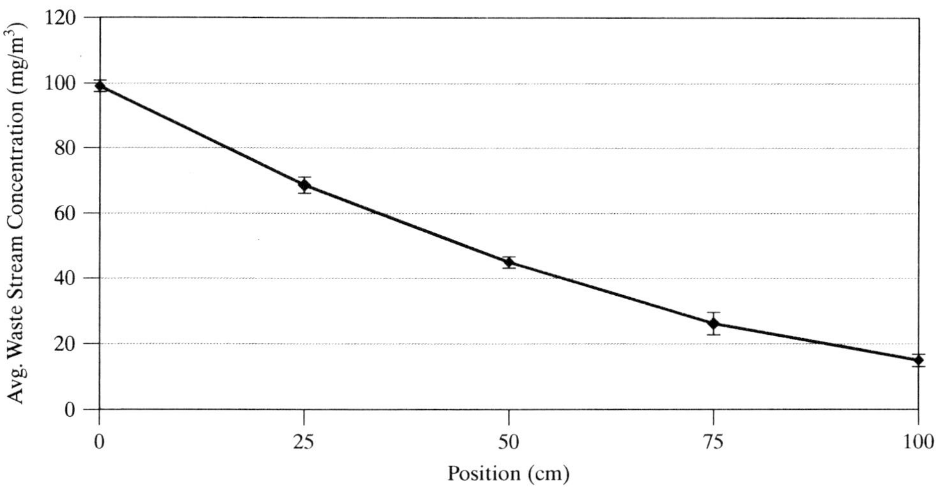

Figure 5. Relationship of position in the biofiltration column and pollutant concentration in the gas waste stream [showing standard deviation of data at each point].

TABLE 2 Comparison of Three Air Filters for an Aquarium

Filter	Calibration Factor	Flow Rate (gallons/hour)	Efficiency
Power	4–5	1000	Least
Canister	1–2	250	Moderate
Wet/Dry	3–4	700	Highest

numbered consecutively—Table 1, Table 2, etc.—and each has an explanatory title and is referred to in the text. Sometimes the numbering of tables is done in Roman numerals (e.g., III, IV, and V) instead of Arabic numerals, to differentiate the tables from figures.

Figures

Figures are what we call every other sort of graphical representation: bar charts, line graphs, scatter plots, schematics, plan views, photographs, pie charts, etc. In written documents, all figures are numbered consecutively and have explanatory labels and titles. A few books and journals still call certain illustrations (especially drawings and photographs) "plates" instead of "figures." All figures (and plates) are also explained and referred to in the text.

Here are descriptions of the more common types of figures.

Figures: Graphs

X–Y graphs are most engineers' favorite way of displaying data. They are particularly useful for showing trends in data over time and changes in the relationship of two variables. The independent variable is usually plotted on the x-axis and the dependent variable on the y-axis. In Figure 6, the independent variable is time (in seconds), and the dependent variable is position, measured in meters from the starting point (0). The graph shows how far a ball travels when it is tossed into the air until it starts to fall. In this graph, the slope of the curve mimics the physical reality that the graph expresses; however, graphs do not usually try to look like what they are depicting.

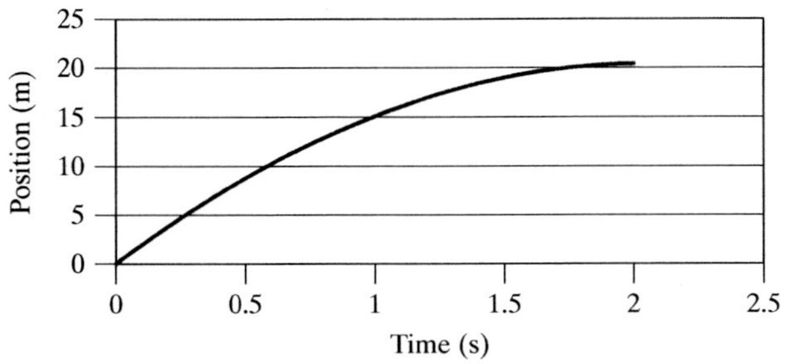

Figure 6. Distance a ball travels when tossed into the air.

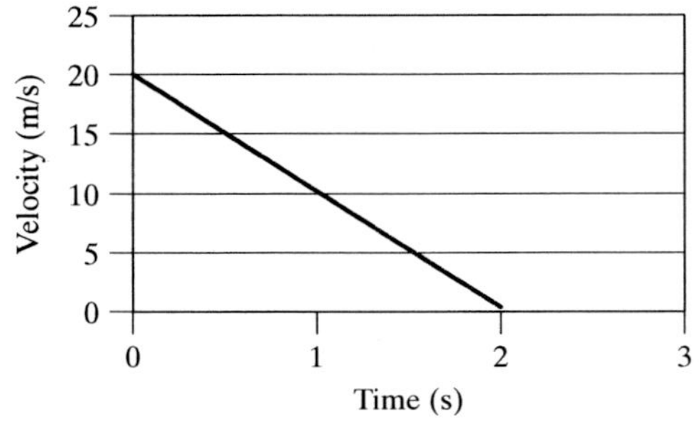

Figure 7. Velocity of ball tossed into the air (at constant acceleration of - 9.8 meters/second^2).

TABLE 3 Data Set for Figures 6 and 7

INITIAL POSITION	0		
initial velocity	20		
time (s)	**position (m)**	**velocity (m/s)**	**acceleration (m/s^2)**
0	0	20	- 9.8
0.5	8.775	15.1	- 9.8
1	15.1	10.2	- 9.8
1.5	18.975	5.3	- 9.8
2	20.4	0.4	- 9.8

In Figure 7, the slope of the curve actually looks like the opposite of what is happening. This figure displays the velocity of a ball as it travels upward through the air to the point where it would start to fall and actually show a negative velocity.

A portion of the data set for Figures 6 and 7 is shown in Table 3. To graph the *distance* the ball travels, you would plot the values for the independent variable in column 1 (time) and the dependent variable in column 2 (position). To graph the *velocity* of the ball over time, you would plot columns 1 (time) and 3 (velocity). These data can be generated by the formulas you enter for velocity and by the ranges you specify for time and position. Since acceleration is a constant governed by gravity (g), you probably would not want to graph it, but you should include it in your explanation of the figure and, whenever possible, show such relevant data on the figure itself.

Line graphs are useful to show changes in one variable over time or differences in values for certain items. You use line graphs when the x-axis is showing labels or increments, not numerical values. Thus, for example, you could show the differences in grades among four students as displayed in Figure 8.

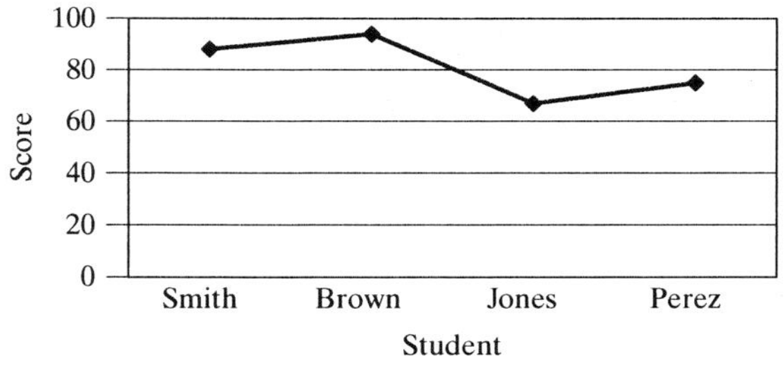

Figure 8. Final scores for four students.

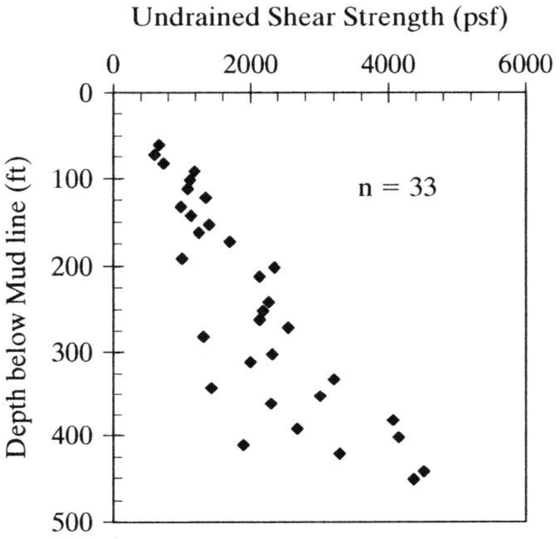

Figure 9. Relationship of undrained shear strength and depth for soil samples taken in the Gulf of Mexico, July 2001. *Graph courtesy of Dr. Robert Gilbert, University of Texas at Austin.*

Most line graphs could also be constructed as bar charts. Do not confuse line graphs with scatter plots that have a line drawn through the data points. Engineers do not use line graphs as often as x–y scatter plots because the former do not plot the relationship between two variables. For line graphs to make sense, the increments along the x-axis must be equal.

Scatter plots show a relationship between variables for a particular phenomenon or quality at precise points. For example, Figure 9 shows measurements of soil samples from an offshore boring that was drilled from an oil production platform. The measurements are of the undrained shear strength of the soil, which is an indication of how strong the soil is. In the graph, the strength is shown in relation to the depth (below the mud line) of the sample taken. The important conclusions from this plot are twofold:

1. The plot shows a trend wherein the undrained shear strength tends to increase with depth.
2. The amount of scatter about this trend increases with increasing depth.

Notice that the labels for the x-axis are on the top of the figure rather than at the bottom and that the independent variable (depth) is plotted along the ordinate (the y-axis) instead of along the abscissa (the x-axis) as is usual. This exception to the rule governing placement of the dependent and independent variables occurs most commonly for graphs that plot depth. Because the concept of depth is inherently visual, we normally plot depth where it makes visual sense—along the vertical axis. Note also that the number *n* of data points is indicated on the figure. Whenever possible, place such helpful information right on the figure itself. If the information is more complex, it may be more appropriate to put it in a caption. If not, you will want to discuss it in your text, along with any other information about the experimental, modeling, or mathematical procedure behind the graphic.

Figures: Charts

Bar charts are probably the most widely used graphic in the popular press, if not in engineering reports, because they are easy to understand and multifunctional. They come in many varieties, too, including histograms and Gantt charts. Comparing the sizes of bars is an easy way to compare differences in values or amounts of particular items. The prevalence of bar charts in annual reports suggests that cost comparisons are a natural subject for these kinds of charts. Figure 10 compares the costs of various air purifiers.

Area charts, or stacked charts, are really a series of line graphs piled on top of each other. They are tricky to read because, beyond the bottom line, you do not read from the beginning of the scale: You separate out the "chunk" that is bounded by two lines and figure out the values within that chunk. Most nontechnical people, and even a lot of engineers, tend to read stacked charts improperly. Consequently, this type of chart should be used only rarely in technical documents. Consider Figure 11 and ask yourself how to determine the number of men who got degrees from this university in 1984–1985.

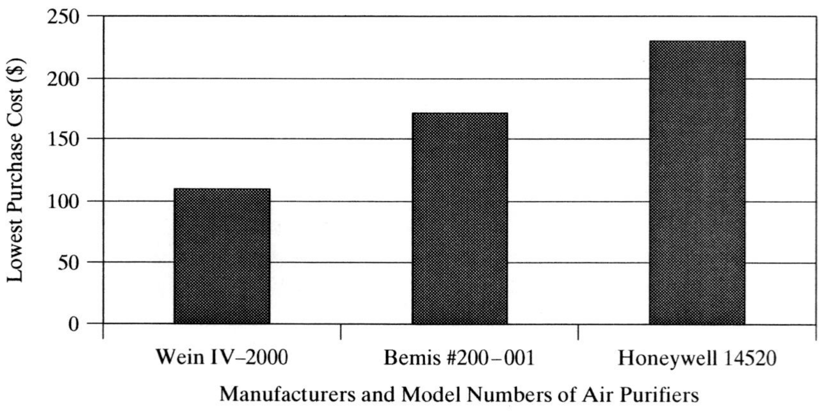

Figure 10. Lowest purchase costs found for Bemis, Honeywell, and Wein air purifiers.

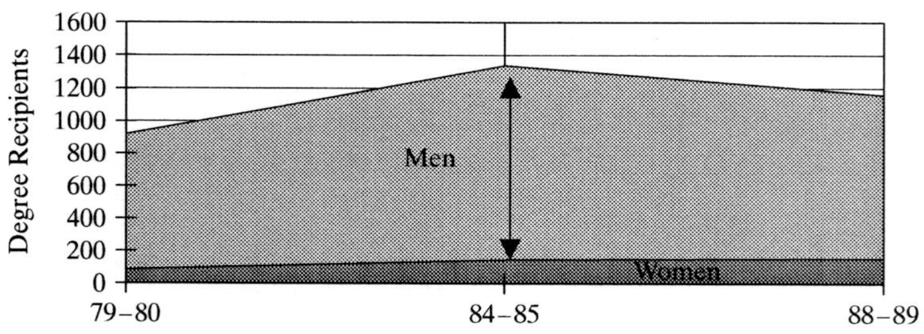

Figure 11. University of Malibu degree recipients by gender [hard to read].

In order to figure out the number of men receiving degrees in 1984–1985, you have to estimate how many are indicated in the chunk delineated by the arrows. Such an estimate is not very precise, and many people would be tempted to estimate the number by starting at the very bottom of the graph instead of at the line at the top of the "Women" chunk, adding about 100 to the correct number. The correct number is 1,240.

Pie charts are often used to depict constituents of a whole. These kinds of charts are seen everywhere, especially in magazines and newspapers, because they are ostensibly easy to read. They *are* easy to read, but only if they are designed and used well. Figure 12 displays the cost breakdown for the air purifiers (discussed previously) as a pie chart. Designing this chart in three dimensions doesn't make much sense, however, and lends a bit of distortion to the figure. The slices facing toward us seem proportionately larger than they are because they (seemingly) show a third dimension.

Pie charts really come into their own when you display the values as percentages of a whole, as in Figure 13. The reader can easily see the pieces as representative of the portion each material contributes to the total construction budget. In this figure, the pieces are "exploded," which emphasizes their relative size (and therefore value) even more.

Be sure that when you use a pie chart, your goal is to focus on the whole of something: a whole year, a whole graduating class, etc. In the pie chart in Figure 14, the importance of the whole—the total number of graduates from this university in the years 1990–1996 is not immediately clear. Why is that number important, especially when it is not even given? It might be important for some analysis of recent alumni, for example, *but the graphic does not clarify this information.*

Flowcharts make visual a process that otherwise would be more difficult to conceptualize and understand. They can depict processes as simple as activities in a workday, or they can depict more complicated processes, rendering them simpler to understand. Figure 15 shows a flowchart depicting the performance-review process in a company.

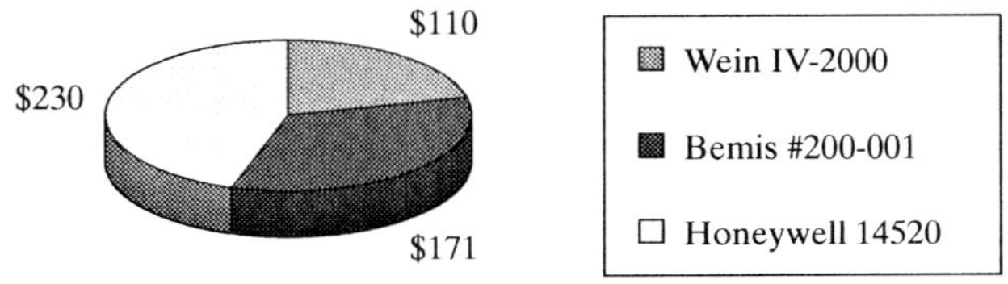

Figure 12. Lowest purchase costs of three air purifiers.

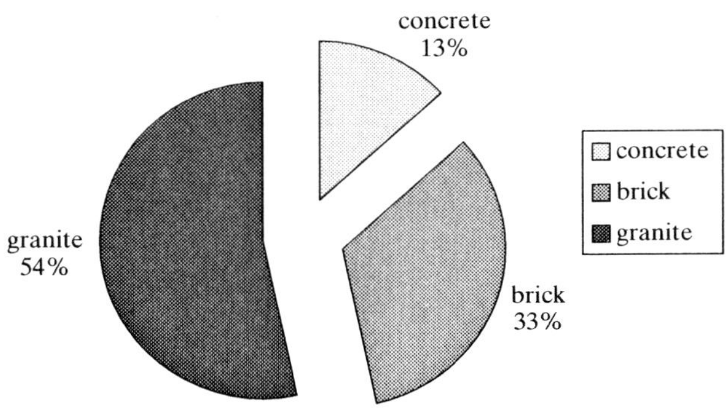

Figure 13. Relative contribution of exterior materials to construction budget.

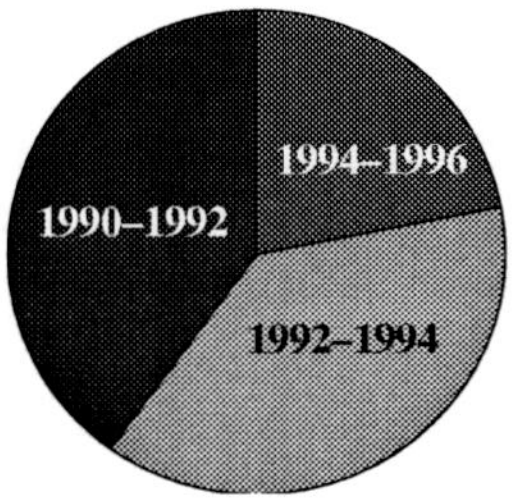

Figure 14. University of Malibu degree recipients, 1990–1996 [inappropriate pie chart].

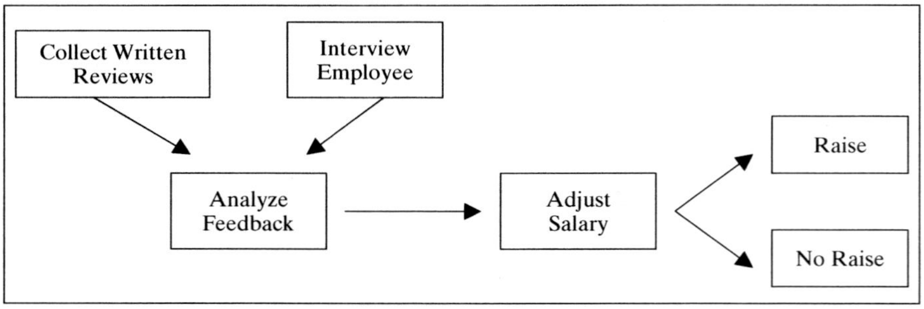

Figure 15. Performance-review process for Integrated Documentation, Inc.

Figures: Drawings and Schematics

Drawings and schematics simplify reality and thereby show the essential elements necessary to understand a particular problem or phenomenon. Figure 16 is a schematic of a bioreactor. The actual column, made from stainless steel, does not look exactly like the schematic in Figure 16, but the drawing shows all that is essential to understand how the experimental work was accomplished using this bioreactor. We see that the contaminated air stream was fed into the column at the top, that samples of air were taken every 25 centimeters, and that the column is 100 cm long. The figure also establishes that, for purposes of graphs showing results of the sampling, "0" is the top of the column, where the air is first fed in.

Figures: Combinations

Some figures artfully combine the attributes of several types of figures, as in Figure 17, a drawing and flowchart depicting the process of pollution in an aquarium. Be careful of trying to combine too many types of figures and thereby failing to help the reader understand information quickly. Figure 17 works because of its elegant simplicity.

HINT

Guidelines for Graphics: *Labels*

- Label each figure or table clearly with a number and a title. For tables, the number and title are centered or left justified above the table. For figures, the number and title go beneath the figure. Your graphics program might have different defaults, but override them.
- Give pertinent details about the experiment either on the figure itself or in parentheses after the title.
- Create a title that draws attention to significant aspects of the illustration. Be more creative than using a simple "y vs. x" title.

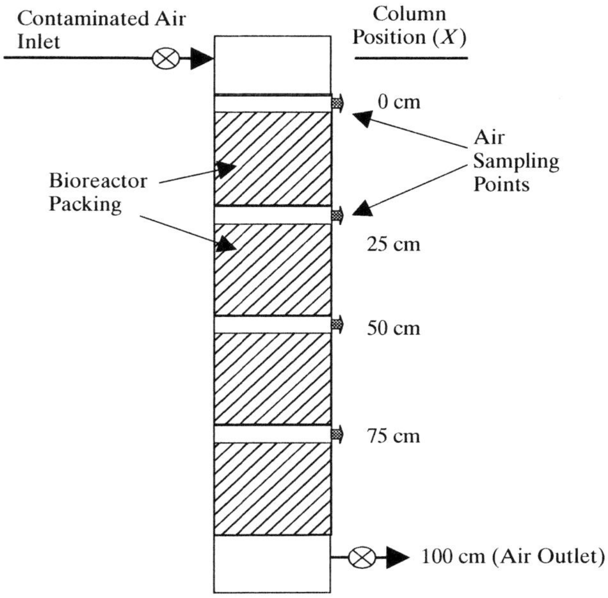

Figure 16. Schematic of an experimental bioreactor.

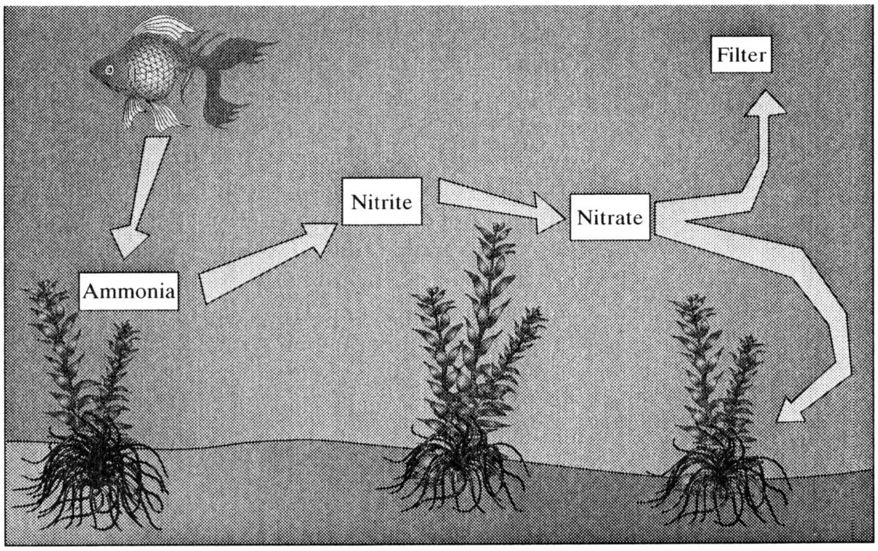

Figure 17. Process of pollution in an aquarium.

PRACTICE QUESTION: LABELING GRAPHICS

What is wrong with this title for a graph?

"Time vs. Temperature"

SUGGESTED ANSWER

The title gives no more information than can be found by reading the axes of the graph. That information is already on the figure, so the title should add more information, such as the material involved (e.g., the temperature of what?).

4 GUIDELINES FOR DESIGNING FIGURES AND TABLES

Because figures and tables are often the heart of an engineering document, your figures must be clear and correct. If they are not, your methodology begins to seem questionable to the reader, and the foundation for your interpretive judgments crumbles. A picture may be worth a thousand words, but a graphic will be worth less than nothing if it represents the data inaccurately or confusingly. How do you construct useful and readable figures and tables?

Graphs

Here are some instructions for constructing a clear, readable graph. The abscissa is the horizontal axis, which typically displays values for the independent variable. The ordinate is the vertical axis, which typically displays values for the dependent variable. The point of origin of both the ordinate and the abscissa is generally zero, although there is a lot of disagreement about the importance of adhering to this guideline. In displaying certain data, zero may not be a meaningful value, and having to begin the ordinate with zero may mean creating a scale that does not capture the relevant data. To ensure that your audience will not misunderstand the graph for such a case (especially when you are unsure about the audience's technical background), consider starting the ordinate values at zero and then inserting a zig-zag break mark to indicate that the scale is being broken for the sake of displaying the relevant data. In Figure 18, all of the relevant values for the dependent variable (the STC values) were between 50 and 58 decibels, so the author started the y-axis at zero but then jumped to 50 to begin the next interval. The break mark indicates this jump. Unfortunately, Excel does not contain a feature that will insert a break mark. Other programs, such as SigmaGraph, do contain such a feature.

Both axes should be properly labeled. These labels should use words and symbols whenever possible and include the units of measurement; using both words and symbols can tie the figure well to the text. Common abbreviations may be used; when in doubt, however, spell out the word. In Figure 18, dB is the common abbreviation for "decibels."

Tables

Tables can be useful for showing comparisons of and changes in data, but they can easily become overcrowded and hard to read. Be sure to highlight important information in tables, such as totals or especially significant numbers or words. Color is effective as a highlighter, but remember that someone may make black-and-white copies of your table. For example, the original version of Table 4 had red numbers indicating unacceptable settlement. But in a black-and-white copy, those numbers are lost in a sea of other black-and-white numbers, and thus the point of this table (that certain sizes of piles allow too much settlement of a structure) is not as clear. Also, the caption has to be

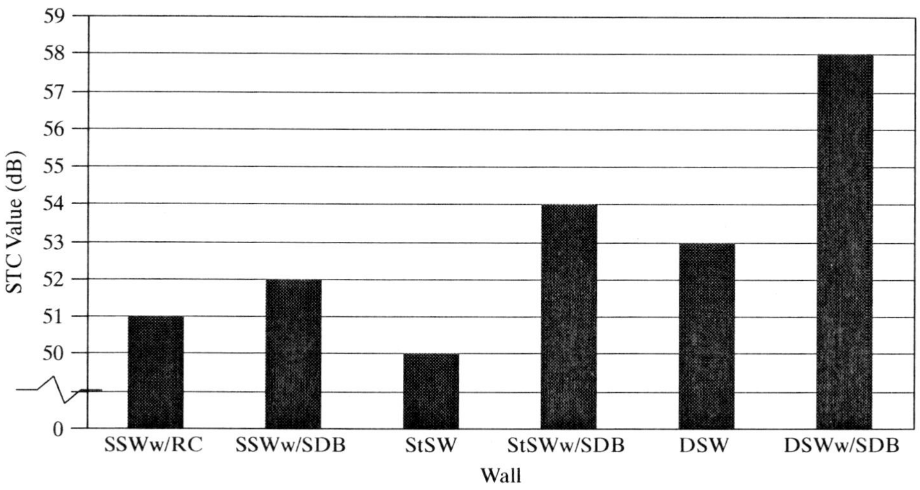

Figure 18. Performance measured as Sound Transmission Class (STC) values for each wall design. *Source:* Simmons, Leslie. *Construction: Principles, Materials, and Methods*. 2001.

TABLE 4 Comparison of Settling for Various Types of Piles

Piles that exceed 0.25 inches of settlement are shown in red. Piles that are shaded in gray were discarded. *[Author's note: Red cannot be seen in black-and-white reproduction.]*

Concrete Pile			Hollow Steel Pile			Concrete-Filled Pile		
Diam (in)	**Len (ft)**	**Sett (in)**	**Diam (in)**	**Len (ft)**	**Sett (in)**	**Diam (in)**	**Len (ft)**	**Sett (in)**
12	50	0.19	12	50	0.07	12	50	0.19
12	55	0.37	12	55	0.18	12	55	0.37
12	60	0.36	12	60	0.17	12	60	0.36
12	65	0.61	12	65	0.28	12	65	0.61
14	50	0.20	14	50	0.10	14	50	0.20
14	55	0.33	14	55	0.16	14	55	0.33
14	60	0.43	14	60	0.21	14	60	0.43
14	65	0.68	14	65	0.26	14	65	0.58
16	50	0.23	16	50	0.10	16	50	0.23
16	55	0.32	16	55	0.20	16	55	0.32
16	60	0.41	16	60	0.42	16	60	0.41
16	65	0.43	16	65	0.43	16	65	0.43

rewritten, since red will not show up in the black-and-white copy. You can use shading to differentiate.

Tables often give a more compete "story" than a line graph (e.g., tables give specific numbers, while graphs typically present only approximate numbers), but generally they are not as effective as more pictorial forms. Consider including tables in appendices and creating line or bar graphs to summarize the same data in the body of your document.

Taking Control of the Design

The default parameters used in standard software for generating figures are not always the best choice for conveying technical information. Figures 19 and 20 contain the same data; however, Figure 20 is much easier to read.

HINT

Guidelines for Graphics: Engineering Conventions

- In graphs, use lines to illustrate predictions of models, and points to indicate data values found through experiment, especially when comparing the two types of information.
- Follow the convention for direction: When tables or figures are printed sideways on a piece of paper, the paper should be turned clockwise for the table or figure to be read).

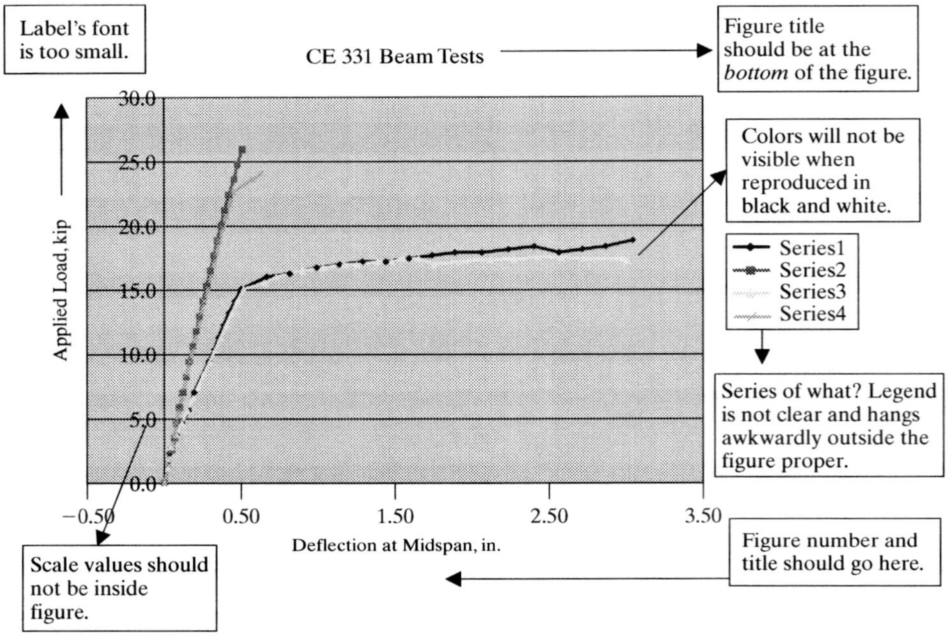

Figure 19. Concrete beam deflections at various loads [badly formatted graph].

The modified graph is easier to read because it uses a larger font and has a larger area in which to display the data. The modified graph also has a figure number and a meaningful title in the proper place.

Note that Figures 19 and 20 seem to flout the rule that the independent variable is plotted on the abscissa: Deflection, or strain, is dependent on the load (or stress) applied in the experiment. In structural engineering, however, stress–strain curves are always plotted such that strain is plotted horizontally and stress is plotted vertically. Be sure to become familiar with the conventions of your particular branch of engineering. A good way to obtain this familiarity is by reading around in the journals in your field. Notice how figures and tables are presented. Do the axes always cross at zero? Are units of measurement always abbreviated? Is the independent variable always plotted on the x-axis? Are tables numbered with a different style than are figures?

Consider the following items when preparing your figures:

- Gray backgrounds, such as that in Figure 19, look nice on the screen, but the technical information tends to be lost when the figure is plotted in black and white.
- Whenever possible, select scales so that the axes cross at (0,0). In all cases, position the axis labels outside the graph area.

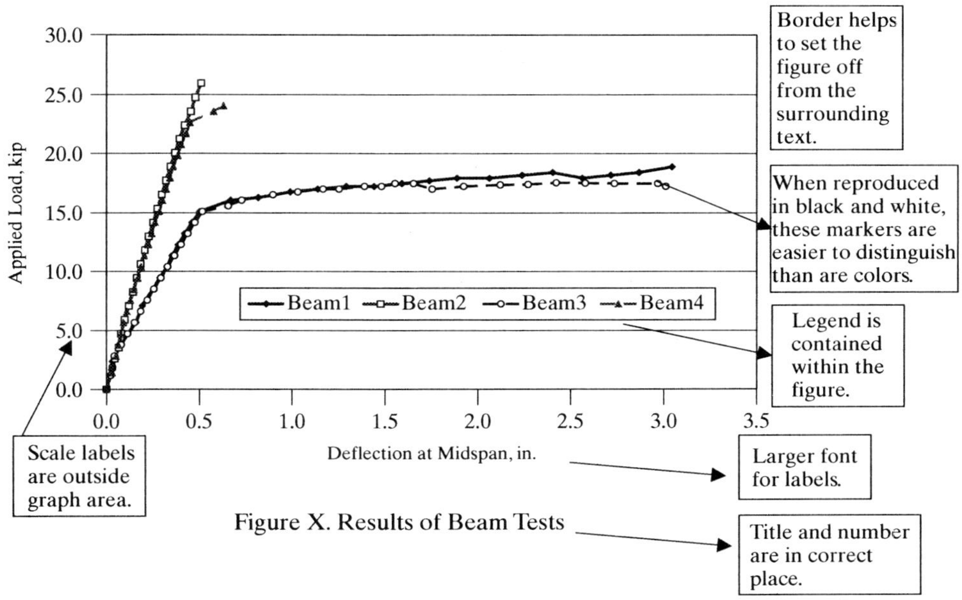

Figure 20. Concrete beam deflections at various loads [well formatted].

- It is not always easy to distinguish among the default colors and symbols when graphs are printed in black and white. Selecting open and filled markers is one way to distinguish one data set from another.
- Default text sizes from graphing programs are typically too small when graphs are imported into a word processor.
- Default line widths from graphing programs are often too thin when graphs are imported into a word processor.
- The location of the legend may decrease the available area for displaying technical information.
- A figure should have a number and a meaningful title placed below, not above, the figure.

Similarly, when designing tables, the default parameters used in standard software are not always the best choice for conveying technical information. Tables 5 and 6 contain the same data; however, Table 6 is much easier to read.

PRACTICE QUESTION: DESIGNING TABLES

What changes make Table 6 easier and faster to read than Table 5?

SUGGESTED ANSWER

- Text and labels are centered.
- Unnecessary horizontal lines are removed.
- Column and row headings are in boldface and single spaced.

TABLE 5 Example of Table that Uses Default Parameters

Column heads are not spaced sensibly and are not in boldface.

Sieve Size	Weight Retained (g)	Percent Retained	Cumulative Percent Retained	Percent Passing
#4	54	10.2	10.2	89.9
#8	78	14.7	24.9	75.1
#16	125	23.5	48.4	51.6
#30	116	21.8	70.2	29.8
#50	88	16.6	86.8	13.2
#100	60	11.3	98.1	1.9
#200	5	0.9	99.0	1.0
Pan	5	0.9	100.0	0.0
Total	531	100.0		

Too many horizontal lines. Our eyes become distracted from the data.

Numbers and text are not centered.

TABLE 6 Example of Table with Modified Parameters

Sieve Size	Weight Retained (g)	Percent Retained	Cumulative Percent Retained	Percent Passing
#4	54	10.2	10.2	89.9
#8	78	14.7	24.9	75.1
#16	125	23.5	48.4	51.6
#30	116	21.8	70.2	29.8
#50	88	16.6	86.8	13.2
#100	60	11.3	98.1	1.9
#200	5	0.9	99.0	1.0
Pan	5	0.9	100.0	0.0
Total	531	100.0		

All of these changes create a graphically simpler table that lets us focus on the data it presents rather than distracting us with meaningless lines and too much white space. Notice that the number and title of tables are placed on top.

HINT

Guidelines for Graphics: Golden Rules

- Position each graphic as near as possible to the text it supports, but always after the first reference to it.
- Keep every illustration as simple and uncluttered as the complexity of your data allows.

A Word about Titles

Titles for graphics are often slapped on at the last minute without much thought. As such, far too often in engineering documents, they are superficial and do not help the

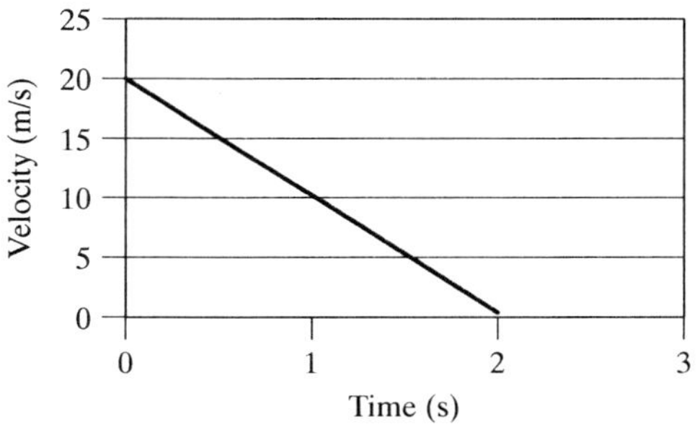

Figure 21. Velocity vs. Time [bad title].

reader understand the significant relationships or phenomena displayed by the graphic. A title that simply repeats the labels of the x-and y-axes adds nothing to the knowledge of the reader. Figure 21 depicts the decreasing velocity of a ball tossed into the air and has the simple title, "Velocity vs. Time." With this inadequate title as a guide, we cannot understand why velocity is decreasing, because we have no idea as to what object's velocity is being measured. Here is a better, more descriptive title:

> **Figure 21. Velocity of ball tossed into the air** (at constant acceleration of -9.8 meters/sec^2).

Another example of an inadequate title is shown in Figure 22, which depicts a graph showing air density as a prediction of the ideal-gas law. This x-vs.-y title simply repeats the labels on the figure without *interpreting* the data at all. A much better title would be the following:

> **Figure 22. Predicted relationship of density and temperature of air at standard atmospheric pressure**

If any of your readers are likely to be nontechnical, you might want to add a caption after the title, such as the following:

> **Figure 22. Predicted inverse relationship of density and temperature of air at standard atmospheric pressure.** Note that density tends to decrease with increasing temperature.

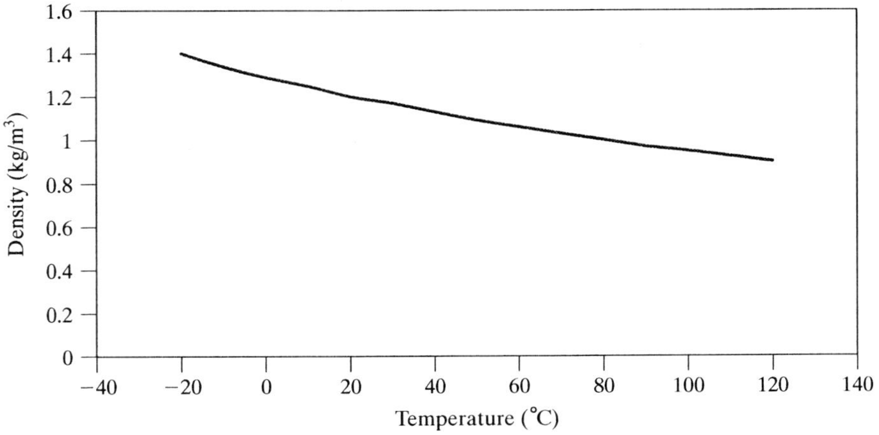

Figure 22. Density vs. Temperature [bad title].

Captions are sometimes used in addition to titles. Captions can help the reader understand the data being presented; they often refer either to the way the data have been shown or to the conditions under which the data were collected. Let us look again at the title and caption from Table 4, which shows settlement data for piles. The caption tells us how to read the table for the greatest understanding. But, of course, this particular caption works only if the table is presented in color. Remember that *you cannot rely on your work always being reproduced in color in hard copy.*

Table 4. Comparison of settling for various types of piles.
Piles that exceed 0.25 inches of settlement are shown in red. Piles that are shaded in gray were discarded.

The punctuation of titles and captions has not been standardized across all engineering and scientific fields. Even the venerable *Chicago Manual of Style* does not indicate a preference for capitalization of titles and captions. Thus, you may either capitalize the first letter in all the words in your figure and table titles (not counting articles such as "a" and "the") or capitalize only the first letter of the initial word. If the figure has a title *and* a caption, the caption is most often treated as a sentence. You may have noticed that, in this chapter, all the figure titles have only the first letter of the initial word capitalized, whereas table titles have all "major" words capitalized [i.e., words other than conjunctions, prepositions, etc.]. In addition, the table titles in this chapter have no period, even if a caption follows, but figure titles have a period at the end, even if no caption follows. These decisions were made by the publisher, not by me. Most journals use the same format for both figure and table titles—the position (below vs. above the graphic), not the title's format, is what distinguishes a figure from a table. I suggest using periods only after the figure number (to separate it from the title) and not at the end of the title, unless it is followed by a caption. If there is a caption, then use a period to separate the title from the caption, for both figures and tables. This latter treatment is recommended by the *Chicago Manual.* The best way for you to find out what to do is to read the journals in your field and notice how they handle figure titles and captions. Also, ask your instructor for guidance.

5 INTEGRATING TEXT WITH FIGURES AND TABLES

The most important point to remember when integrating your text with your figures and tables is this: Each figure or table should be explained and interpreted in your text, and yet each should be able to stand by itself, to make sense even out of context. Your explanation of a figure or table should answer two questions sequentially:

1. What information does the figure or table contain, and (in many cases) how were those results obtained?
2. What does this information mean?

First orient the reader to the particulars shown in the figure or table (experimental details, values obtained, etc.); then fully interpret and discuss all changes, relationships, processes, or phenomena indicated by these data. If your figure or table is partially a repetition of information shown in earlier figures or tables, simply remind the reader of that information; do not repeat all of it. If the previous writing up to the point of any particular figure or table is good, very little new information should be necessary in order to answer the first question. You should focus instead on the meaning or interpretation of the information in the figure or table.

Using Graphics Designed by Others

Sometimes, the graphic you want to use already exists, because the data it depicts did not originate with you. Sketches and photographs especially fall into this category, and

HINT

Guidelines for Graphics: Process

Make your figures and tables right the first time, including the labeling and titling; it takes the same amount of time whenever you do it, and it is easier to write the text when you have final figures.

HINT

When discussing technical graphics, here are some guidelines:

- Do not repeat in the text *all* of the data represented in the graphic (especially a table).
- Do not continually say, "The figure shows. . . ." The emphasis should be on the information, not on the word "figure."
- Do not fail to interpret the graphic and explain its significance.
- Do not use "data" as a singular word. "Data" is a plural word; "datum" is the singular version.

they are readily available on the Internet. May you use a drawing created and published by someone else? For educational purposes (a report for class, for instance), the answer is a qualified "yes," as long as you acknowledge the source of the graphic. Figure 23 is an example of a picture used in a report comparing rocket engines. This drawing already existed on the NASA website and became an important tool in this student's research. There was no point in redesigning this visual depiction of the process by which a rocket engine produces thrust. So, the student provided the source of the drawing directly underneath the title.

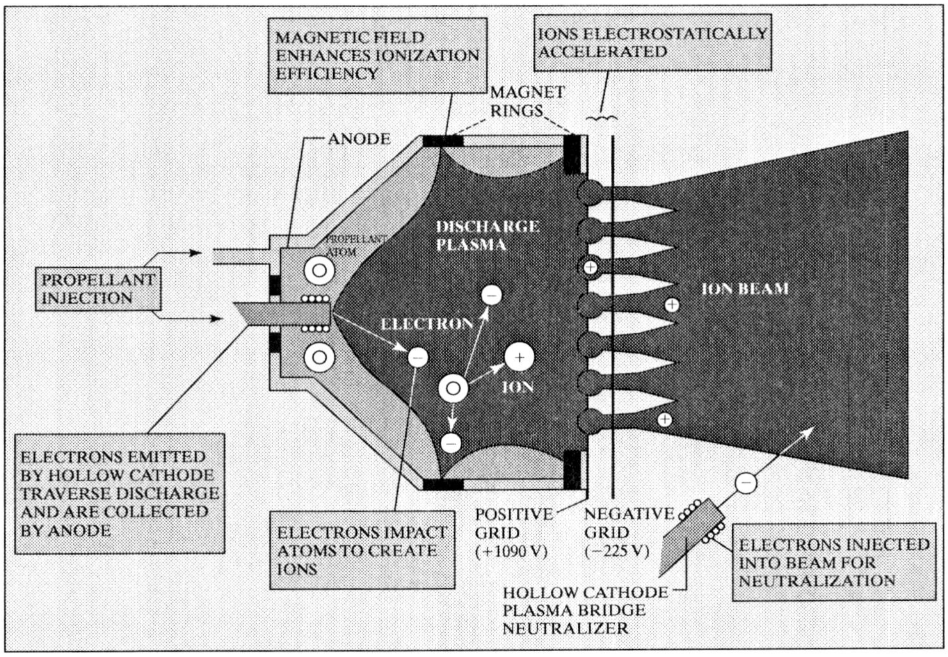

Figure 23. An internal schematic of the Xenon Ion engine. *Source:* NASA Glenn Research Center.

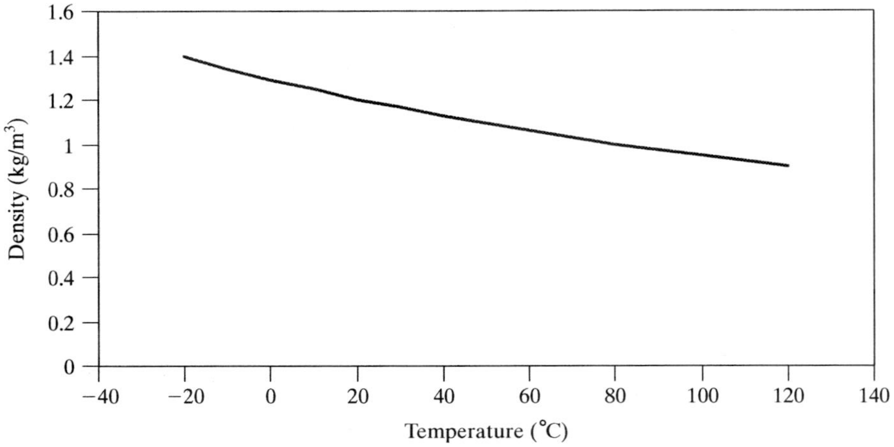

Figure 24. Relationship between density and temperature of air at standard atmospheric pressure. *Source of data:* Crowe, C.T., et al. 2001. *Engineering Fluid Mechanics*, 7th ed.

It is a good idea to provide at least author information in the source citation. In this case, the author is the research center rather than an individual. The full citation should then go in the list of references at the end of the document. In this case, the full citation would include the URL (http://www.lerc.nasa.gov/www/pao/html/ipsworks.htm) and the date the student accessed the site, since the date of publication is not available on the Web page from which this picture came.

The site from which the Xenon Ion engine picture came is maintained by a public agency, so the images are usually considered to be public and therefore usable by the public, especially for educational purposes. The safest plan in borrowing an image, however, is to write to the authors or sponsors of the site (in this case, NASA) for permission to use the picture.

If you have designed your own graphical depiction of some data, but the data themselves came from another source, you do have an obligation to indicate the source of those data. For example, if you create an x–y plot to show the correlation between density of air and temperature at successive measurements, and the values for the plot came from a textbook, you should cite that book right under the title of your figure, as in Figure 24.

You do not have to cite all of the publication details under the title if those details are included in a reference list in your document.

SUMMARY

- Spreadsheet programs help you organize and synthesize numerical and other data that you collect. Use such programs to perform calculations on your data and, ultimately, to represent the data in chart form as a table or figure.
- If you experiment with the different chart forms available in Excel and other spreadsheet or graphing programs, you will learn which forms represent different kinds of data best. Bar charts and line graphs are good for showing changes in one variable over time; x–y and scatter plots are the best choices

for showing changes in the relationship between two variables. Engineers use the latter types of graph most of the time.

- Readers need help in understanding the significance of the information in your graphics. Interpret and explain your graphics in words.
- Give proper credit when you use a graphic created by someone else. If possible, ask permission first.

Problems

1. Record the temperature in your bedroom over a 48-hour period. Then, make an x–y graph showing relationship between time of day/night and temperature. Answer these questions, and justify your choices:

   What decisions do you have to make about displaying the data and labeling the axes?
   Which kind of x–y graph did you choose (with line, with trendline, etc.)?

2. Take the same data on temperature of your bedroom as in #1 and make another kind of graph or chart (other than an x–y graph). Try at least two other types, such as these: a line graph, a pie chart, a bar chart, an area chart. Experiment to see which type of visual display looks best and displays the information best. Which type(s) do you think do *not* work well?

3. Test the graphs you produced in #1 and/or #2 on at least one friend. Which type of visual display works best and why? Ask your friends to explain their preferences.

4. Read an engineering article on a topic you are studying for one of your classes. Study the graphics and write a brief description of their effectiveness by answering the following questions:

   Are the graphics easy to decipher and understand? Why or why not?
   Do the graphics convey the information discussed in the text of the article, or do they introduce new data/information not completely discussed in the text?
   Does each graphic contain enough details such that it could stand alone, without the accompanying discussion?

5. Look again at the graphics in the article you read in #4. Which single graphic (figure or table) seems the best to you, and why? Discuss its virtues—what makes it clear and full of useful information? Can you point to design elements that help make this figure or table clear and useful?

6. Looking at the graphic of the Xenon Ion engine in Figure 23, write a paragraph or two of description of how the engine works. Does the graphic give you enough information to do this, or do you have to fill in some gaps?

7. True or false? Area charts (stacked charts) are the easiest to read of all types of charts.

8. True or false? Using color is always an effective way to highlight numbers on a table.

9. True or false? An acceptable strategy for presenting a set of data in reports is to include a selection of the data in a graphic in the body of the report and the full set of data in a table in the appendix.

10. Write at least one paragraph defending *one* of these positions:

    a. The y axis should always start at zero.

    b. Under certain circumstances, the y axis does not have to start at zero.

FURTHER READING

Cleveland, W.S. 1994. *The Elements of Graphing Data*. Hobart Press: Summit, NJ.
This comprehensive work covers every aspect of understanding how best to graph data. The discussion is high level, but easy to follow. There is more information here than most undergraduates will need, but everyone from beginning students to working engineers will benefit from the substantive discussions of how to make your data stand out and how to avoid ambiguity. Use the index to target answers to any questions you have about specific graphical elements. Error bars, for instance, can actually be ambiguous if the graph does not make clear whether they show the sample standard deviation (one use of error bars) or an estimate of the standard deviation of the standard mean (another use of error bars; see pp. 60–61).

Tufte, E.R. 1983. *The Visual Display of Quantitative Information*. Graphics Press: Cheshire, CT.
This beautifully produced book revolutionized the study of information display and raised the level of awareness of how important graphics are in understanding scientific and technical concepts. When Tufte could not get this book printed properly, he started his own publishing firm in Connecticut. Every page of this book has nuggets of gold, but you should look especially at Part I and the "Chartjunk" chapter in Part II.

Tufte, E.R. 1990. *Envisioning Information*. Graphics Press: Cheshire, CT.
This second major work by Tufte expands the variety of graphics examined, looking at every type of information display, including words themselves. There is a wonderful reproduction of a three-dimensional model published by Euclid in 1570.

Valiela, I. 2001. *Doing Science: Design, Analysis, and Communication of Scientific Research*. Oxford University Press: Oxford, U.K.
The first four chapters of this much-needed book cover how to obtain and analyze scientific information, as well as how to design one's research. Chapters 5–11 offer good guidance and examples for communicating those data effectively and accurately so that everyone, even the general public, can think critically about scientific information and make well-founded decisions.

11

Revising: When Will I Ever Be Finished?

1 INTRODUCTION

The *hardest* part of writing is creating the first draft. When you initiate a writing project—whether it is a business letter or memo, a research proposal, a lab report, or a feasibility analysis—you are creating something out of nothing but the information you have gathered and the synthesizing powers of your own brain. This act of creation is exhausting, demanding, and, finally, rewarding when you have a draft in hand. But the *most important* part of writing is *rewriting*. Rewriting is the multiphase sequence through which you strive to get your writing right, to make it meaningful to the reader and to make it accurately reflect your work and your thoughts. Your reader will never know all that you went through to produce any final document. And that is fine, because a good piece of writing is transparent. Its meaning just shines through the words, and the reader is not even aware of the craftsmanship. But to produce that transparent, seemingly effortless piece of writing, you usually have to craft many drafts. The trick is to make each revision count, so that you can see and track your progress in clarifying and completing the meanings you intend. And to track that progress, you often need the help of others.

Most people use the terms *rewriting, revising, editing*, and even *proofreading* to mean the same thing. Let us use *rewriting* here to mean the entire process of moving from first draft to finished product. *Revising* then means changing your document mainly for content and organization. *Editing* is a two-step process: correcting for logic and flow and improving your language sentence by sentence (this latter step is sometimes called "copyediting"). And *proofreading* is the final step in preparing a document, in which you correct mechanical errors.

SECTIONS

1 Introduction
2 Writing Is Rewriting
3 Organizing a Peer Review
4 About Grammar Checkers
5 Revising First for Structure and Content
6 Editing for Logic and Flow (Mostly Paragraph Level)
7 Editing for Sense and Meaning (Mostly Sentence and Word Levels)
8 Proofreading for Final Copy

OBJECTIVES

By reading this chapter, you will learn the following:

- the importance of reserving time for revision;
- how to develop peer-review groups;
- the benefits and shortcomings of grammar checkers;
- how to revise in manageable stages and correct for organization and content;
- how to edit paragraph by paragraph and correct for logic and flow;
- how to edit sentence by sentence and correct for grammar and style;
- to use a checklist for finalizing and proofreading your written document.

2 WRITING IS REWRITING

Take a look at the following passage, and determine whether it needs rewriting of some kind:

> Lilliputian Engineering gathered information about rope strength from several sources. First, Dr. Swift, a professor at Baker University, was consulted about the differences between ropes woven out of seaweed and grass. Then, we performed numerous lab experiments in which we determined the tensile strength of the individual fibers of the S-37, S-40, and J2000 ropes. **Resistance to discoloration was determined through phone interviews with company representatives**. After this, Lilliputian Engineering quantified internal rope stresses by hanging loads onto ropes of various diameters and measuring rope stretching.

This passage is not an example of terrible writing, but it is definitely writing that needs revision and editing. What are some of the problems? Well, for one thing, the sequence of investigation seems slightly askew. Surely, the researchers performed experiments on tensile strength *and* on internal stress before moving on to the interviews about discoloration. This report on rope strength seems to have been written in the order that the author remembered doing things rather than the actual, logical sequence of investigation or the order of importance of tasks. In addition, the sentence about phone interviews (in bold type) cries out for a citation (i.e., which company representatives were consulted?). And what about the imprecise nature of the words "numerous" (how many experiments is that?) in the third sentence and "this" in the last sentence? Also, do you not find it disconcerting that the sentences alternate between passive and active voice? In the third sentence, the writer states that "we *performed*" experiments, but in the fourth sentence, he or she says that discoloration "*was determined*." There is no rule about sticking with one verb form or the other throughout an entire paragraph, but the flip-flopping here seems unnecessary and awkward. In addition, if you really look closely, you will be bothered by the comparison in the second sentence: Are all the ropes woven out of both seaweed and grass, or are some ropes woven out of just one material?

So how do we fix all these errors? Well actually, most workplace documents undergo several rounds of editing and revising before they are considered final documents. Most technical editors advocate what is called "levels of edit" to break up the rewriting process into manageable chunks of effort and to ensure that any document works on all levels: content, organization, logic, and sense. This procedure may mean that you end up writing three or four drafts before achieving the final version. Sounds like a lot of effort, but remember that each draft is faster to produce, with fewer changes, and the result—a crisp, convincing, and thoughtful document—is well worth the effort. That document will get you what you want, whether it is funding for a new project, recognition for research well done, or a satisfied client.

Here is a checklist that presents important writing guidelines to help you revise and edit several times before handing in, circulating, or publishing any document.

Rewriting: Guidelines for Revising, Editing, and Proofreading Your Writing

Whenever you finish a draft, put it aside for *at least a few hours*. Then follow this sequence:

1. Check the structure and content of your document:
    - Is the structure what the audience expects in a document of that type (report, proposal, etc.)?
    - Have you included all necessary information, and are your presentations of data as complete as you can make them? Have you indicated and justified omissions where you simply cannot be complete?

- Do any passages include facts or figures that did not come from your original research? If so, would you consider that information to be general knowledge (in which case you do not have to cite references)? Or must you insert citations to outside sources of information?

2. Check the logic and flow of your ideas and information:
   - Is the logic based on the reader's frame of reference or yours?
   - Are there any passages that do not seem to follow logically from the preceding passage?
   - Does each paragraph begin with a topic sentence whose content is supported and expanded (through examples, description, etc.) throughout the rest of the paragraph?
3. Edit line by line for correct grammar and style: Check your sentences for sense, clarity, and precision.
4. Proofread for punctuation, spelling, words left out, spacing, and consistency of headings. Try reading the document aloud to "hear" any remaining errors.

The foregoing steps are generally sequential—step 1 usually comes before step 4—but they are also iterative. Proofreading, for instance, may reveal a missing piece of data that you thought you had included. When you make a substantive change (such as adding another piece of information), you will need to return at least to step 2 to check logic and flow, and then you will have to run through the remaining steps once again.

3 ORGANIZING A PEER REVIEW

Now, here is the $64,000 question: Are you able to catch all the errors when revising something you wrote? If you are really honest, you will most likely answer, "No, probably not." It is extremely difficult for *any* writer to see all of his or her own errors. Because those sentences have come out of your head, you often continue to see what is in your head rather than what is on the page. Recent studies on human perception have shown that we often see what we expect to see even when it is not there or when something very unusual has taken its place. In an article in *The New Yorker* (June 11, 2001), writer Malcolm Gladwell gives an example from a psychological study (by Daniel Simons and Christopher Chabris) in which 50% of the observers of a videotape showing basketball players making passes failed to see the woman dressed up like a gorilla walk into the middle of the action. Those observers were so busy counting passes that they simply did not see a person in a gorilla suit standing in the middle of the basketball court. Gladwell's point in this article, about highway safety, is that research is beginning to show an astonishing insight: ". . . it is not just that our memory of what we see is selective; it is that seeing itself is selective" (p. 54). So it is no wonder that we do not see our misspellings or mistakes in grammar, but do see in our mind's eye that word we actually left out.

There are some proofreading strategies you can use to fool your mind into catching your own errors, but those strategies do not usually catch the more substantive mistakes that should be corrected. The best way to catch your errors in structure, logic, completeness, citation, and flow is to ask someone else to read and comment on your draft. Whom should you ask? Not your boss or your teacher (unless he or she asks to read your draft), but a peer—a fellow student or engineer. Preferably, your reviewer should be someone who is not familiar with the work you are writing about; that way, your reviewer will not substitute what he or she already knows about the subject for the information that is actually there on the page.

Now, it is not easy to ask someone to read a piece of writing that you know needs work. But think of the alternative: sending an unedited, incomplete, or just poorly written report or proposal to a decision maker (e.g., your boss, your teacher, or your client). It is better to risk embarrassment with a peer than run the risk of not getting a job or a decent grade. Besides, you can always offer to read your peer's drafts in return.

You can form a peer-review group with just two people or with many, at school or at work. Make a pact with each other that you will take the time to read carefully and comment on each other's drafts. Here are some guidelines to follow when you ask for a review of your work:

HINT

Getting the Most out of a Peer Review of Your Draft

- Ask whether your reviewer wants to write down suggestions directly on your draft, to talk his or her comments through with you, or both. Ask whether he or she is more comfortable reading and commenting on the electronic or the hard copy.
- Come to an agreement with the reviewer on how much time he or she needs to review your draft; gently remind him or her if the deadline passes.
- Call your reviewer's attention to any critical elements on which you want him or her to focus.
- Do not expect a complete correction of all grammatical and sentence-structure errors. Your reviewer is probably not a grammarian, and what you really need are comments on the substance, completeness, and flow of your information.
- Receive all comments, whether written or verbal, in a positive, accepting spirit. Do not be defensive; think about how lucky you are to get this free critique that is bound to make your document better, if not perfect.
- Thank your reviewer, and always offer to reciprocate.

Here are some guidelines to follow when you review someone else's draft:

HINT

Giving the Most when Reviewing the Writing of Others

- Ask for the document in the form you are most comfortable working with: electronic or hard copy. If you know how to use the editing functions in your word processor (such as the Track Changes function in Microsoft Word), use them to make suggestions for revision. Consider learning about these editing functions if you are not already familiar with them.
- Try to comment on what is good as well as what could be better.
- Do not use language that is personally critical: "I think there are a lot of sentence-level errors in this passage," *not* "Your writing really stinks here."
- Focus on the big picture—the clarity, structure, and flow of the document—rather than on catching every word error. You do not want this task to be so much work that you end up hating the draft *and* its author.
- Whenever possible, offer to meet with the writer and go over the document together. It is easier to explain some things in person than in written comments.
- Use a positive, upbeat tone in your verbal comments, even if you see lots of errors. Be patient in your explanations. Writers are very sensitive about their own creations and often have a hard time seeing another way of saying something.

Remember that in the workplace, peer review does not necessarily take the place of a final review by a professional editor or technical writer within the organization. The better the draft you give the technical editor, the more he or she can focus on polishing your writing instead of trying to make substantive changes in subject matter on which he or she is not an expert.

Some school or work groups use guided questions to stimulate and standardize the peer-review process. Rather than asking reviewers to completely edit your work, it is a good idea to ask them to focus on certain issues that tend to arise in your particular document. For example, say that you have written a research proposal for your company. In this company, proposals of this type usually outline methods for conducting a feasibility study of solutions to the client's need or problem. Budgets included in these types of proposals should make clear that the costs are for conducting an investigation of possible solutions, not for implementing a chosen solution. In the following set of questions for peer reviewers of your proposal, the words with initial capital letters are headings of required sections in the proposal:

Sample Peer-Review Questions for a Research Proposal

1. **Does the Summary seem complete? If the Summary you are reviewing seems incomplete or unpersuasive, let the author know, and offer a suggestion for improvement.**
2. **Does the Problem Definition present the client's problem in enough detail?**
3. **Does the Scope of Project state the specific objective(s) of this project? Does it outline the alternative solutions that will be investigated and the methods of study that will be used to investigate and evaluate them? Does this section say what the deliverable product of this project will be?**
4. **How specifically has the author defined the solution criteria? If the criteria are not very specific, does the author indicate how he or she will go about defining them during the project?**
5. **Has the author indicated that certain criteria are more important than others? If the criteria are not prioritized, can you suggest how they might be?**
6. **Are the steps outlined in the Proposed Procedure specific enough? Does each step follow logically from the others?**
7. **Does the Schedule section reflect the tasks outlined in the Proposed Procedure? Do you have any suggestions for revision?**
8. **Does the Budget seem realistic? Does it reflect the cost of the author's conducting the study, *not* the cost of implementing a solution to the client's problem?**
9. **Do you think the proposal is persuasive? Do you think the proposal would convince the primary audience (the client) to fund this project? Why or why not?**
10. **Can you see any evidence in this proposal that it is written also for the secondary audience, those other interested readers (such as advisor or owners or financial managers)? Why or why not?**

As an individual, you can also, of course, ask your reviewer to focus on certain issues that tend to arise in your own writing. For instance, if you know that you have trouble getting to the point, be sure to have someone check your statements to make sure you have clearly explained at the beginning of the document what the project is all about and what the document intends to do (propose something, conclude something, etc.). In other words, you want to ask your human reviewers to do what computer-based grammar checkers cannot do: check for clarity, logic, consistency with company standards, and completeness of information.

4 ABOUT GRAMMAR CHECKERS

The grammar-checker function included in most word-processing software is, on the whole, a boon to clear writing. Grammar checkers highlight for us the phrases and sentences that may contain errors in grammar or style. Figure 1, for instance, is the screen that pops up when I do a grammar check in Microsoft Word on a funky sentence I have written critiquing another author's writing.

I knew the sentence was "funky" because Word had underlined it in green. To find out more specifically what is wrong with it, I chose the Spelling and Grammar function from the Tools menu. The checker considers my question to be a fragment for a couple of reasons, probably: The question starts with "and," and I have buried a question within a question (in parentheses). The first problem is arguable: Some grammarians believe it is never proper to start a sentence with "and" or "but," and some think that it is OK to do so. In fact, the 15th edition of the *Chicago Manual of Style* (2003) says it is perfectly fine to start a sentence with a coordinating conjunction ("and" and but" are the most common), but traditional composition guidelines die hard. Without going into more details, I suggest a middle-of-the-road course that limits your use of "and" or "but" as a sentence starter to those infrequent occasions when the link (similarity or dissimilarity) to the previous sentence is more important than any other meaning in the current sentence.

The second problem—burying a sentence within a sentence—is a more serious issue and should never be done in formal or even semiformal technical writing. I chose to leave that sentence as it is because I want to be informal in the style of this book. Also, when I am picking apart a passage to explain stylistic nuances, I have to work harder to get your attention. A question buried within a question may have gotten your attention. In any case, I considered what the Word grammar checker indicated and then used my own judgment to decide what to do.

As long as you use your own judgment in conjunction with your word processor's promptings, grammar checkers are helpful in catching those grammatical and stylistic errors we tend to make all the time (such as overuse of passive voice and disagreement of subject and verb). In addition, the spell checker is good at catching those words we tend to misspell. I can never remember, for instance, whether "attendence" or "attendance" is correct. But Word's spell checker tells me immediately that the first spelling is incorrect by underlining it in red. However, there are many misspelled words that Word cannot catch because *Word*

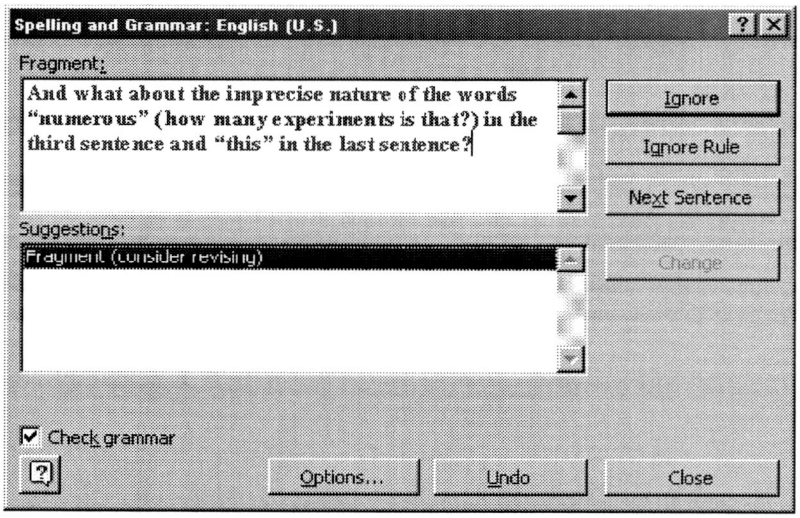

Figure 1. A Sample screen from the grammar-checking tool in Microsoft Word.

does not read for meaning. Word processors thus far do not have artificial intelligence, so they cannot read for sense; for example, they cannot know whether the correct word for you to use in a particular sentence is "from" or "form." Both spellings are words in the dictionary, and thus only a being with a certain level of cognitive reasoning skills can tell you whether you want to write, "The data were extrapolated form many sources" or "The data were extrapolated from many sources." That being is usually you, unless you have a careful peer reviewer or the luxury of a technical editor. My word processor did not underline "form" in any color in the previous sentence, even though it was incorrect in the context.

PRACTICE QUESTION: GRAMMAR CHECKERS

Type this sentence into your word processor, and then run the grammar checker on it:

> The director explained the environmental process review guidelines.

Did the checker make any suggestions? If not, can you pinpoint at least one problem with the readability of this sentence? Is anything unclear?

SUGGESTED ANSWER

The noun phrase, "environmental process review guidelines," is rather difficult to understand precisely. Is the director explaining a process for reviewing environmental guidelines? Or is she explaining the guidelines for reviewing environmental processes? If your grammar checker did not ask for clarification here, then you will have to catch these confusing noun phrases for yourself.

5 REVISING FIRST FOR STRUCTURE AND CONTENT

Revising for structure and content is often difficult to separate from editing for logic and flow (the next phase). After all, if you discover that some content is missing (some specific test results, for instance), then your conclusions will not flow logically from the data presented; there is a problem with content *and* with flow, and of course you will have to restructure that section. Do not worry about trying to separate out the two phases; just proceed from the macro to the micro as you revise and edit. Start by looking at the entire document (remember to check your table of contents for consistency with headings in the body of the document), move fairly quickly from section to section, and keep in mind the big picture. Ask yourself these questions as you read:

1. Does this draft conform to standards and expectations for this type of document in my company, agency, or classroom?
2. Have I highlighted the really important points, even if that means I repeat them in different places in the document?
3. Do I introduce the document and each section properly so that the reader knows what to expect next?
4. What do I want the reader to get out of this particular section?

To answer the first question, make sure that your draft includes the sections (with appropriate headings) expected by your client, boss, or teacher. If you do not know what is expected, go and find examples of similar documents. Certainly, you may have to add or subtract certain sections as you realize that they do or do not apply, respectively, to the particular project you are writing about, but make these changes carefully and justify them to yourself. For documents that are longer than four pages, adding a summary at the beginning is always a good idea. For longer reports, this section is called an executive summary. For most documents, a conclusion, whether labeled as such or not, should summarize and highlight important points for the reader.

HINT

Checklist for Revision of Structure and Content

1. Ensure that the text conforms to structural standards for the particular type of document.
2. Check that the document highlights important points (perhaps with graphics).
3. Make sure that the document introduces each section such that the section's purpose is clear.
4. Ensure that each section fulfills a purpose for the various readers of the document.

Let us examine an example summary for a proposal:

Proposal to Evaluate Wall Designs for Apartment Sound Control

Summary

ERC Properties, an Austin apartment management company, is searching for a wall design to control sound in their new apartment complexes. In their existing complexes, residents have complained about the insufficient sound control provided by the common wall between two apartments. These high levels of noise have convinced some residents to move to other apartment complexes, forcing ERC Properties to search for replacement tenants. To avoid losing residents in their future complexes, ERC has determined to improve the sound control. Therefore, they have asked Firm Ideas to evaluate different wall types and materials that would provide better sound control in the new complexes.

> checklist item 2: The most important points in this proposal have to do with the *problem* of controlling sound between apartments and with the various *solutions* that involve redesigning the walls. Repeating the words "sound control" emphasizes those points.

Firm Ideas has begun the process of choosing and evaluating *wall designs* for *controlling noise* between apartments. This proposal outlines the *initial designs* and the methods for evaluating them. These six initial designs include two types each of single-, staggered-, and double-stud walls. Each *wall design* will be evaluated based on the wall's performance, material and labor cost, construction time, and usage limitations. After the evaluations are complete, Firm Ideas will recommend the *most suitable wall design* to ERC Properties.

> checklist item 4: Methodology statements convince the *advisor* reader.

The wall-design evaluations are expected to take 10 weeks. A progress report of this project will be available on March 31, 2003, and a final report will be submitted to ERC Properties on May 1, 2003. The estimated cost of Firm Idea's services on this project is $1,750.00.

> checklist item 4: Cost and time estimates are important for the *decision-making* reader.

Most proposals start with a summary, so the reader's expectations are being fulfilled here (item 1 of the foregoing checklist). Notice that this summary also answers the most important questions for the client and thereby fulfills items 2 (highlighting important points) and 4 (educating readers). Of course, different readers of this proposal will have somewhat different concerns: The *decision-making* reader will want primarily to know how quickly and at what cost the sound-control problem can be solved. The *advisor* reader will want to know how carefully the writers evaluated various solutions. At the beginning of a report or proposal, however, all readers ask themselves the following questions:

What is the purpose of the document I am reading?
What is the purpose of the proposed project?
How long will this project take, and how much will it cost?

The proposal summary answers those questions in three short paragraphs. In fact, the summary has fulfilled item 3 in the foregoing checklist (letting the reader know what to expect in the section) by letting the reader know what to expect in the entire document. For other sections in this proposal, how do you introduce them so as to preview their content for the reader? Here is the beginning of a section called "Solution Criteria":

> Each of the six wall designs will be evaluated according to four criteria: performance, cost, time to construct, and usage limitations. These criteria are explained next in prioritized order.

Those two sentences tell us exactly what to expect next and, indeed, from the whole section.

6 EDITING FOR LOGIC AND FLOW (MOSTLY PARAGRAPH LEVEL)

Making sure that you let the reader know what to expect next is the task that connects reading for content with reading for logic and flow. Since the most important location in any document is usually the beginning, be sure to check the first section (whether a summary, an introduction, or just the first paragraph of a letter) most carefully for logic and flow. Ask yourself if the text clearly conveys the purpose of the document and what you want the reader to get out of the section. Make sure the opening statements follow a logical order and present the most important information first. The following paragraph, from the summary of a proposal, follows a sequence that moves from company *need* to underlying company *problem* and *source* of that problem:

ERC Properties, an Austin apartment management company, is *searching for a wall design to control sound* in their new apartment complexes. In their existing complexes, residents have complained about *the insufficient sound control* provided by the common wall between two apartments. These *high levels of noise* have convinced *some residents to move* to other apartment complexes, thus *forcing ERC Properties to search for replacement tenants.* To avoid losing residents in their future complexes, ERC has determined to improve the sound control. Therefore, they have asked Firm Ideas to evaluate different wall types and materials that would provide better sound control in the new complexes.

The client's *need* (what this proposal is responding to).

Source of noise problem.

Source of three problems for client:
1. Tenants are moving out.
2. ERC must search for new tenants.
3. Future tenants may move, too.

After checking the beginning section carefully, read your document paragraph by paragraph. Reading at the paragraph level often becomes intermixed with reading at the document level, but try to focus on flow and transitions, and do not worry so much about the big questions of purpose and content. Let us look at an *earlier* version of the paragraph we just examined. Do you see places where the logic is not as tight as it could be, where a step seems to be missing in the explanation presented?

> **First Draft of First Paragraph of a Research Proposal**
>
> ERC Properties, a central Texas apartment management company, is searching for a wall design to control sound transmitted through shared walls in an apartment complex. ERC Properties has received complaints from residents of transmitted noise through adjacent units in several of their apartment complexes. *These high levels of noise have convinced some residents to move to other apartment complexes, thus forcing ERC Properties to search for better sound control.* Therefore, they have asked Firm Ideas to evaluate different wall types and materials that would provide better sound control in the new complexes.

Take a close look at the third sentence. Is the cause-and-effect relationship presented there entirely clear? Actually, there are two cause-and-effect relationships—an implied one in the first half of the sentence (residents are moving because of the "high levels of noise") and a direct one in the second half (residents are moving; "thus," ERC is searching "for better sound control"):

Implied cause and effect

These high levels of noise have convinced some residents to move to other apartment complexes, thus forcing ERC Properties to search for better sound control.

Stated cause and effect: residents' moving means ERC must search for better sound control

But does it necessarily follow that residents' leaving means that ERC must search for better sound control? There is actually a step missing: Residents are leaving; thus, ERC is forced to find new tenants, and *in order to find and keep new tenants*, ERC must improve sound control. You help the reader understand more quickly if you break causal relationships down into the simplest chain possible, which may mean adding more links to the chain. When you add more links to a cause-and-effect chain, you usually need to keep sentences as short as possible; thus, you often end up crafting more sentences to clarify the cause-and-effect relationships. Check out this revision, in which the original sentence is now broken into two sentences:

> These high levels of noise have convinced some residents to move to other apartment complexes, thus forcing ERC Properties to search for replacement tenants. To avoid losing residents in their future complexes, ERC has determined to improve the sound control.

In the first sentence, the cause-and-effect chain has two large links (or steps): The high levels of noise cause residents to move, and residents' moving causes ERC to search for replacement tenants. Notice the clarification here: ERC must "search" for "replacement tenants," and that search has led them to search for ways to improve sound control in future buildings. In the second sentence, the chain moves backward, presenting the desired effect first—to avoid losing tenants in the future—and the cause second—ERC's projected improvement of sound control. This way of explaining things has the added benefit of presenting ERC's search for sound control as the cause of the current proposal writers' investigation as well. The next sentence practically writes itself (as yet another effect of the cause presented in the previous sentence):

> Therefore, they have asked Firm Ideas to evaluate different wall types and materials that would provide better sound control in the new complexes.

What else needs clarifying in this first draft of the proposal's first paragraph? If ERC wants new tenants to rent their apartments, and they have determined that inadequate sound control is the problem that keeps tenants away, then we would expect that ERC would try to renovate their existing apartments to provide better sound control. But our expectations are wrong; we are misled for most of the paragraph. We learn only in the last sentence, second-to-last word, that ERC is not in fact going to renovate existing apartments: They are planning on incorporating better sound-control materials in their *new* apartment complexes. The edited version clarifies this difference between existing and new apartments simply by using the words "existing," "new," and "future" earlier in the paragraph.

First Paragraph of Research Proposal Edited for Logic and Flow

ERC Properties, an Austin apartment management company, is searching for a wall design to control sound in their *new* apartment complexes. In their *existing* complexes, residents have complained about the insufficient sound control provided by the common wall between two apartments. These high levels of noise have convinced some residents to move to other apartment complexes, thus forcing ERC Properties to search for replacement tenants. To avoid losing residents in their *future* complexes, ERC has determined to improve the sound control. Therefore, they have asked Firm Ideas to evaluate different wall types and materials that would provide better sound control in the *new* complexes.

HINT

Checklist for the *Logic-and-Flow* Level of Editing (Paragraph Level)

1. Ensure that the document uses organizing patterns such as cause-and-effect or problem–solution structures.
2. Within that pattern, each sentence of the paragraph flows logically from the previous sentence.
3. Make sure that links in cause-and-effect chains are clearly delineated. There should be no more than two links per sentence.
4. Ensure that the topic sentence introduces the theme of the paragraph and that it makes a logical link with the final sentence.
5. Make sure that no sentence in a paragraph is more general than the paragraph's topic sentence.

7 EDITING FOR SENSE AND MEANING (MOSTLY SENTENCE AND WORD LEVELS)

After we edit a document for logic and flow, we should take one more look at it in order to concentrate on editing it at the sentence level, checking for sense and meaning and making sure that every word is as precise as possible and the grammar is correct. Consider the following paragraph, from the summary of a proposal, where each sentence has been numbered for the purposes of our discussion:

> **[1] ERC Properties, an Austin apartment management company, is searching for a wall design to control sound in their new apartment complexes. [2] In their existing complexes, residents have complained about the insufficient sound control provided by the common wall between two apartments. [3] These high levels of noise have convinced some residents to move to other apartment complexes, thus forcing ERC Properties to search for replacement tenants. [4] To avoid losing residents in their future complexes, ERC has determined to improve the sound control. [5] Therefore, they have asked Firm Ideas to evaluate different wall types and materials that would provide better sound control in the new complexes.**

When you are editing sentence by sentence, you need to slow yourself down, but keep reading for sense. (In proofreading, you focus more on mechanics than sense.) One way to slow down and really see what is written is to read the text out loud to yourself. This technique has worked for many writers. The trick is to keep reading aloud and not lapse into reading silently. The technique depends on your ear hearing what your eye no longer sees (e.g., a word left out, an overly wordy sentence, or an ambiguous pronoun).

A possible problem in the foregoing paragraph occurs in sentence #1 with the pronoun "their." "Their" refers to ERC Properties, a company, which is a *collective* noun, meaning it can be either singular or plural, depending on the meaning. Here, the company is presumably acting as a single entity, and indeed the singular verb "is" describes what the company is doing ("searching"). Why, then, does the writer switch and use the plural pronoun "their" in the second part of the sentence? Well, the singular form of the pronoun that would apply is "it," since we do not usually refer to companies as "he" or "she." But "it" does not seem quite right either, since we know that the company is composed of human beings making decisions and doing the sort of cognitive work that an "it" cannot do. So, we often cheat a bit and use the plural pronoun—a form of "they"—to indicate a singular entity composed of more than one human. The practice is so common by now that it does not seem worth worrying about, and the fixes are also awkward; it is easy enough to say "ERC Properties Á *are* searching," but it is a bit more difficult to use a plural verb with the shortened form of the company's name, as in, "ERC *have* determined to improve." In sentence #1, every reader will know whom you are indicating by "their," but be careful when there is more than one plural noun in the sentence. Look at sentence #2: "*their* existing complexes" could refer to "residents," since that is the plural noun closest to the pronoun "their." You would be wise not to use "their" there. So, let us change that sentence by removing "their" and repeating a shortened version of the client company's name:

> In existing **ERC** complexes, residents have complained about the insufficient sound control provided by the common wall between two apartments.

You might also want to change the "their" to "the" in sentence #4.

Unnecessary wordiness is a common problem with almost everyone's writing. The paragraph here has become quite streamlined and focused through the process of editing for flow, so we do not see a lot of extra words. But how about the word "thus" in sentence #3? It is a common and useful transitional word, but we do not really need it here because we have the participle "forcing" to explain the consequence of residents' moving out. Rereading this paragraph multiple times enables us to see another wordiness issue: The word "complexes" is used in every single sentence! It is an important word to use, because "apartments" or "buildings" does not mean quite the same thing, and we want to use consistent terminology, especially when focusing on the client's problem. If we examine each sentence, however, we can see that "apartment complexes" is probably not necessary—or even accurate—in sentence #3. We can just say "convinced some residents to move out." In addition, we may not need the introductory phrase "In existing ERC complexes" in sentence #2; we could just make that information part of the main clause: "Residents of existing complexes have complained. . . ." That way, we can remove a comma and get right to the main subject of the sentence: "residents."

So here is our new, streamlined version of the paragraph:

> ERC Properties, an Austin apartment management company, is searching for a wall design to control sound in their new apartment complexes. [2] Residents of existing ERC complexes have complained about the insufficient sound control provided by the common wall between two apartments. [3] These high levels of noise have convinced some residents to move out, forcing ERC Properties to search for replacement tenants. [4] To avoid losing residents in planned future complexes, ERC has determined to improve the sound control. [5] Therefore, they have asked Firm Ideas to evaluate different wall types and materials that would provide better sound control in the new complexes.

Actually, the previous version of the paragraph that had been revised for logic and flow had very few sentence and word-level issues. This degree of polish is a common reward for paying attention to each level of edit as you work you way through the rewriting process.

At this sense-and-meaning level of edit, though, you want to remember what you know about your own stylistic and grammatical foibles. What have writing teachers commented on in the past? Do you tend to use a lot of passive voice? Are your sentences described as "wordy" on returned papers? If punctuation is a particular problem for you, you will need to pay attention to it at every level of editing, because every time you rephrase a sentence, the punctuation has to be rethought as well. You can use the Find function of your word processor to help target word and punctuation mistakes you tend to make. For example, because the apostrophe is usually used to indicate the possessive, many people use "it's" mistakenly. "It's" means "it is," not the possessive of "it," which is indicated by "its," as in "its function" (the function of "it"). So, I usually set my word processor to look for both "it's" and "its" so that I can check for correct usage. Similarly, if the semicolon is still an untamed beast to you, search for words such as "however" and "therefore." If one of those words occurs in the middle of a sentence, see whether each set of words on either side could form a complete sentence. If so, you probably need to put a semicolon before the "however" and a comma afterward.

HINT

Checklist for *Sense-and-Meaning* Level of Edit (Sentences and Words)

1. Ensure that sentences are linked together with transitional words where necessary.
2. Make sure that words are not ambiguous in meaning
3. Check that technical terms and keywords are used consistently and not changed just for variety's sake.
4. Consider suggestions by your computer's grammar checker.
5. Ensure that sentences have been checked for these common problems:
    - too much passive voice
    - ambiguous pronoun reference
    - subject–verb disagreement
    - wrong word: form/from, their/there/they're, cite/site, to/too/two, etc.
6. Check that figures and tables have meaningful titles and adequate labeling.

Use the computer's "Find" function to search for your common errors

PRACTICE QUESTION: COMPUTERS AND REVISION

If grammar checkers are still so unreliable, how do you use your word processor to pinpoint problems at the sentence level?

SUGGESTED ANSWER

Use the Find function to search for particular words and phrases that you know you tend to misuse. If your instructor is a stickler for using "data" as a plural word, then search the document for "data" and make sure every verb connected to that noun is plural.

8 PROOFREADING FOR FINAL COPY

At this final phase in the process of improving your document, you are checking for mechanical errors, which include the following:

- spacing problems (between both lines and words)
- mistakes in numbering (sections, headings, pages, figures, and tables)
- missing words and marks of punctuation (especially the right-hand quotation mark or parenthesis)
- misspellings

Some writing experts suggest reading backward in order to proofread without getting caught up in the sense of the sentences. You could read the entire document backward, word for word, or you could read each sentence forward, starting with the final sentence and moving backward sentence by sentence through the document. I have to say that I find this method unbearably tedious and cannot make myself do it. I read forward, but very slowly, paying little attention to sense and lots of attention to the look of the words on the page and to sequences of any kind: numbering, pairs of punctuation marks, and all those little details that are usually too boring to pay much attention to.

Whatever proofreading method you use, make sure that you allow time after editing for flow and sense before you attempt final proofing. Of course, if you catch a problem with sense or flow during proofing, you will want to correct it, but you are likely pretty tired at this point, and you may no longer be the best judge of what flows or makes sense. If you have already edited for sense and meaning, then clean up the mechanical errors of punctuation, numbering, spacing, and spelling, and you will have a professional-looking document, even if a few language errors remain. Chances are good, in any case, that those remaining errors are matters of style and therefore questionable and not necessarily rule bound.

Speaking of getting tired, remember the reality of the writing process: Writing is tiring. When you find a particular mechanical error (such as no right-hand quotation mark), you are likely to find others nearby. Simple mistakes tend to cluster in our documents at the point at which we got tired, and therefore sloppy.

HINT

Suggestions for Powerful Proofreading

- Allow time after editing.
- Check numbering.
- Check whatever comes in pairs (e.g., quotation marks and parentheses).
- Check numerical calculations.
- Remember that mistakes tend to occur in clusters.
- Turn the document upside down to check spacing.
- Consider reading backwards.
- Use the spell checker!!!

PRACTICE QUESTION: PROOFREADING

Proofread the following passage to see how many errors you can catch and correct:

> **Everyone had their own idea about how the software would be developed, what platform would run on best, and what hardware would make it most scalable. When the company folded, the former CEO says "what should have been a brilliant new product for pricing derivatives is only a half completed bundle of program code. The CEO is now seek funding for a new start-up venture.**

SUGGESTED ANSWER

Does your proofread version look like this?

> **Everyone had *his or her* own idea about how the software would be developed, what platform *it* would run on best, and what hardware would make it most scalable. When the company folded, the former CEO *said*, "What should have been a brilliant new product for pricing derivatives is only a *half-completed* bundle of program code." The CEO is now seeking funding for a new venture.**

SUMMARY

- Revising your writing may be the most important part of the writing process. Approach revising in stages, focusing first on rewriting for structure and content.
- Edit also in stages, focusing first on examining paragraphs and then on correcting sentences for grammar and clarity.
- Do not rely on electronic grammar checkers to catch your sentence-level errors. Use the checkers with caution, and make sure you have a good style guide at hand.
- Asking others to review your writing before it is finalized is a great idea. Form a peer-review group with other students or coworkers. Receiving a peer review, however, is no substitute for doing a final edit yourself.
- Learn to leave time for proofreading, and adopt the strategies that work best for you. Not everyone has the patience to read backwards.

Problems

1. Take a draft of a current writing assignment you have been given and edit it yourself. Time yourself and see how long it takes you to produce a second draft. Write a brief explanation of how long the second draft took to produce, how difficult the edit was, and whether you think the assignment is now in final form. Could you feel good about handing in the second draft?

2. Take the same draft as in #1 (the first draft, not the revision) and give it to a friend to edit. Ask your friend to read as carefully as possible, not to be afraid to criticize, and to write down all the errors he or she can see. Ask your friend to indicate where the writing gets at all confusing.

   - Now, if you have already edited the draft yourself, compare the two edits—yours and your friend's. How are they different? Write a brief description of the differences.
   - If you have not already edited the draft yourself, write a list of the mistakes and problems your friend encountered in editing your draft. What surprises you?

3. Ask the friend who edited your draft in #2 to talk to you about his or her suggestions. Is it easier to understand someone's criticism when they explain it to you verbally? On a scale of 1 ("fabulously helpful") to 10 ("a total waste of time"), rate the usefulness of having someone else edit your work. Which was more useful: the written comments or the verbal comments?

4. Given the way most people read, the most important part of a document is which part?

   The beginning
   The middle
   The end

5. Under what circumstances would you want to present background information about a problem before stating the solution? Under what circumstances would you want to present the solution before the details of the problem?

6. True or false? In the stages of rewriting, editing for sense and clarity should come before editing for structure and content.

7. True or false? Once I become a good enough writer, I will be able to write two-page (or even longer) documents in one draft, without revising.

8. If I don't have enough time to finish editing a document, I should at least check which one of the following? Choose only one and explain your choice.

   The figure/table numbers and titles
   The first paragraph's flow and accuracy
   The last paragraph's flow and accuracy
   The headings' format and wording

FURTHER READING

Chicago Manual of Style, 15th ed., 2003. University of Chicago Press: Chicago.

This is the mother ship of style manuals. This big volume contains more than you'll ever want to know about manuscript preparation, documentation, and the publishing process, but it will also have definitive answers for you on all those niggling little questions about style (do you write out the number twelve or use the Arabic numeral?). The sections on "Grammer and Usage" and "Punctuation" are a terrific reference tool for engineers. Most discipline-specific style manuals in engineering are based on the *Chicago Manual.*

Elbow, P. 1998. *Writing with Power: Techniques for Mastering the Writing Process*, 2d. ed. Oxford University Press: New York.

Chapter 3 of this influential book on the writing process describes the critical importance of "sharing" as a prerequisite for good writing.

Gladwell, M. 2001. "Wrong Turn: How the fight to make America's highways safer went off course," *The New Yorker*, June 11, 2001, pp. 50–61.

This article's discussion of how we literally misperceive reality is focused on how such misperception causes automobile accidents, but the evidence presented from research studies of perception seem very applicable to other situations as well. When we edit our own writing, we may well not see what is really there, because we have the "right" text visible in our mind's eye.

Samson, D.C. 1993. *Editing Technical Writing*. Oxford University Press: New York.

This book is aimed mostly at professional writers and editors and focuses on editing as a collaborative process. But three of the chapters are relevant for engineering students. The two chapters on editing text and editing graphics offer guidance and examples on editing with audience and purpose in mind. And the chapter "Degrees of Edit" gives some very practical advice for doing the most with the time you have. The reality of professional life is that you simply don't always have time to do a heavy edit. But that's no excuse for not doing at least a light edit. The examples presented are clear and useful.

12

Speaking: Do I Really Have to Stand Up and Talk in Front of All Those People?

1 INTRODUCTION

Engineers have to communicate information orally all the time—sometimes very informally through one-on-one conversations, sometimes a little more formally through project briefings in a conference room, and sometimes much more formally as a speaker in a big auditorium. In all these situations, you want to get comfortable with using body language, voice quality, and visuals to your advantage. You can use those tools to help deliver the message of your talk, whether you are making recommendations to a client, requesting funds for a new project, or giving the results of a research study. It is all about interacting with your audience. If the members of your audience (whether composed of 1 or 100 people) feel you are speaking directly to them and would be willing to *listen* to them in return, you will have succeeded. After all, your audience probably prefers hearing and watching your presentation than having to read a long report on the subject.

HINT

Most people would rather listen to someone explain technical information than read about those same concepts. After all, it is more fun and memorable to look at pictures of butterflies and flowers than to read about the pollination characteristics of butterflies.

SECTIONS

1 Introduction
2 Planning the Content of a Presentation: Outlining
3 Planning the Visuals
4 Planning the Delivery
5 Controlling Nervousness
6 Handling Questions
7 Preparing for a Team Presentation

OBJECTIVES

By reading this chapter, you will learn the following:

- how to plan and outline your presentation;
- how to plan content and visuals for your talk;
- how to use voice quality and body language to deliver your message;
- how to use the most appropriate delivery mode for your talk;
- how to use your nervousness to prepare for and deliver a talk;
- how to handle questions from the audience;
- how to prepare for a team presentation.

Of course, most people are nervous about speaking in public, especially in a professional situation. But controlled nervousness can be used to your advantage, to give you energy and motivation. In this chapter, we will look at techniques for controlling nervousness. The most important strategy is to be prepared. There is no substitute for planning and rehearsing your talk in advance.

2 PLANNING THE CONTENT OF A PRESENTATION: OUTLINING

The first step in planning a good presentation is usually to make an outline of the contents. The next steps in developing your presentation would be to create the visuals and then rehearse. These steps are not always linear (you might, for instance, start by preparing a graph of test results), but you want to allow time to accomplish all three steps.

Here is a plan for step one, outlining your talk. Consider these items as you outline:

- Know your **audience**.
- Choose **three to five main points** to **highlight**.
- Divide your talk into **three sections:**
  - **Repeat** the main points in each section.
  - Decide where **visuals** will be useful.

As the talks you give become more complex, you may find that you need to make more than three to five main points, but this rule of thumb is useful to keep in mind throughout your career, especially when talking to a less technical audience.

Audience

You never know who will come to a presentation you are giving. Even if the audience is supposed to be just your boss and a couple of other project managers, your boss might rush in at the last minute and tell you he or she is bringing a couple of visitors from corporate headquarters to your talk. Or it might turn out that a couple of inspectors from the regulatory agency want to hear what you have to say about the air-pollution studies you have been conducting. Even at technical conferences, you cannot be sure these days that everyone has the same technical background you do. Interdisciplinary projects and studies are becoming increasingly common, and you have to be sure that everyone understands your information. Sometimes, this diversity of audience means that you must add more background information and explanation than you might think is necessary. Remember, too, that we all hear differently then we read. When we read, we can go back and reread information we do not quite understand. Your listeners cannot do that, so you have to enable them to understand your presentation from the beginning.

Highlighting

When you are presenting, time tends go to by faster than you thought it would. You have to build in time for pauses and (sometimes) more lengthy explanations than you had originally thought. So, do not try to cover too much material! (Some of your professors have probably given you an unintentional lesson in cognitive overkill during a lecture.) You cannot begin to cover all of the material that is in your report, thesis, or proposal. So, choose three to five main points to highlight, and build your talk around them.

HINT

You only have time to present and support three to five main points!

Three Sections

That old chestnut about an essay having three main parts—beginning, middle, and end—is really true for a talk. Listeners need a lot of structure to keep their mental place in listening to a talk. Studies have shown that even the most well-intentioned listeners retain less than one-third of the information given in a talk. If you think about how *you* listen, you will realize that you fade in and out of attentiveness and do not even hear everything that is said. Also, you may not understand all of a speaker's information or be able to put it in a meaningful context. Help *your* listeners remember the important points of your talk by repeating the points throughout the presentation. After all, if you state a critical point three times, you are pretty likely to get close to a 100% chance that the listeners will remember it ($3 \times \frac{1}{3} = 1$, which you can think of as 100%).

While you are making your outline, you can also be thinking about where you will need to show visuals and which of these visuals should convey technical information and which should keep listeners on track by indicating what sort of information the talk is about to cover (conclusions, recommendations, etc.).

Plan for a 15-minute talk and you cannot go wrong. Most individual student presentations cannot be longer, and most individual presentations at conferences or meetings are a similar length. At conferences, there will probably be a session moderator keeping track of time, and your talk most likely will be sandwiched in with at least two others during a 60-minute session. Allowing for questions after each talk, this situation means 15 minutes maximum for your talk itself. If you are told that you have more time than that, terrific. If you planned for only 15 minutes, then you can always add a few more details. (Somehow, there is usually more to say about a subject!)

Here is a checklist for the first section of a typical 15-minute presentation:

Introduction (2–3 minutes):

- Introduce yourself.
- Motivate the audience to listen to you.
- Explain the purpose of your work and of the presentation.
- Preview the main ideas.
- Establish key concepts.

Beginnings are important. The impression you initially make will stay with your listeners. Begin in an upbeat manner; remember that you are meeting and greeting this audience. Try to use your voice and movements to convey some excitement about the project you are about to discuss. If you do not sound excited about the work you have done, why would an audience want to listen to you for upwards of a quarter of an hour? In your introduction, preview the main ideas you will discuss and mention any key concepts you want folks to keep in mind while they listen to you. One example of such a key concept might be to remind them that your test results are preliminary and need to be confirmed in the next phase by field studies. Do not forget to introduce yourself if no one does it for you: Who are you, where are you from, and why are you talking to these people?

You do not want to memorize your talk word for word (because you would sound too stilted), but you might want to memorize the first few lines of your own introduction. You will probably be pretty nervous, and if you can begin smoothly, you will become more confident as you go along.

The following is a checklist for the second section of the presentation:

Body (8–10 minutes):

- Use an easy-to-follow organization.
- Use transitions.
- Provide supporting data.
- Bring in key concepts.

Keep the body of your talk well organized with good graphics, and you will probably keep people awake. That is no mean feat in this age of overwork and multitasking! Be sure to discuss your topics and key concepts in the same order that you introduced them in the introduction. Keeping the same sequence is a mnemonic device that helps your listeners remember the information. Especially in the body of your talk, where you present a good number of details to support your main findings or points, you must focus on logical, repetitive organization and repetition of keywords.

Here is a checklist for the third section of the presentation:

Conclusions (2–3 minutes):

- Provide a clear wrap-up.
- Comment on the significance of the results.
- Do not just stop in your tracks!
- Ask for questions.

Never just stop and say, "That's all." Plan a clear ending to your talk that emphasizes the most important fact or conclusion of your work. Remember that many in your audience will probably remember best the last thing you said (other than, "Are there any questions?"). Especially if your talk has been fairly technical, you must use the ending to underline the significance of what you did, found, or concluded. For talks that give instructions of some kind, finish with the most important fact, whether it is a safety warning or a statement of how easy these instructions will make the job or how important the job is. Repetition is your friend here. Do not fail to say something just because you have already said it.

Of course, you always end any talk with a request for questions or comments from your listeners. Your job is to make sure your message came through, and, fortunately, presentation allows you to know for sure whether you succeeded. Asking for questions gives you another chance to be clear if anyone is confused or in doubt about something you have said or done.

Here is a sample outline for a 15–20-minute talk given to the U.S. Environmental Protection Agency. The talk was a progress report on a research project testing a new material for possible use as a landfill liner. You can use this outline as a guide while preparing a talk of your own. The wording you use in your outline may find its way onto your slides, but keep in mind that *these outline sections are not yet slides*! The outline is as follows:

Sample Presentation Outline

INTRODUCTION

Project

Designing and conducting bench-scale hydraulic conductivity tests of manufactured bentonitic blanket materials for landfill liners and caps.

Purpose of Project: To provide data to EPA on whether these new materials are as effective in preventing leachate from escaping into soil (and rain from getting into landfill) as are current liners (three feet of compacted clay).

Purpose of Talk

To update audience on testing and results so far. To solicit feedback on test design and methods and on preliminary conclusions.

List of Main Topics (what I will discuss)

Background on material
Testing equipment
Testing procedure
Current status of testing

List of Key Concepts

Designers in a hurry to use new material
Test designed only for ideal lab conditions
Further testing needs to be done

BODY

Main Points

1. Background on liner material
composition, cost, ease of installation

2. Testing equipment
9 steel tanks, gravel, water, etc.
collection dish under drain

3. Testing procedure and duration
3 sets of tests for each tank
each set takes two months
readings taken weekly
hydraulic flow calculated: $Q = kiA$
test stopped when steady flow is reached

4. Current status
first set of tests complete
hydraulic conductivity values calculated
second set of tests set up

CONCLUSION

Tentative Conclusions

1. Conductivity is within range published by manufacturers of three brands tested.
2. Conductivity is comparable to (actually slightly lower than) that of clay liner.
3. Further testing is needed—e.g., simulating actual landfill conditions.

3 PLANNING THE VISUALS

When we get information in both words and images, we understand better. A picture may be worth a thousand words, but we still need words to give us the context for the picture. Be sure that your words and your visuals are in synch all the time: Do not leave a visual up while you talk about something else. Also, it is OK for the presentation screen to be dark or blank sometimes. Remember that *you* are giving the talk, not your fancy pictures.

HINT

Words + Images = 

Do not try to use too many visuals; it is better to explain a few things thoroughly than to have to skim through and skip visuals because you are running out of time. You have to allow time for pauses and for potentially more-thorough explanations than you had anticipated because you can see the audience is confused.

HINT

Creating Visuals

- Begin planning visuals as soon as the outline of your presentation is completed.
- Plan on using 10–20 visuals for a 10–15-minute talk.
- Consider all delivery methods:
  - Computer projection is the most professional.
  - There may be reasons to consider other methods as well.

Here is a checklist for designing visuals:

- *Simplify* any hard-copy graphics (graphs, tables, etc.)
- Think big.
- Include mapping visuals.

The following subsections explain these three points.

Simplify

A long table or a complex graph that works in a report will *not* work well in a presentation. The audience does not have enough time to process and understand visuals that are very complex and dense. Remember that each visual will probably be shown for less than one minute. Consider Table 1.

Table 1 does not make a good presentation graphic. There are too many numbers to absorb and understand quickly, and the significance of those numbers is not made instantly visually clear. The numbers showing diameter, for example, look as important as the numbers showing settlement (presumably the critical information). But the main problem with this visual is the small-sized and wordy caption; the labeling is left over from the hard-copy version. Presentation visuals do not need to be numbered, and the titles should be very simple and short, usually without captions. In any case, the caption here is meaningless in the hard-copy, handout version of your slides.

For visuals that are mostly words, remember again that an audience does not have time to read and process information the way a reader of a book can. You must use fewer

TABLE 1 Comparison of Settling for Various Types of Piles. The piles that exceed 0.25 inches of settlement are shown in red. The piles that are shaded in gray were discarded. [*This table is too crowded to make a good presentation visual. Titles for graphics should not have captions.*]

Concrete Pile			Hollow Steel Pile			Concrete-Filled Pile		
Diam (in)	Len (ft)	Sett (in)	Diam (in)	Len (ft)	Sett (in)	Diam (in)	Len (ft)	Sett (in)
12	50	0.19	12	50	0.07	12	50	0.19
12	55	0.37	12	55	0.18	12	55	0.37
12	60	0.36	12	60	0.17	12	60	0.36
12	65	0.61	12	65	0.28	12	65	0.61
14	50	0.20	14	50	0.10	14	50	0.20
14	55	0.33	14	55	0.16	14	55	0.33
14	60	0.43	14	60	0.21	14	60	0.43
14	65	0.68	14	65	0.26	14	65	0.58
16	50	0.23	16	50	0.10	16	50	0.23
16	55	0.32	16	55	0.20	16	55	0.32
16	60	0.41	16	60	0.42	16	60	0.41
16	65	0.43	16	65	0.43	16	65	0.43

words and allow more white space between the words. Here is a general rule of thumb for text on slides:

- No more than seven or eight lines of text per slide.
- No more than seven or eight words per line.

Think Big

One of the prime benefits of using a software presentation-package such as Microsoft PowerPoint is that the default text sizes are always large enough to be seen easily. Because we are so used to judging visual ease of reading from a small screen or from a hard copy, we forget that projecting text and images over distance necessitates a larger scale if the visuals are going to be read easily. If you customize the default font settings for any particular design template in PowerPoint, make sure that you maintain a font size that can be read easily from the back of a large room. If you are using an overhead projector, always enlarge the hard-copy versions of your tables and graphs. Use these minimum font sizes for presentation graphics:

- 18-point minimum font for labeling axes
- 32-point minimum font for titles
- 24-point minimum font for first-level bullet points
- 20-point minimum font for bullet points and main text (only for second- or third-level bullet points)

Use fonts larger than these wherever you can. A good test for readability of your slides is to print them out in slide view, place each one on the floor, and stand back about a foot. If you can read the slide easily, then it is probably large enough for your audience to see easily when it is projected on a screen in front of a large room.

Include Mapping Visuals

You are probably aware of the importance of content visuals in conveying information on the subject you are discussing. For example, the graph in Figure 1 plots the theoretical

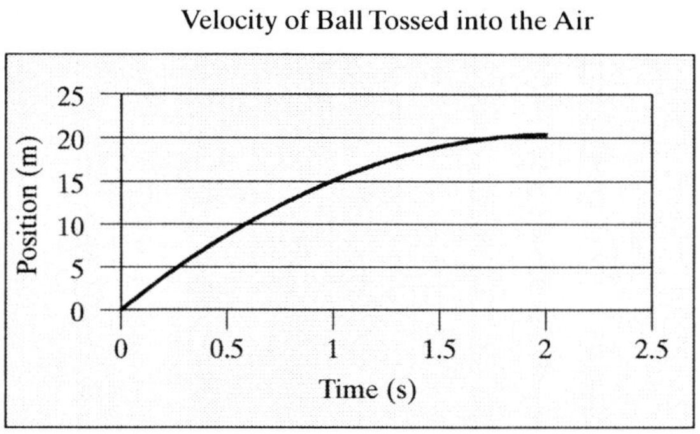

Figure 1. Example of content-oriented slide for a presentation.

HINT

Content-oriented slides display graphs, bar charts, etc. —mostly images.

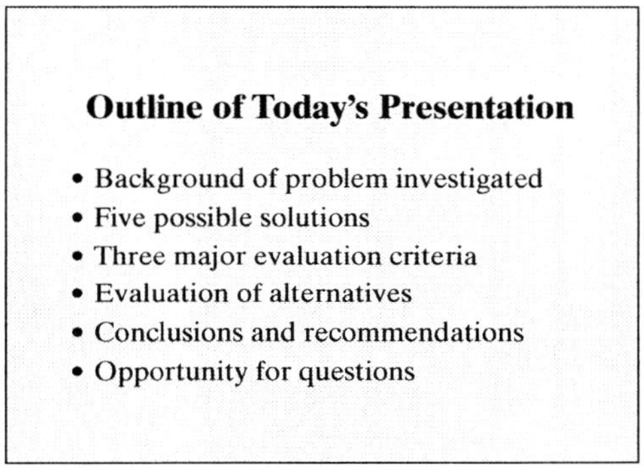

Figure 2. Example of a mapping slide for a presentation. You should use both content-oriented and mapping visuals.

HINT

Mapping slides map the structure of the talk—mostly words.

speed and position of a ball. The visual shows the exact relationship faster than words alone could explain it. But you may not be aware of the importance of visuals that map your talk and emphasize certain key points (e.g., how the theoretical information was used to solve a problem). You want to tell your audience where you are in the talk and what to expect. A mapping slide that outlines the content and purpose of the talk, as shown in Figure 2, prepares the listener for what is to come, thereby enhancing his or her understanding of topics and ideas as you explain them in more detail.

The first mapping slide should always be your title slide. Tell the members of your audience both orally and visually who you are, where you are from, why you are talking to them, and what is the general subject of your talk. In Figure 3, the font sizes are Arial 40 for the title, Arial 32 for the presenter's name, and Times 24 for the speaker's affiliation and date. Arial is a sans-serif font (meaning that it has no little "feet" at the edges of the letters), so it projects very well. Times is a serif font that should be used carefully for presentation graphics. Use Times for information that is somewhat less important than other information, or else make sure you never use less than a 24-point size.

Following the title slide, an introduction slide should outline in a bit more detail what information you will impart. (See Figure 2.) Similarly, labeling the conclusions

Figure 3. Sample title slide for a presentation.

slide "Conclusions" and writing the conclusions out in simplified form tells the listener that the talk is almost at an end. If the listener's mind has been drifting, he or she knows that now is the time to listen attentively and get the talk's most important points.

Modes of Delivery

Computer projection has become the professional standard for delivery of most presentations. There are tremendous advantages and a few potential problems with using computer projection. If you have designed your slides using presentation software, you can update them even at the very last minute. There is also no chance of getting the slides out of order (as with transparencies or 35-mm slides). You can also use the full range of multimedia options available to you through your computer and the projection capabilities at the presentation site. To gain these advantages, however, you must understand and be confident about the equipment you will be using. When problems arise, it is generally because you have not checked out the equipment beforehand, especially when you are off site, away from school or the office. It is possible that the projector will not be compatible with your computer or that the right cables will not be at hand. If you have your presentation on disk, the borrowed computer may not have the right software to run your files. Or, if some of your presentation depends on the Internet or other network connections, the correct hookups may not be available. If at all possible, you should rehearse or at least visit with the equipment to be used. Go there the day before and assess the equipment.

At this point in the early 21st century, the most commonly used presentation software is probably Microsoft PowerPoint. This package is very user friendly and very powerful. I have seen students learn to create decent slides within 10 to 15 minutes and then learn the finer points of designing with PowerPoint within the next hour. All you have to do is open the program and start experimenting. Use the templates to give yourself ideas on design and format, but remember that others may be using those very same templates. Your talk should be unique, because both you and the way you present your particular information are unique. PowerPoint is very customizable, so take the trouble to experiment with design and format.

Packages such as PowerPoint help convey information in manageable chunks. This sort of software has revolutionized business and technical presentations by forcing the

HINT

Suggestions for Designing with Presentation Software

- Use the design templates rather than the genre templates ("sales pitch," etc.). Design templates offer a consistent format for slides (font size and type, placement of titles, etc.), but do not try to dictate content.
- Be careful about colors: Use very dark text on a very light background or vice versa.
- Avoid red texts, as it is difficult to read from a distance, no matter how large the text.
- Either keep slides consistent in design or vary the design according to subject matter. For example, if you are introducing a multiphase project, you might make the introduction slides slightly different in format from the body slides. Conclusion slides might then repeat the introduction slides' design.
- Experiment with animations and dimming, but consider your audience before using these effects: A semitechnical audience may appreciate "entertainment" more than would the members of your dissertation committee.

presenter to distill his or her thoughts and information for greatest simplicity and clarity. Bullet points or talking points can be created effortlessly in PowerPoint, and the product has a large selection of clip art just a click away. See how the slide shown in Figure 4 clearly introduces a case study on environmental contamination, using a piece of clip art to render the subject matter more memorable in the mind of the listener.

You can use presentation software not only to create the visuals for your talk but also to organize your thoughts (and therefore your talk) and to create handouts for the audience. Once you start using the software, you will realize how the necessity to type your thoughts in logical order, using short phrases or sentences, helps to organize your thoughts in the first place. You may have thought you knew what you wanted to say about the difficulty of monitoring a site suspected of being contaminated with benzene, but chances are that if you just started extemporizing (speaking off the cuff), you would say a lot more than what's contained in the second bullet point on that slide. Yet, for a general audience, that bullet point contains a good summary of the problem.

Be careful, however, of letting your bullet points speak for you. Those bullets are simply the tip of the iceberg; it is up to you to provide the connections between the bullet

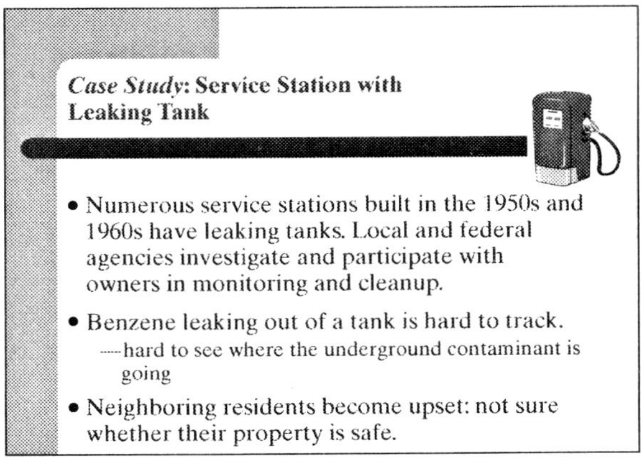

Figure 4. Sample slide for introducing a case study.

points that make them come alive. You want to create *meaning*, not simply a *list*, out of your data and your analysis. *Never stand in front of your audience and read your slides.* Use the slides for your own visual reference, and repeat keywords, phrases, and critical pieces of data, but always add more information, further examples, or a story that illustrates the point indicated by the bullet.

HINT

Do Not Let Your Bullet Points Speak for You

- Your job is to create meaning so the audience understands the significance of your work.
- Do not simply read your bullet points.
- Explain: Provide "connective tissue" between bullet points.

Never let your visuals upstage you or the content you are presenting. A package such as PowerPoint supports all sorts of multimedia applications (including animation, video, audio, and other special effects) that can be used to make your talk understandable, memorable, and entertaining. But as you plan your visuals, always consider first the audience to whom you will be presenting. The more technical knowledge the audience members have about your subject matter, the fewer bells and whistles they will need. And remember that bells and whistles should never be used for their own sake, but always in the interest of stressing or explaining a particular point that you are making. For example, the text on the PowerPoint slide shown in Figure 5 does not match the illustration.

There are situations in which computer projection may be unavailable or may not be the best choice for your talk. If you are showing photographs that were not computer generated, you may want to use 35-millimeter slides to achieve the best resolution. If you are leading an interactive meeting or workshop, you may want to use overheads (transparencies) to capture the ideas or modifications suggested by participants. Sometimes, the overhead projector will be the only equipment available to you.

Whatever your mode of delivery, keep these guidelines in mind as you develop your visuals: Make sure that each visual makes the right point. Do not get carried away with the possibilities offered by clip art and by all the many pictures you can download from the Web. Let your talk's content guide you, and then try to build in a little gentle humor here and there.

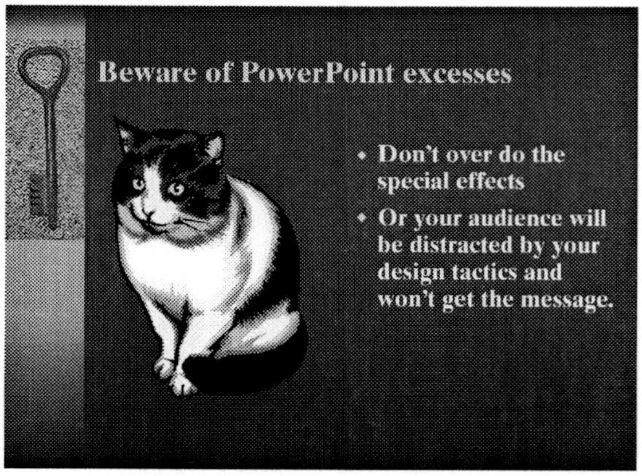

Figure 5. Example of a possibly frivolous slide.
[If this slide is not meant to illustrate something about cats or catlike qualities, then find another picture. Never use a picture just because it is "cool" or you like it.]

HINT

Guidelines for Designing Visuals in All Modes

- **Design each visual to make one main point.**
- **Two visuals may be better than one.**
- **Be sure each visual makes the right point.**
- **Leave a visual up only as long as you are talking about it.**
- **Remember that *you* must still be the focus, not your visuals.**
- **Avoid clutter. On graphs, be sure to omit needless gridlines, tick marks, etc.**
- **Check your visuals for distortion, spelling errors, and computational mistakes.**
- **Think big. Follow software guides. Most text should be in at least 24-point font. The bigger the presentation room, the bigger the font size should be.**
- **Use few words, and separate them with lots of white space. The benchmark maximum lengths are as follows:**
  - **Seven to eight lines of text per slide**
  - **Seven to eight words per line**

PRACTICE QUESTION: DESIGNING SLIDES

How could the slide shown in Figure 6 be improved? Identify specific problems with it.

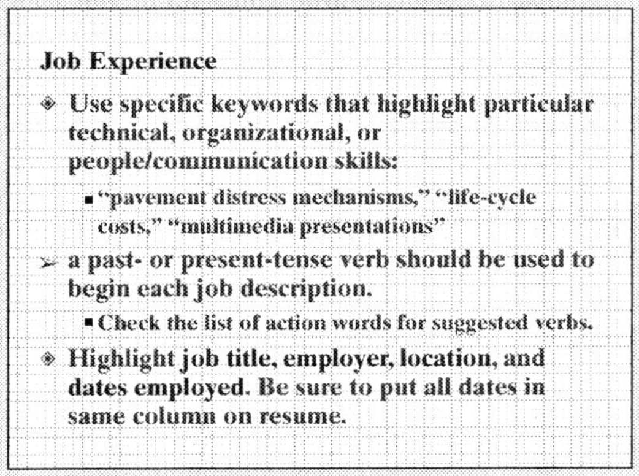

Figure 6. Example of badly designed slide.

SUGGESTED ANSWER

Some of the problems with the slide include the following:

- too much text
- awkward placement of text
- inappropriate background for a text-laden slide—the lines are distracting
- the size of text doesn't always reflect the level of bullet
- the bullet design is not consistent
- capitalization is not consistent

Figure 7 shows a suggested revision.

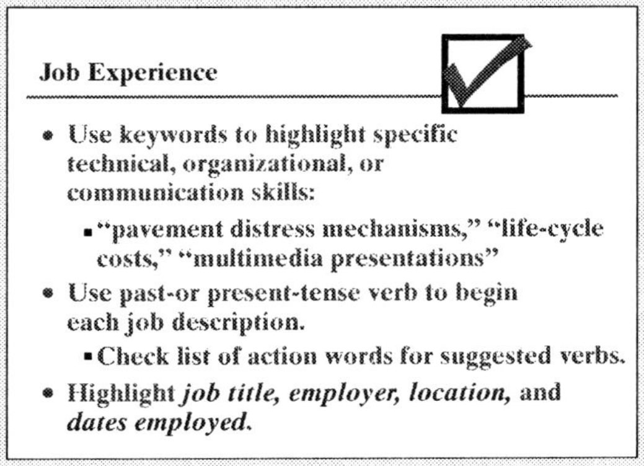

Figure 7. Example of improved slide design.

4 PLANNING THE DELIVERY

Remember that your audience will need your help to understand clearly what you are saying. After all, readers can read at their own pace and stop whenever they want to.

Listeners, on the other hand, are at the mercy of the speaker's pace and level of detail. Remember this factor when you give a talk. Plan not to talk too quickly, and keep an eye on the audience members to see if they are understanding you. For many people, it is more difficult to retain large amounts of information when listening than when reading. When we listen to a talk, we are also listening to what is in our own heads—all that inner conversation that most of us have going on almost all the time. Use voice quality, body language, visuals, and transitions—the tools described here—to help keep your listeners with you during your talk.

You have some powerful tools to use when giving a talk. Aside from the words you speak and the visuals you show about the topics indicated in your outline, you have your own voice quality and body language. You can use these tools to convince an audience of your credibility, intelligence, and enthusiasm for the work you have done. Enthusiasm may be the most important ingredient of your talk: If you show that you care intensely about the subject and the work, chances are your audience will also care. Conversely, if you speak in a monotonous, flat tone and if you stand in one place or slouch, the audience members will get the message that you do not care that much, so *neither will they*. The best way to improve your delivery strategies is to harness the enthusiasm that should already be there inside you (though perhaps rather buried). Positive energy communicates best. But first you need to get feedback on how you come across to an audience. You can use rehearsing, videotaping, and anticipating questions as strategies for eliciting feedback.

Rehearsing

The best way to improve the delivery of your talk is to rehearse in front of someone. Rehearsing in front of at least one other person enables you to get constructive feedback—on what you are doing right and what you are doing wrong—that will help you communicate effectively with your audience. Ask your classmates or coworkers to watch you run through your talk, and offer to do the same for them. Rehearsing at least three times is a good idea for less experienced presenters: Rehearse once on your own, once standing up in front of someone, and once (ideally) with your audiovisual equipment in the space where you will be giving your talk. Ask your surrogate audience members to critique your delivery. Could they hear and understand everything you said? Could they follow the talk, and did you make it interesting? The HINT box contains a suggested checklist of questions for your reviewers to answer:

HINT

Questions for Your Rehearsal Reviewer to Answer about You

Is the structure of the talk evident?
Do the various sections hang together and form a whole?
Does the pace seem right?
Is the subject matter clear and compelling?
Are the visuals clear and compelling?
Does the presenter sound enthusiastic?
Does the presenter speak to the point without using verbal fillers, such as "um" and "you know"?
Does the speaker maintain eye contact?
Does the speaker move naturally, without fidgeting?
Could I hear every word of the presentation?
Does the talk hold my interest?

Voice Quality

Tone of voice is so very important in any interpersonal communication. Think of how many times you may have been misunderstood because your listener thought you were being sarcastic or bored when you really were not—your voice just somehow sounded that way. A good voice is an enormous personal asset. Your career will truly benefit if you can learn to control, modulate, and use your voice to be persuasive, informative, or inspiring, as the occasion demands. In a presentation, you have to ensure first of all that you are speaking loudly enough for everyone in the room to hear you—everyone, even those people in the very back rows. Every speaker can learn to speak more loudly. Try the exercise explained in the upcoming box to increase the volume of your voice. The trick is not to put more pressure on your throat—that is not where your voice originates—but rather to apply pressure to your diaphragm, where your voice originates. Your diaphragm is like a bellows that can heat up your voice to a roaring fire.

PRACTICE EXERCISE: INCREASE VOICE VOLUME AND RESONANCE

- Stand with your feet shoulder width apart and your hands on your diaphragm, between your belly button and breastbone. Open your mouth and emit a constant sound at the same pitch. Press on your diaphragm and see how your voice volume increases without your even trying. The harder you press, the louder is the sound.
- Remember that your voice does not come from your throat; it rises up from your diaphragm and gains power from the air in your lungs.

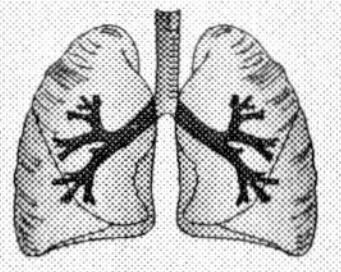

Second, you want to work on the richness, pleasantness, and variety of your voice. These qualities are controlled by what sound experts call *tone*, *pitch*, and *dynamics*. Some of us naturally have a rather monotonous voice, one that stays at the same pitch, tone, and dynamic. In any talk, we want, however, to use our voice as an instrument that conveys emphasis, importance, humor, and sometimes even emotion. *Tone* is the resonance of your voice. If you have a squeaky or thin voice, check out that voice exercise in the previous box for a method of strengthening your voice so it sounds richer and more resonant. *Pitch* is the highness or lowness of your voice. Try to vary your pitch to emphasize important points. Remember that your audience understands meaning by the way that your voice rises and falls. A rising voice at the end of a sentence is the way that we understand you are asking a question, for instance. Do not make or let your voice rise at the end if you are not asking a question. However, do not let your voice fall so low at the end of a sentence that the listener cannot catch the last few words. Be careful not to run out of energy (and therefore volume) at the end of your sentences. *Dynamics* refers to loudness and softness of a voice. Again, you can emphasize important points by raising your voice volume a little.

Also be careful not to speak too fast. Nervousness makes us speed up. On the other hand, if the pace of your talk is slow, it may seem that you are unsure of what you are going to say next. Strive for a medium pace, one that carries listeners along without losing any information because you are whizzing over the words.

Body Language

You want to show that you care by leaning slightly forward, toward the audience, and using your hands appropriately to emphasize points and to include the audience in your

HINT

Voice Quality

- Speak loudly enough for all to hear.
- Make sure your voice is varied in pitch and dynamics.
- Speak in a resonant tone.
- Make your voice rise and fall appropriately; don't run out of steam at the end of a sentence.
- Speak at a medium pace. Generally, speak more carefully and slowly than you normally do.
- Speak *toward* the audience. Glance toward slides only briefly.

thoughts. If you stand with your arms crossed, lean away from the audience, or slouch against the table or podium, your gestures are excluding the audience and creating a barrier between you and those to whom you are speaking. Any audience member wants to feel as though you are speaking directly to him or her. If you stand in one place all the time, your audience will become bored, and chances are you will become increasingly nervous. Moving around helps control nervousness, because movement relieves stress. Be careful as you move around, however: Move naturally, and do not develop annoying repetitive movements without realizing it. I once watched a young speaker drive her audience crazy by continually scratching behind her left knee. She had no idea she was doing this, but the audience was mesmerized by this repeated nervous motion and consequently could not pay attention to the admirable content of her talk.

The other way to distance yourself from and alienate an audience is by failing to make eye contact. In the United States, if someone does not look at us, we either lose interest in what he or she is saying or feel that the speaker is untrustworthy. The same reaction may not occur in other cultures however; in fact, if you are from a non-Western country, you may be uncomfortable looking directly at your audience. Nevertheless, if you are speaking in the United States, you should know that looking someone in the eye is considered to be a trait of truth and strength. If you gaze over the heads of your audience, down at the floor, or, worse, out the window, your audience will probably simply stop listening to you. If you are not looking at your audience because you are afraid of what you will see, relax. Most listeners are glad to be hearing your information delivered in an accessible and visually appealing way. They are on your side, and if you look out at that sea of faces, you will probably see a lot of friendly, positive expressions. You can get a nice charge of energy by looking at your listeners and seeing someone nod as though he or she understands or even smile as though he or she is happy to have this information. Do not cut yourself off from this positive energy by failing to look out at your audience. On the other hand, do not stare at just one or a couple of people. Practice a sweeping motion, resting your eyes for just a couple of seconds on each face. Try to include the entire room in this sweeping motion.

HINT

Body Language

- Use appropriate movement. Move and position your hands naturally.
- Use inclusive gestures (no crossed arms).
- Maintain good posture (no leaning against the podium or shoving both hands in pockets).
- Do not make distracting tics or noises.
- Maintain good eye contact.

Stand and face the audience!

Using Visuals

It is easy to use too many visuals, now that they are relatively easy to create. Check yourself during rehearsal by timing your talk *while using* visuals. A good rule of thumb for the number of visuals is to use no more than 1 per minute, or 10–20 for a 15–20-minute talk. Be careful of a natural tendency to include more visuals than you really need to make your points. Ask a reviewer whether your visuals support your points well, raise further questions, or seem rather empty of content. Practice displaying the visuals using whatever technology you have chosen; the last thing you want to do is to have to fiddle with the equipment during the presentation. Learn how to use a laser pointer so that you do not flash it around needlessly.

Using Transitions

Moving a reader along from point to point is difficult to do in writing; in a presentation, you have more tools—pitch and tone of voice, gestures, etc.—to help. And you have another tool that you also use in writing: transitional words and phrases. These words and phrases can be more direct and informal, however, in speech than in writing. Rhetorical questions work really well to sum up one section of information and point to the next section: "So, what did we find once we had completed the lab experiments?" "What did our studies show us about the load-bearing properties of this material?"

Remember that in speech, you will want to use active voice—e.g., "we found deformation"—much more than passive voice—e.g., "deformation was found." Active voice will help your speech be clearer, more concise, and more direct.

Videotaping

An excellent way to improve your presentation skills is to have someone videotape you so that you can see for yourself what your strong and weak points are. As you watch the tape later, ask yourself these questions:

- What is my greatest strength as a speaker?
- What is the weakest part of my presentation? Why?
- What, specifically, can I do to improve? Should I do voice exercises? Practice more? Reorganize?
- What should be the result of my improvement efforts in my next talk?

Always notice what you do right as well as what you do wrong. Some students realize, for instance, that even though they were very nervous giving the talk, the tape shows a person seemingly in control, with few signs of nervousness. This is great information to have: You do not *show* your nervousness nearly as much as you *feel* it. Knowing that should give you more confidence the next time around.

When it comes to improving certain aspects of your presentation (increasing voice volume, for instance), practice strategies for planning, rehearsing, and answering questions and see what difference the strategies make the next time around. Set realistic goals for yourself. Your voice does not have to be broadcast quality, like your favorite news reporter's. It just has to be loud enough and resonant enough to get your meaning across. Listen closely to your tape, and pretend to be someone in an audience listening to you. Take notes on your performance. When you tape yourself again, compare the new notes with the old. *Any* improvement is a significant improvement.

Anticipating Questions

The question-and-answer (Q&A) session that inevitably follows almost every presentation is an important mechanism for exchanging information. The audience gets to clarify or learn more about your subject matter, and you get to discuss some details that had to be left out of your talk or to correct any misunderstandings that may have arisen. You, as well as the audience, can learn a lot during this session. Unfortunately, many speakers are afraid of Q&A and do not even want to think about it beforehand. That is unfortunate, because preparing for Q&A will give you a good level of confidence. How do you prepare?

You can ask your rehearsal reviewers questions about the content of your talk: What is not quite clear to them, or what do they want to know more about? The first type of question will lead you to reorganize or clarify some part of your talk, and the latter type will give you a heads-up on possible questions your audience will have following your actual talk. You cannot prepare information on every topic related to yours, but being aware of future directions or similar work in other areas would enable you to answer very nicely such a "tell me more" kind of question.

As you look over your outline and rehearse your talk, think to yourself about questions that might arise from your content. Jot them down, and spend a little time thinking about them. Don't spend too much time—you cannot prepare for every type of question, nor should you have to. Your job is not to know everything, just to have done your homework on the subjects you are presenting.

5 CONTROLLING NERVOUSNESS

Everyone gets nervous at some point before giving a talk. The trick is to use your nervousness as a source of energy without becoming overly anxious. Let your nervousness give you energy. Being nervous means at least you are not apathetic; after all, people who are bored and indifferent are going to *sound* boring. But do not let nervousness overwhelm you. Look over the upcoming strategies for coping with nervousness before and during your talk, and see which ones might work for you. Different strategies work better for different people. The goal is to keep your nervousness within reasonable limits at the same time that you harness it for energy, motivation, and the willingness to go the extra mile in preparing.

Before Your Talk

Deep breathing is something you can do right before your talk or anytime you feel anxious. Simply breathe in slowly on a count of, say, four, and then breathe out on double that amount, or eight counts. Counting slows your breathing down, and soon your body will feel more relaxed. When the body relaxes, the mind tends to follow suit.

Mental imaging is a technique used widely by coaches for teams and individual players to buttress their confidence. For example, a kicker might use this technique before being brought in to attempt the extra point in a football game. He would create a mental movie that shows him moving his foot back, swinging forward, and connecting with the ball at exactly the right point. He would watch mentally as the ball arcs up into the air and sails over the goal right between the goalposts to thunderous applause from the spectators. In your case, imagine yourself getting up to give a talk, beginning confidently, and speaking at exactly the right pace, with people looking pleased and affirming by nodding their heads. Your mental movie would include your finishing on time with good energy, a definite conclusion, and perhaps a smile as you ask for questions. Remember also to picture the audience—a group of people with glints in their eyes of understanding and even admiration for you and the work you have done. Such a mental movie builds confidence, and confidence enables you to make the movie come true.

Mind games can help divert your mind from focusing on your own fears and anxieties. Count the chairs in the room, or say people's names backward, for example. Bring a crossword puzzle to occupy you during the time before you get up to speak. Or really concentrate on the other talks and think about their content instead of on your own talk. Last-minute cramming and memorization usually does not work and ends up making you feel more nervous by the time you actually come to the podium.

HINT

Tips for controlling nervousness before your talk:

- Try exercising.
- Avoid caffeine the day of your talk.
- Replace negative self-assessments with positive self-talk.
- Check all equipment in advance.

Just Before and during Your Talk

Here is a strategy for settling yourself down right before your talk: As you come up to the podium, speak to a few folks along the way. If you know someone in the audience, be sure and say hi to them as you move to the front; if you do not know anyone, speak to the moderator or to the speaker who went before you—tell her what a good job he or she did. You could even admit to the moderator that you are nervous; chances are he or she will say something reassuring. The point is that by carrying on these small conversations, you ready yourself psychologically to behave during your talk as though you are simply continuing a conversation, only now with lots of people. Conversing with people is different than thinking of them as some sort of enemy from whom you would like to hide. This strategy should make facing the audience less intimidating and put you in the right frame of mind to relax, focus on what you want to say, stay connected with your audience, and even have fun.

Once you are at the podium, take a moment before speaking to get relaxed and adjusted to your audience. Think of the audience as a group of friends, even if you have never seen any of the people before. As you talk, keep an eye on the audience members' reactions, and slow down or ask for questions if some of them are looking puzzled. This checking takes practice; do not let the reactions of a few folks alarm or throw you. You can hardly ever make *everyone* understand *all* of your points; do not even try. There are always some people who will look bored or distracted for particular reasons. Think about that 8:00 AM class you took; you were probably not the most lively looking listener.

Remember, your audience always wants you to succeed! They will give you the benefit of the doubt because they are happy sitting there being informed and, perhaps, even slightly entertained.

PRACTICE QUESTION: HARNESSING NERVOUSNESS

In what ways is nervousness a useful quality for you as a speaker?

SUGGESTED ANSWER

Nervousness gives you energy. It makes you more alert. Use your nervousness as a source of strength and enthusiasm. Nervousness encourages you to rehearse more.

6 HANDLING QUESTIONS

Audience questions can give you, the speaker, very valuable feedback on the direction and methodology of your work. Here are some strategies for handling questions as they come at you:

Tips for Answering Audience Questions

1. Remain standing at the front of the room following your presentation. Do not run off or start heading to the edge of the speaking area. Convey confidence and a willingness to entertain questions.
2. Be aware of your body position; the audience is watching you. Avoid playing with your clothing, hair, notes, or writing implements. If your partner is answering a question, pay respectful attention.
3. Try to select questions from different parts of the room. In most Q&A situations, it is best to try to include the entire audience by using this method. Even if a particular person does not get a question answered, he or she won't feel ignored.
4. Repeat the question before you answer it, so everyone can hear it. Paraphrasing serves three crucial functions: (1) It allows everyone in the audience to hear the question; (2) it gives you time to think of your response; (3) it ensures that you understand exactly what is being asked of you.
5. Answer each question clearly and then move on. Provide an efficient and accurate response, and move to the next question. Try not to ramble or wander into a related topic.
6. Signal when you will accept one last question. To avoid the perception that you are trying to escape a particularly nasty question, warn the audience in advance that your time for questions is quickly passing. In a classroom setting, however, this role usually falls to the instructor.
7. Brainstorm possible questions in advance. The best way to prepare for a Q&A period is to think of possible questions ahead of time. You may not accurately guess what your audience will ask, but you will feel more confident. In addition, you may very well guess accurately, and then your answer will sparkle with professionalism.
8. Do not assume that tough questions are hostile questions. You may have simply gotten someone really thinking! If a question is long, and complicated, or you just do not understand it, do not try to bluff your way through an answer. Admit that you had not thought of that or studied that as part of your work, and *offer to speak with the questioner privately* after the session.
9. Practice answering questions as a part of rehearsal. If you practice with a potential audience member (and you should), have that person ask you questions about the presentation. This practice will help simulate the actual presentation experience and will boost your confidence.

Remember that the Q&A session can be the most valuable and the most enjoyable part of the presentation for you. You can learn new perspectives and ideas, and you will get feedback on the direction of your work or approach to a problem. Stay upbeat and positive, even when you are confused by a long, complex, or unforeseen question. Most listeners are not trying to test your knowledge; they really want to learn from your talk. But they have their own individual agendas, and their concerns and backgrounds are often not the same as yours. So each listener is trying to fit your information into the framework of what he or she knows and works on. Sometimes the fitting process takes some work on the part of speaker and listener, but especially in science and engineering, listeners and speaker usually share a strong bond: Everyone wants to see progress in solving that particular problem or in advancing the body of knowledge about that particular field. And when someone says publicly (as will happen to you sometime) that he or she really enjoyed your presentation, you will feel *great*.

7 PREPARING FOR A TEAM PRESENTATION

Many presentations in engineering practice are given by teams of engineers or by multidisciplinary teams that may include engineers, architects, planners, marketing experts, salespeople, attorneys, academic researchers, regulators, and other types of professionals. In many college courses, presentations of research are also done by teams—usually, teams of students with varying backgrounds and areas of expertise. So, learning to present well involves learning how to coordinate with other speakers as well as how to plan and deliver your own material.

All of the same strategies you might use for planning, rehearsing, and controlling nervousness apply to team presentations as well as individual ones. What changes a bit is the delivery. You have to consider not only where you will stand and how you will move and run the equipment, but also where your teammates will stand and move, etc. And you have to plan what you will do when you are not the one talking. At many presentation venues, the entire team will either sit or stand at the front of the room, so you will need to practice keeping your hands calm and trying not to fidget when you are not talking. Because a team presentation has to be even more "choreographed" than a solo presentation, rehearsing with the rest of your team, probably several times, is even more important for success.

What, then, are some special strategies for preparing a team presentation? First of all, you must be sure to share contact and schedule information with each other at the beginning of the project. Meet as often as you can to plan and rehearse the talk. Second, divide up the responsibilities and the work in such a way that draws on each member's particular strengths. Each of you will most likely be responsible for a particular part of the presentation's content, but it might be best to have one person creating the slides, so that they are uniform in design. If you have the choice, you will want to select the person with the most outgoing, upbeat personality to introduce the talk (and possibly also to conclude it). It may be best to have this person act as a sort of emcee, directing audience questions to the most appropriate speaker, for instance.

During rehearsals, be supportive, but constructively critical, of each other's content and delivery style. Sometime before the dress rehearsal, go out into the room, sit at the back, and see whether you can hear everything each speaker is saying. Pretend you are an audience member who does not already know a lot about the project. Do the slides and the flow of information make sense? Are the points easy to understand? Does each speaker provide transitions between important points? At the dress rehearsal, of course, the entire team should be up front, and you should all be practicing how to "pass the baton" to each other. Remember to remind the audience of the name of the speaker who will be speaking after you; no one is going to remember an entire list of names on a title slide. Enjoy the synergy of presenting with a group. You will find that the audience will also enjoy that energy. Your tone with your teammates should be natural and relatively informal, and small touches of humor may be used here and there. Ask at least one non-team member to watch a rehearsal and comment about the humor to ensure that it is appropriate and relevant.

HINT

Tips for Presenting with a Team

Preparation

- Share contact information at the first meeting. Stay in touch!
- Divide up the preparation according to individual strengths:
  - Who is the best slide designer?
  - Who knows the best sources of information for each topic?
- Divide up speaker roles according to individual strengths:
  - Who knows most about each section or topic?
  - Who is the most upbeat?
- Rehearse often, and seek constructive criticism.
- Work on transitioning smoothly between sections and speakers; do not get in each other's way. Does it make sense to have one person running the computer projection?
- Practice answering questions the team has brainstormed.

Delivery

- "Pass the baton" to each other by mentioning the next speaker's name. Consider previewing what he or she will discuss.
- Answer those audience questions with which you feel most comfortable. If no one can answer a particular question, offer to find out the answer and get back to the questioner later.
- Stay upbeat and enjoy the synergy of presenting with colleagues!

SUMMARY

- Planning an effective presentation means thinking about your audience and their level of familiarity with and interest in your subject matter. You should design visuals and adjust your language to make technical concepts and data clear to that audience.
- Testing your presentation on a sample audience is a great way to prepare. Ask for feedback from the test audience-members on your voice, delivery, and visuals. Was your talk interesting to them, and did they understand everything?
- Visuals for a talk should generally be bigger, with less text, than you think, so that people can read them easily from the back of the room.
- You can use nervousness as a source of energy to make your presentation more vivid and keep yourself alert and on the ball.
- There are many strategies for controlling the level of nervousness that threatens to overwhelm you. Learn which strategies work best for you, and practice using them before and during your delivery.
- Preparing for a team presentation requires rehearsal with your teammates, not just by yourself.

Problems

1. Along with several other students (if possible), attend a talk given at your college or in the community. Take notes on the speaker's delivery, the visuals shown, and your general reaction to the tone and content of the talk. Answer the following questions:

   Was the talk interesting? Did it hold your attention?
   Did you understand all the information?
   Did the visuals enhance your understanding? If so, how?
   How did the speaker's voice and body language enhance or detract from your understanding or your interest?

2. If you attended the talk in #1 with at least one other student, compare your notes with the other students'. Answer the following questions:

   Did you all agree on which aspects (visuals, language, body language, voice, etc.) were most effective?
   What differences do you note in your reactions to the talk? Can you postulate any reasons for those differences?

3. Make an outline for a presentation you will deliver in one of your classes. Begin the outline by summarizing in 30 words (or less) your objective in giving the talk. Phrase your objective using an active verb, e.g.: "In this talk, I want to *demonstrate/persuade/inform/show* my audience *something*." Show the outline to your instructor or another student for feedback.

4. Pick your favorite design template in MS Powerpoint (or another presentation software package) and create a slide (not the title slide). Try to use a projector to look at the slide on a big screen (rather than on your computer screen). Ask a friend to look with you. Then, answer these questions:

   If your slide has text on it, does the design get in the way of reading the text?
   If your slide has a visual, does the design get in the way of seeing and understanding the visual?
   If you use this design, which slide formats or layouts will you need to avoid?

5. Using MS Powerpoint or another presentation software, create a slide with nothing but red, 24-point text. Then project the slide on a big screen in as big a room as you can find. How close to the screen do you have to sit to be able to read the red text?

6. Along with a friend, find a big room or, preferably, an auditorium somewhere on campus or in your town. Ask your friend to sit in the back of the room while you stand in the front or on the stage. Talk in a normal voice. If your friend cannot hear you very well, ask him or her to keep moving closer to the front. How close to the front does your friend have to move to be able to hear you easily?

7. True or false? I should work hard to eliminate all traces of nervousness in myself before I give a talk.

8. The most important quality in giving a successful presentation is which *one* of the following?

   Showing only one visual per minute (on average)
   Never saying "uhm"
   Being enthusiastic
   Always remembering to step to the side of the big screen

FURTHER READING

Gurak, L.J. 2000. *Oral Presentations for Technical Communication*. Allyn & Bacon: Boston.

This is probably the best book available that is devoted entirely to technical presentations; most publications on presentations are overgeneralized and geared to the sales pitch or after-dinner situation. Presentation of complex technical information requires much more thought about the background of your audience and the best way to display data. Chapters 16, 17, and 18, on visuals in presentations, are particularly helpful.

Houp, K.W., Pearsall, T.E., Tebeaux, E., and Dragga, S. 2002. *Reporting Technical Information*, 10th ed. Oxford University Press: New York.

This venerable textbook, now in its 10th edition, covers all aspects of writing and presenting technical information. The chapters on techniques and applications of writing are excellent, and of all the many technical-communication textbooks out there, this one does the best job on presentations. Make sure to read Chapter 19, "Oral Reports."

Tufte, E.R. 2003. *The Cognitive Style of PowerPoint*. Graphics Press: Cheshire, CT.

This pamphlet argues that overuse of PowerPoint has led to a sort of dumbing down of business and professional audiences. Relying on bullet points to convey complex messages, Tufte argues, allows the speaker to avoid his or her responsibility to analyze and describe complexity—the "connective tissue" between those bullets that is the real information.

13

Producing Engineering Documents: The Rubber Meets the Road

1 INTRODUCTION

As an engineer, you will produce many different kinds of documents, including working drawings, specifications, and technical instructions. But the majority of documents you will produce in your career falls into the category of reporting on results or on plans for obtaining results. These sorts of documents generally use sentence-and-paragraph format to convey information to a variety of readers, including your boss, your boss's boss, upper management, clients, subcontractors, suppliers and vendors, public agencies, and sometimes the press. Understanding the expectations of those readers about format as well as content will be critical to success. This chapter presents samples of these most common types of engineering documents, with annotations that explain why each is a good example of its type. The focus here is on the big picture: the way the document is designed and put together to convey the information.

This chapter also describes the document design and production process, for which you may eventually have abundant help in the form of technical writers and editors, but which may at times need your guiding hand. Every technical professional should know how to set up documents for publishing in various forms (hard copy, soft copy, typical digital forms, etc.) and should have some idea of the time and resources required to publish and then disseminate documents.

2 DESIGNING PROFESSIONAL DOCUMENTS

Once you have done the research, gathered and organized the information, written (probably) several drafts and revised several times, you are ready to produce the document in hard copy or digital form (often both). This section offers general guidelines for designing readable pages. Most of

SECTIONS

1 Introduction
2 Designing Professional Documents
3 Sample Correspondence
4 Sample Research Proposal
5 Sample Progress Report
6 Sample Feasibility Report

OBJECTIVES

By reading this chapter, you will learn the following:

- principles for designing professional-looking documents;
- standard formats for the most common engineering documents;
- how to write correspondence;
- how to write a research proposal;
- how to write a progress report;
- how to write a feasibility report.

You will also be able to study annotated samples of the following types of engineering documents:

- memos;
- research proposals;
- progress reports;
- feasibility reports.

these guidelines apply to both hard and soft copy, although the end of the section outlines some of the design differences to keep in mind.

You do not need to think of yourself as a graphic designer to realize that two elements are common to all document pages: text and white space. The way that you arrange those two elements makes some pages more attractive and easy to read than others. White space is the most underappreciated design element. And yet, using white space is the easiest way to set off important words, phrases, graphics, and beginnings of sections—any textual or graphical element of importance.

HINT

Use white space to set off important or "different" items:

- figures and tables (including titles)
- titles of sections
- headings and (some) subheadings

Using text as a design element means considering and making decisions about justification, indentation, line spacing, and the size and type of font. Those are some of the decisions that affect the look and readability of your text. When choosing a type of font, remember that the commonly used fonts break down into serif and sans-serif fonts. Serifs are the little "feet" appended to the edges of some letters. Times is an example of a serif font. Sans-serif fonts do not have these little marks; the letters are more blocklike. Arial and Helvetica are common sans-serif fonts. Figures 1 and 2 display the same two paragraphs with two different sets of guidelines for spacing and fonts. In Figure 1, the text is fully justified, a line is skipped between paragraphs, and a sans-serif font is used (Arial). Both samples have the same size font (10) and use $1\frac{1}{2}$-line spacing.

In Figure 2, the text is left justified (with a ragged right margin), paragraphs are indented (so no extra line is skipped between them), and a serif font is used (Times New Roman).

Which scheme is better? The answer usually depends on what the reader needs to do with the document. For text that needs to be projected (as in a presentation) or read from some distance or under unusual circumstances (such as out in the field), a sans-serif font works best. For text that is long (more than a page) and meant to be read up close, some research suggests that a serif font is best, because readers can distinguish more

> The scope of my project will involve comparing different wall types, including single-, staggered-, and double-stud walls, constructed with a combination of building and sound-control materials. These materials will include gypsum board, resilient channels, wallcavity sprays, thin polymers, insulation, and sound-deadening boards. The different wall types are shown in Figure 1.
>
> I will not consider noise exterior to the apartment. I will assume that the apartment complexes are two to three stories and of a generic design, as given in Figure 2. In addition, I will only be analyzing the section of wall given in Figure 2 to evaluate the different wall types (called the "party wall," through which most interior noise passes).

Figure 1. Paragraph using full justification and Arial 10 font.

The scope of my project will involve comparing different wall types, including single-, staggered-, and double-stud walls, constructed with a combination of building and sound-control materials. These materials will include gypsum board, resilient channels, wall cavity sprays, thin polymers, insulation, and sound-deadening boards. The different wall types are given in Figure 1.

I will not consider noise exterior to the apartment. I will assume that the apartment complexes are two to three stories and of a generic design, as given in Figure 2.
In addition, I will only be analyzing the section of wall given in Figure 2 to evaluate the different wall types (called the "party wall," through which most interior noise passes).

Figure 2. Paragraph using ragged right margin and Times New Roman 10 font.

quickly between the letters. Similarly, some research shows that fully justified text, while it looks very professional, is not as easy to read in longer documents. Readers who turn away from a page can find their place again more easily, it is thought, with a ragged right margin. For two-column formats, however, full justification makes for a much cleaner look.

HINT

Visual Elements of a Page

Text
Headings and titles
Graphics
Simple separators like lines and boxes
White space

Get to know your word processor well so that you can control elements such as spacing, placement of page numbers, and size and placement of headings. If you learn to set up style sheets in your word processor, you can let your computer "remember" what size fonts to use for text, titles, and headings for each document; where to put the page numbers; how much white space to put around headings; etc. In Microsoft Word, the Style command is in the Format menu.

HINT

Designing Pages: A Few Tips

- Use the same design for all pages that contain the same kinds of information (e.g., beginnings of sections).
- Use a maximum of two typefaces. Serif fonts such as Times are slightly easier to read in long documents.
- Text with a ragged right margin is slightly easier to read in long documents than is text with a fully justified margin.
- ALL CAPS ARE HARD TO READ; USE THEM SPARINGLY.

Headings play a key role in helping the reader identify information quickly. Use plenty of white space around headings (more space for bigger headings), and check carefully that you have kept levels of headings consistent. After the title of the document and the titles of sections, the major headings are level #1, the subheadings are level #2, etc. Be sure to make the look of each heading reflect its level. Thus, if "Introduction" and "Conclusion" are both major headings, their size, font, and format should be exactly the same. Figure 3 shows some examples of different options for keeping headings consistent.

Headings

Option One: Use Size and Position to Differentiate

Heading Level One [14-point Times New Roman]

Heading level two [12-point Times New Roman]

Heading level three. Might be run in with text. If you number the heading, do not use bullets points.

Be consistent with capitalization.
Run-in headings usually need to be in boldface or italics to set them off from the text.

(a)

Headings

Option Two: Use Numbering to Differentiate

1. Heading One

1.1 Heading Two

1.1.1 Heading Three

This example uses position, as well as numbering, to differentiate the headings.

(b)

Headings

Option Three: Use Font Styles to Differentiate (Example #1)

Heading One [Times New Roman 14 bold]

Heading Two [Times New Roman 12 unbolded]

Heading Three. Italics help differentiate heading from body text.

This option also generally uses size as well as style to differentiate between heading levels.

(c)

Headings

Option Three: Use Font Styles to Differentiate (Example #2)

Introduction [Tahoma 12]

This is the introduction, etc. [Times New Roman 10]

You might use one font type for headings and another for text. *Never use more than* ***two*** *font types* in a technical document (except for manuals).

(d)

Figure 3 (a–d). Different options for keeping headings consistent in format and level.

Style Sheet Checklist

† **Document**
- Paper type and cover page
- Single sided or double sided?
- Margin width

† **Sections**
- Start on right-facing pages?
- Use line or other graphic device to divide?

† **Headings and Subheadings**
- Position
- Font: size and characteristics (e.g., bold, italics?)

† **Main Text**
- Font (serif or sans serif)
- Paragraph indentation, or skip a line between left-justified blocks?
- Line spacing (single, 1½, or double)

† **Page Numbering**
- Where on page?
- What font or type?
- Different for different sections?

† **Tables and Figures**
- Integrated with text?
- Where do labels go?

† **References**
- Citations in parentheses in text?
- Reference list goes where?

† **Abbreviations, Acronyms, Equations**
- Include glossary? Where?
- Acronyms spelled out where?
- Mathematical variables explained where?

Figure 4. Standard checklist for a document's style sheet.

Since size is the most intuitive reflection of importance, choose size first as a way to indicate level of heading or importance, and then add on other differentiating characteristics. Do not use bullet points in headings.

To avoid last-minute chaos, make decisions about headings and text design early in the process of writing. Learn to set up style sheets in your word processor so that all headings of a certain level will be consistent. If you are writing as part of a team, all members will need to agree about these stylistic decisions. The style sheet checklist in Figure 4 should help you remember all of the style issues that will need resolution (e.g., Where should page numbers go?). If you resolve these issues sooner rather than later, you save yourself headaches in the production, as well as the writing, process. If each person is supposed to write 8–10 pages for his or her section, for instance, what happens when one writer is using 8-point Times and therefore ends up producing lots more text than the person using 12-point Times? You have a much more difficult editing and cutting job—that is what happens.

HINT

More Tips for Designing Pages

- Find out whether your organization has a style guide. If so, use it!
- The team producing a document needs to create style guide if one does not already exist.
- For journal papers or conference proceedings, check the Information for Contributors or Call for Papers sections (usually in the front of the journal) for style guidelines.

Standard fonts for engineering reports are Times and Times New Roman 11 or 12. Some companies prefer Palatino or other fonts. Always check with your professor, company, or agency to see whether a style guide (even an informal one) exists.

Disseminating Documents Digitally

If you need to disseminate digitally a word-processing document that you do not want altered (such as a final report or perhaps your résumé), it is generally a good idea to convert the document to PDF, or portable document format. Some word-processing programs, such as Microsoft Word, allow you to convert documents to PDF; you can also convert documents using Adobe® Acrobat® software. Your file can then be opened using Adobe® Reader®, which is available free of charge at www.adobe.com. PDF has become the standard format for exchanging documents, since PDF files cannot be altered as easily as files in other formats, such as Microsoft Word.

ASSIGNMENT

Take two pages of a document you have already written, and try setting up styles for it. In Word, define the styles for the different levels of your headings: Choose the type and size of font (bold? italics? underline?).

Choose the spacing before and after the headings. Define the style of your "normal" text. How does your document look when you apply those styles? Now change some of the styles and see which ones you think make the document look better. More importantly, show a couple of different styles to a friend, and ask which one he or she likes better.

FURTHER READING ON DESIGNING DOCUMENTS

Williams, R. 1994. *The Non-Designer's Design Book*. Peachpit Press: Berkeley, CA.
This little book is the best guide to document design for folks who are not graphic designers.

Williams, R., and Tollet, J. 1998. *The Non-Designer's Web Book*. Peachpit Press: Berkeley, CA.
A follow-up to the previous reference, this volume provides helpful guidance to the novice about how to create and post a website.

3 SAMPLE CORRESPONDENCE

This section presents an internal memo from the head of quality control (QC) at a wood-truss manufacturing company to a supervisor who is not an engineer. The company has just received a shipment of wood from the supplier, and, uncharacteristically, the wood has a moisture content above the fiber-saturation point. The supervisor (Ellen Austin) majored in business and believes the wet wood would be fine to use for the current construction job. George Lumberman, the QC engineer, must explain the technical consequences of using the wet lumber.

A Word about Using the Sample in This Section

This section presents a model of a good letter for you to use. The model may not, however, fit the purpose or the requirements of a document you are called upon to write. Always ask your boss or instructor for samples to look at. Don't then copy them slavishly by using a fill-in-the-blanks approach, but rather notice the type of information presented in each document or section and decide whether your information fits that type.

LONGHORN LUMBER INTRAOFFICE MEMO

DATE: 11/18/2003

TO: ELLEN P. AUSTIN, MANAGER

CC: MARIA G. JONES, PRESIDENT

FROM: GEORGE G. LUMBERMAN, QUALITY CONTROL

RE: NUECES CONSTRUCTION PROJECT

It has come to my attention that the Southern pine lumber we recently received from Peachtree Lumber Supply has a moisture content above the fiber-saturation point. I was quite surprised, as Peachtree has an excellent track record of shipping well-seasoned, high-quality lumber. I have contacted Peachtree, and they regret the mistake (citing a recent roof leak in their warehouse) and have offered to send a new shipment free of charge, contingent on the return of this one. This is a very generous offer, but somewhat inadequate. It would take 10 days for the new shipment to arrive, and this lumber was intended for use in the Nueces construction project. As you know, the Nueces project is already behind schedule. The workers will be ready to install the trusses tomorrow, and no further work can be done until they are in place. However, it is highly preferable to halt construction for 10 days and await the new shipment than to use the lumber we currently have on hand.

Using lumber with a moisture content above the fiber-saturation point would be catastrophic for the structural integrity of this building. Wood is made of thin-walled tubes, which are also called cellulose cells or fibers. When wood absorbs water, these fibers become saturated with water, swell, and move apart. The fibers give wood its mechanical strength; as the fibers move further apart, the density of the fibers decreases and the wood loses strength and stiffness.

The relationship between the strength of wood and its moisture content is demonstrated by a test that I performed on the lumber we received this week and some well-seasoned lumber left over from a previous project. Each beam was placed between two supports,and a load was applied at a constant rate to the center. The test results are tabulated in the upcoming table. In both tests, the dry lumber sustained significantly larger loads before failure. The structural design for the Nueces project demands that the trusses be much stronger than the wet beams we tested. Furthermore, the displacements recorded for the wet beams are much larger than for the dry ones. The design cannot accommodate such large displacements.

Sample description		Maximum load (lbs.)	Displacement (in.)	Strength ratio (dry/wet)
Test 1	Dry	2557	1.14	1.707
	Wet	1498	1.60	
Test 2	Dry	1905	1.08	1.922
	Wet	991	1.50	

Therefore, my recommendation is to *halt construction* until the new lumber shipment arrives. An alternative you may have considered is to let the wood we have dry out. However, this process is likely to take longer than 10 days, given the current damp weather conditions. Another alternative is to investigate other suppliers who could fill our need more quickly.

Under no circumstances should we use the wet wood. Not only would using this wood cause an unacceptable safety hazard, it would be catastrophic to our reputation.

NOTES ON SAMPLE MEMO

Notice that in correspondence (especially in internal memos), personal pronouns can be used. Because all correspondence is written from one person to another, these pronouns are acceptable, except in the most formal, legalistic letters. Notice also how the author, who is an engineer, gently educates the recipient, who is *not* an engineer, about the structural integrity of wet vs. dry wood. It is not unusual for middle managers in engineering companies to have business backgrounds rather than engineering degrees. Even with engineering degrees, some supervisors have been away from hands-on engineering long enough that they do not remember or have not kept up with technical work in their field. Understanding your audience is an important part of communicating technical information effectively. Even when the audience is your boss, do not assume that person has the same background that you do, and think about what technical explanations may help him or her make good business *and* engineering decisions.

In this memo, the most important points are made at the beginnings and endings of paragraphs. The fact that the wood is wet is given in the very first sentence. The first paragraph then ends with the information that halting construction is preferable to using the wet lumber for the current project, and the next paragraph begins with a firm statement about the unacceptable risk to structural integrity of using the wet wood. The word "catastrophic" should not be used very often, but when you need to use it, do not bury it in the middle of a paragraph. The final paragraph begins with the recommendation not to use the wet wood, and the memo ends with a warning of the consequences of not following that recommendation. The final paragraph has only two sentences—an uncommon occurrence in technical writing, but justified by the critical nature of the information imparted.

Figures and tables are handled differently in correspondence than in other technical documents. Notice that the table has no title or number. Generally, the few graphics in letters or memos are explained so thoroughly that they do not need formal labeling, except if they are included as attachments and therefore might be photocopied or read separately.

Generally, memos and letters are single spaced.

4 SAMPLE RESEARCH PROPOSAL

This section presents a sample research proposal for a consulting project evaluating different materials for soundproofing an apartment building. The comments in boxes highlight the function of each section within this proposal. See Section 6, Sample Feasibility Report for more information on structuring longer documents.

A Word about Using the Sample in This Section

This section presents a model of a good research proposal for you to use. The model may not, however, fit the purpose or the requirements of a document you are called upon to write. Always ask your boss or instructor for samples to look at. Don't then copy them slavishly by using a fill-in-the-blanks approach, but rather notice the type of information presented in each document or section and decide whether your information fits that type.

Relyon Engineering, Inc.

Proposal to Determine Improved Methods of Sound Insulation

Summary

> **Clear and specific statement of the problem**

The residents of Gaylord Apartments in central Austin have complained to the owner, Louise Smith, about the lack of proper soundproofing in the building. Ms. Smith is aware that several tenants intend to relocate because of high noise levels, and she is concerned about the possibility of also losing new business if she does not address the problem.

> **Statement of purpose**

> **Possible solutions have been narrowed down**

> **Criteria for evaluating solutions described**

Ms. Smith has requested the services of Relyon Engineering, Inc., to propose and evaluate alternatives for reducing noise between apartments. These alternatives include improvements to insulating materials, alterations to flooring materials, or some combination of the two. We will evaluate potential solutions based on their effectiveness in reducing sound transmission, as well as their cost per apartment, construction time, and aesthetics. Upon completing our research, the Relyon team will present the findings to Ms. Smith and recommend the most appropriate solution to the noise problem.

> **Answers to managers' questions: How long will project take? How much will it cost?**

The estimated time frame for research and evaluation of the sound problem is 11 weeks. We will deliver a progress report to the client on March 24, 2003, and a final report containing the findings and the recommended solution by May 4, 2003. The estimated cost of Relyon Engineering's services for this evaluation is $1,400.00.

Definition of the Problem

> **Expansion of statement of the problem**

In recent months, a large number of residents have complained about the level of noise coming from other apartments in the building. Although noise control is adequate between adjacent apartments on the same floor, excessive sound is transmitted through ceilings and floors and creates a major disturbance for the tenants. In the past year, approximately 20% of the building's residents did not renew their leases, and 50% of those tenants cited high noise levels as a factor in their decision.

Potential causes of this problem include inappropriate choice of flooring materials, a lack of proper insulating materials in apartment ceilings, and structural damage to the ceilings. Hardwood floors, which are in all of the building's apartments, transmit more sound than carpeted floors. Various types of insulating materials can be placed inside or on the exterior of ceilings to help absorb noise; the absence or underutilization of such materials could be a factor in the excessive noise transmission. Structural damage such as small cracks can also significantly affect the ceiling's ability to act as a sound barrier; in an older building such as Gaylord Apartments, such damage would not be unexpected.

In order to improve the quality of life for the remaining residents, as well as to increase the market value of these apartments, Ms. Smith feels it is necessary to implement a solution before July, when most leases are up for renewal.

Scope of Project

Our team will consider two basic methods to increase the sound insulation in the apartment: carpeting the floor in the apartments and adding insulating materials to the ceilings. Possible solutions could also include a combination of these techniques. Although we have also discussed structural damage as a potential cause of the problem, any repairs of this nature would be too time consuming and disruptive to residents and therefore are beyond the scope of this investigation.

Expanded description of possible solutions

The general scope of the project will be to more fully assess the problem, determine which specific materials (flooring and/or insulation) we will consider using to solve the problem, and analyze each possibility against our criteria, which are described in the next section. Our research will include sound measurements inside the building to help us quantify the current noise levels and resident surveys to gain more information about what improvements they want or expect to see. Consultation with an acoustical consulting firm and companies that produce insulating materials will provide a basis for more specific alternative solutions. We will then gather additional information from textbooks, journal articles, and product specifications to help us determine which solution to recommend to the client.

Methodology used to study problem and evaluate solutions

Our final report will recommend the strongest solution of those studied. We will present these solutions to the client in both oral and written form. The oral presentation is scheduled for April 23, 2003, and we will submit the final report on May 4, 2003.

Reiteration of project's time frame

Solution Criteria

The alternative solutions will be evaluated according to the criteria outlined next. These criteria are listed in order of priority, from highest to lowest, as designated by the client.

Whole section is a detailed description of criteria to be used to evaluate possible solutions

Effectiveness in Reducing Noise Transmission

The client has asked that we determine an appropriate threshold for noise transmission, based on the preferences of residents. We will measure each solution in terms of noise reduction (NR), a quantifiable term expressed as a function of the transmission loss (TL) of a barrier (defined as the ratio of the sound reradiated by the barrier to the sound absorbed by it, in decibels), the area of the barrier, and the ability of the receiving room to absorb sound.

Cost of Implementation

The owner would like to spend no more than $1,500 per apartment to implement the solution. This cost includes materials and labor, but is separate from the consulting fees for Reylon Engineering's recommendation report.

Time of Construction

Most new resident apartment leases begin July 1, and the client has specified a construction completion date of June 15 in order to prepare the apartments for their new tenants. Since vacating tenants must leave by June 1, renovations must be completed in the two-week period from June 1 to June 15.

Aesthetics

We will consult residents as to their aesthetic preferences among the various solutions involving carpeting and ceiling insulation. We will also survey residents to determine how important aesthetics are relative to effectiveness in reducing noise, using a simple numerical scale to rank their preferences. These rankings will help us determine how much consideration to give aesthetics in our final recommendation.

Step-by-step outline of tasks necessary to accomplish goal of project: recommend best solution to the noise problem

Proposed Procedure

We will begin our research by investigating the apartment building itself. We will obtain the original building plans and determine what materials are presently acting as sound barriers in the ceilings, if any. We will also visit the apartments and determine the amount of sound being transmitted between apartments through the ceilings and floors, using standard equipment for measuring decibel levels. This process will help us determine the current noise reduction (NR) between apartments and provide a baseline with which we can compare the effectiveness of potential alternatives. Surveys of the residents will provide us with information about what constitutes an acceptable level of noise, as well as what aesthetic preferences they have, if the ultimate solution involves altering the flooring material. Based on these data, we will set specific targets for the NR and aesthetics criteria.

Once we have clarified these goals, we will consult professionals currently working in the area of residential sound insulation, including the acoustical consulting company Nelson Acoustical Engineering, Inc., and Victor Gomez from Flour Daniel, Inc., to help us develop a more specific list of possible solutions. We will also contact the insulating-materials companies Netwall Noise Control and Building Green Products to gain information about particular products that could aid in improved insulation and noise reduction. In all of these cases, correspondence will be mainly via telephone and in-person office visits.

After finalizing our list of possible solutions, we will consult the appropriate technical journals, textbooks, and product specifications to estimate each alternative's effectiveness in reducing transmitted noise (NR), as well as the material and labor costs involved. To evaluate the amount of time necessary to implement each solution, we will consult area contractors who have experience with the specific renovations we are considering. Finally, survey data gathered from residents at the beginning of our project will allow us to factor the aesthetic criterion into our recommendation.

All feasible solutions will be presented directly to the client in an oral presentation, followed by a written report of the results of the study, which will include material layouts and a cost breakdown, as well as cost estimates for the entire project.

A page break is inserted here so as to fit Table 1 on one page. Never break a table over two pages!

Schedule

The proposed work schedule for the project is as follows:

Table 1. Project Schedule

Activity	2/22	2/26–3/05	3/05–3/12	3/12–3/19	3/19–3/26	3/26–4/09	4/12–4/24	4/23–5/04
✓ Analyze building plans	←→							
✓ Perform sound measurements		←→						
✓ Survey residents		←→						
✓ Consult sound engineers			←→					
✓ Consult materials companies				←→				
✓ Evaluate solutions					←→			
✓ Deliver progress report					*3/26*			
✓ Prepare report draft						←→		
✓ Submit report draft to supervisor						*4/09*		
✓ Review draft and prepare oral report							←→	
✓ Deliver oral report to Harriet Romo							*4/24*	
✓ Prepare final report								←→
✓ Submit final report								*5/04*

Table title in correct location

Budget

The estimated total cost of our team's services is $1,400.00. This total includes all research, funding for collaboration with other firms, and preparation of reports, as shown in the following cost breakdown:

Narrative justification of budget

Table 2. Project Budget

Activity	Number of Hours	Hourly Rate	Cost
Research and analysis	15	$35	$525
Correspondence with other firms	5	$35	$175
Cost estimating and comparisons	5	$35	$175
Presentations and reports	15	$35	$525
		Total Cost	**$1,400**

Table gives quick visual overview of budget

Qualifications

The project manager has a good working knowledge of the fundamentals of architectural, structural, and mechanical engineering. She also has experience working in a large civil-engineering firm.

Jane Daniels

Education: *Bachelor of science in civil engineering*, the University of Texas at Austin, to be awarded December 2005.

Experience: *Intern*, Turner Collie and Braden, Inc., June 2003 to August 2003.

Qualifications of researcher doing the work

Sources of Information

- Stein, Benjamin, and John S. Reynolds. 2000. *Mechanical and Electrical Equipment for Buildings*. 9th ed. New York: John Wiley & Sons, Inc.
- *Journal of Sound and Vibration*, London.
- Victor Gomez, P.E., Flour Daniel, Inc., Austin, TX.
- Gina Davis (interview, March 23, 2002), Nelson Acoustical Engineering, Inc., Austin, TX.
- Netwall Noise Control, Austin TX.
- Building Green, http://www.buildinggreen.com/index.html (accessed March 2002).
- Louise Smith, owner, Gaylord Apartments.

This reference list is not formatted according to any accepted conventions; it is simply an indication of where this student will look for information. Ask your instructor whether you should use a formal references list.

5 SAMPLE PROGRESS REPORT

All progress reports provide a bridge for readers between past, present, and future. These reports explain what has happened on a particular project, what is happening now, and what is projected to happen in the future. They also provide a bridge for management readers not only between different phases of a project, but between different projects, each of which is part of some bigger plan or process. For some projects, you will report periodically, as often as once per week; for others, you may write only a midproject report. In all cases, you are helping the reader look ahead toward the successful (you hope) conclusion of the project, whether it is a building being constructed, a manufacturing process being implemented, or a series of experiments testing some new material.

There are many good reasons for taking the time to stop what you are doing and write a project report. Here are a few reasons:

- to reassure clients that you are making progress
- to provide clients with a brief look at some findings
- to give clients a chance to evaluate your work and request changes
- to give you a chance to discuss problems with the project
- to force you to adjust your work schedule to conform to reality

Very often, progress reports are compared with a project's original proposal to see whether the project is keeping to budget and schedule. And, very often, engineering projects deviate somewhat from what is expected. So, a major function of this type of report can be to explain and justify necessary changes and to show how problems (many of them unforeseeable) are being handled. Writing this type of report and presenting it to the client also gives you the chance to solicit help in dealing with problems or with changes requested by the client. The sample progress report shown here has no such problems, but many projects need more money or resources than originally thought.

Here is a sample report on the status of a project to examine various orbital configurations of satellites around the moon in order to help NASA establish continuous communication with the far side of the moon. See Section 6, Sample Feasibility Report, for more information on structuring longer documents.

A Word about Using the Sample in This Section

This section presents a model of a good progress report for you to use. The model may not, however, fit the purpose or the requirements of a document you are called upon to write. Always ask your boss or instructor for samples to look at. Don't then copy them slavishly by using a fill-in-the-blanks approach, but rather notice the type of information presented in each document or section and decide whether your information fits that type.

Wentogle Incorporated
Sending You to Outer Space

Progress Report on the Evaluation of Orbit Configurations That Provide Continuous Communication with the Moon for NASA

Project Description

NASA has requested the development of a system to establish continuous communication with the far side of the moon for future scientific missions. Currently, Earth-based communication with the back side of the lunar surface is impossible because of transmission blockage caused by the mass of the moon. On September 13, 2000, NASA requested that Wentogle Incorporated perform an evaluation of possible satellite orbital configurations capable of providing the communication link.

Wentogle's analysis focuses on the optimization of satellite orbital configurations rather than the type of communication satellites to be used for creating the communication network. Three alternative solutions are currently under investigation: two configurations of satellites in circular orbit around the moon and one configuration of satellites in a lunar halo orbit around the gravitational equilibrium point L1.

Complete communication coverage of the back side of the moon is the primary criterion for selecting a configuration. Other selection criteria include the satellite's orbital stability and lifetime, as well as an overall configuration cost estimate, which will be based on the number of satellites required to provide continuous communication.

The project was initiated on October 1, 2000, by John Carson and Andrew Giacobe of Wentogle, Inc., with a budget of $2,000.

Work Accomplished

The following work on this project has been completed thus far.

Satellites in Circular Orbit around the Moon

The geometrical analysis of the circular orbits was performed on systems of satellites in two perpendicular orbit planes. Two separate cases were considered: three satellites per orbit and six satellites per orbit. The configuration for these orbits, shown in Figure 1 and Figure 2, displays the two perpendicular orbits with three satellites per orbit and six satellites per orbit, respectively. Lunar communication coverage provided by the circular orbiting satellites was determined using a geometric analysis based on the altitude of the orbit. For the analysis, the satellites were assumed to be equidistant from each other, thus forming equilateral triangles.

Figures are referenced and explained in the text.

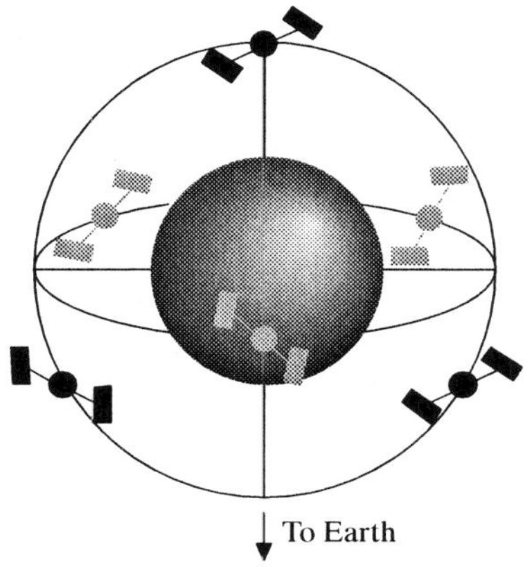

Figure titles in correct place.

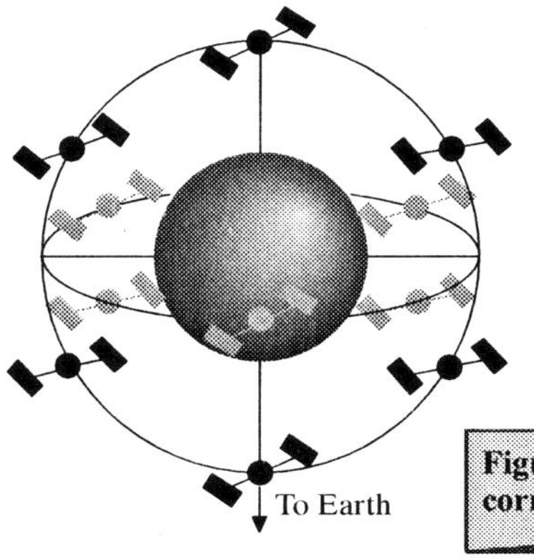

Figure 1. Two groups of three satellites in perpendicular orbits

Figure 2. Two groups of six satellites in perpendicular orbits

Wentogle discovered that the three-satellite configuration provided complete communication coverage when the orbital altitude was a minimum of 3,480 kilometers, which is twice the radius of the moon. The high altitude of the three-satellite system may result in orbital instabilities, so Wentogle decided to research a configuration using more satellites at a lower orbit. The six-satellite configuration required half of the altitude needed for the three-satellite system. The six-satellite altitude is consequently equal to the lunar radius of 1,740 kilometers. The maximum altitude of both orbits is equal to the SOI (Sphere of Influence), which is located at an altitude of 64,380 kilometers above the moon by definition. Above the SOI, Earth's gravitational attraction on the satellites is stronger than the moon's, and the orbits would not remain around the moon.

Satellites in Halo Orbits around the Moon

The communication coverage provided by satellites in orbit around the equilibrium point L1 was determined using a geometrical analysis. The halo orbit of interest is perpendicular to the line between Earth and the moon, as shown in Figure 3. From the analysis, Wentogle discovered that three satellites forming an equilateral triangle in the halo orbit provide communication coverage when the orbital radius is a minimum of twice the lunar radius, or 3,480 kilometers, as shown in Figure 4.

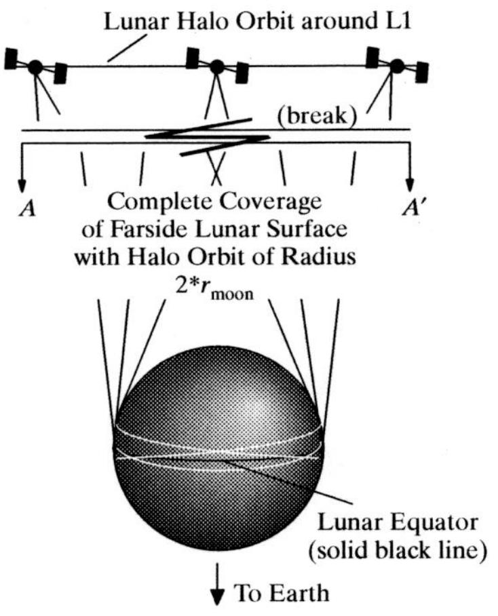

Figure 3. Lunar Halo Orbit with three satellites provides complete communication coverage

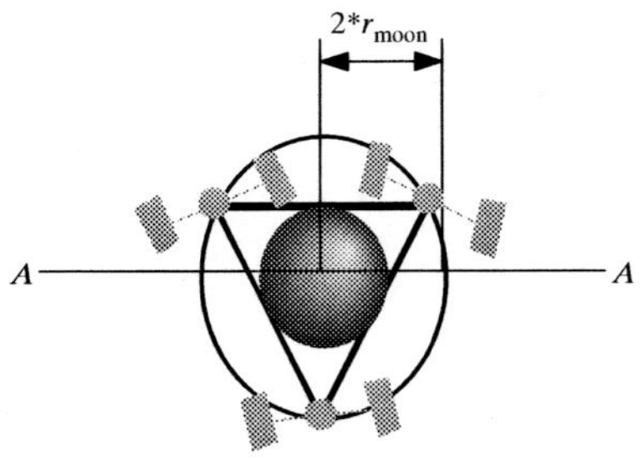

Figure 4. Lunar Halo Orbit with three satellites (view of far side of moon, looking toward Earth)

The geometrical analysis was based on a halo orbit with an altitude of 61,500 kilometers above the center of the lunar equator. This altitude corresponds to the distance of the gravitational equilibrium point L1 from the center of the moon. The maximum halo-orbit radius depends on the stability of the orbit and will not be computed until the stability analysis has been completed.

Work Remaining

Based on the geometrical analysis, both the three-satellite and six-satellite configuration of the circular orbit provide complete communication coverage of the lunar surface. However, accurate orbit simulations will be required to determine whether the selected orbit altitudes are stable. Using the models currently under development and performing a geometric analysis, Wentogle will determine the orbital stability and lifetime of the three-satellite and six-satellite configurations. Similarly, once the halo-orbit simulations are complete, their stability will also be analyzed using the computer models. Then, the stability and lifetime data will be compared with the results from the lunar-circular-orbit analysis.

Finally, we will determine the overall maintenance and implementation costs of the configurations. The cost will be determined for the three-satellite and six-satellite configuration, and a comparison will be made between these two options and the option of satellites in halo orbit around the L1 gravitational-equilibrium point. This information will allow us to select the most suitable option for providing continuous communication with the back side of the moon.

No foreseeable schedule delays or budget changes are expected. The project will be completed on time and within budget by December 7, 2000, when the final report will be submitted to NASA.

ORGANIZATION OF A PROGRESS REPORT

The sample report uses standard sections to outline progress: Project Description, Work Accomplished, and Work Remaining. Here are the pieces of information each of these sections would typically cover or the questions they would answer:

Project Description

- purpose of project
- specific objectives
- scope (limitations, if any)
- alternative solutions being evaluated
- criteria used for evaluation
- dates project began and budget approved
- people working on project
- client (people for whom work is being done)
- other necessary background information

Work Accomplished

- What specific tasks have been accomplished?
- Does the progress of work actually match the schedule in the proposal?
- Which scheduled tasks are still outstanding?
- Which tasks are new?
- Make clear how you have dealt with problems and changes, if any. (Discuss more fully than in The Project Description.)

Work Remaining

- What specific tasks are left to do?
- How will you use the remaining tasks to make a recommendation to the client?
- What are the final deliverables (often a report or a presentation)?
- Is the project on schedule and within budget?

6 SAMPLE FEASIBILITY REPORT

This document reports on a study conducted for the owners of a set of apartment complexes in a medium-sized American city. Tenants have been complaining about noise from neighboring apartments, so the owners have hired a consulting engineering firm to investigate possible cost-effective means of controlling the transmission of sound between apartments. The student consultant has spent the semester defining the problem, defining a scope of work, and evaluating chosen solutions on the basis of a set of criteria worked out with the client. The report was produced for a Technical Communication class, but the methodology and analysis were very professional. And the client was real.

The organization of this report is typical of the way feasibility reports are put together: sections defining the problem at hand, describing how the list of feasible solutions was narrowed down, explaining the criteria used to measure the feasibility of various solutions, presenting the findings of the investigation, applying a method for final evaluation of feasibility, and finally recommending a course of action.

A Word about Using the Sample in This Section

This section presents a model of a good feasibility report for you to use. The model may not, however, fit the purpose or the requirements of a document you are called upon to write. Always ask your boss or instructor for samples to look at. Don't then copy them slavishly by using a fill-in-the-blanks approach, but rather notice the type of information presented in each document or section and decide whether your information fits that type.

FINAL REPORT

Evaluation of Wall Designs That Control Airborne Sound Transmission

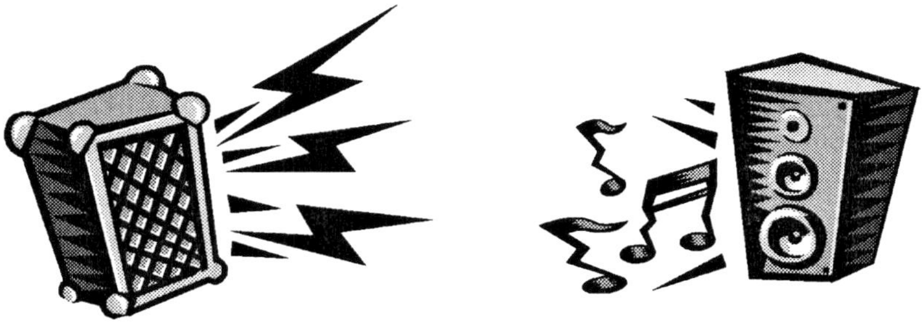

Firm Ideas

FRONT MATTER: COVER PAGE

A report's *cover page* should be designed for ease of reading and simplicity of concept. Like the cover of a book, a report's cover should give some idea of what is inside, without presenting too much detail. Unlike a book's cover, the covers of most technical reports do not name an author, but rather include the logo and name of the company or agency that produced the report (in this case, Firm Ideas). This sample cover has just three pieces of information: the type of report, the title, and the name of the authoring company. Many agencies have coding and numbering systems that require certain numbers to be on the cover of every report; at the very least, you want to let the reader know what type of document he or she is looking at. For example, is it a proposal, a progress report, or the final report on a particular project?

The cover page in this example also has a graphic depicting the subject matter of the work reported, but many technical reports use only a logo on the cover, reserving graphical display for conveying the technical content inside the report. Check with your instructor or company before designing your own cover. Very often, you will simply use a standardized cover and insert your own title.

The title, of course, is critical to the reader's understanding of the scope of what is reported here. Make sure the title on the cover page is exactly the same as on the title page (which comes a little bit later). Be straightforward and direct in phrasing the title, but offer as much detail as a reader will need in order to understand the scope of the work reported. Compare this sample report's title,

> Evaluation of Wall Designs That Control Airborne Sound Transmission

with another version of it:

> Evaluation of Six Wall Designs That Are Intended to Control Airborne Sound Transmission between Apartments via a Common Wall

The second title probably offers more information than is necessary to understand the scope of work. On the other hand, the following title does not offer enough information about this study as done for a particular client with particular sound-control problems:

> Controlling Sound Transmission

The implied scope is much too large here: There are many ways other than designing a wall to control sound. And there are ways to transmit sound other than exclusively through the air. So we don't really understand the scope of this particular study from the title alone.

FIRM IDEAS

100 East 51st Street Austin, TX 78705 (512) 400-3333

May 1, 2004

Mike Parker, Manager of EBC Properties
1000 Summit Lane
Austin, TX 78705

Dear Mr. Parker:

The attached technical report contains Firm Ideas' evaluation of wall designs that control airborne sound transmission. This report includes the descriptions of each wall design, the evaluation process, and the wall design that Firm Ideas recommends EBC Properties use in their new apartment complexes. We used the solution criteria established by EBC Properties and Firm Ideas to evaluate six wall designs:

- Single-Stud Wall with Resilient Channel
- Single-Stud Wall with Sound-Deadening Board
- Staggered-Stud Wall
- Staggered-Stud Wall with Sound-Deadening Board
- Double-Stud Wall
- Double-Stud Wall with Sound-Deadening Board

Firm Ideas recommends that EBC Properties use the Double-Stud Wall in their new apartment complexes. First, the Double-Stud Wall provides moderate performance with an STC value of 53 decibels. Second, the Double-Stud Wall has a relatively low material cost of $11.98 per linear foot and a relatively low labor cost of $8.74 per linear foot. Finally, the Double-Stud Wall can support the highest mountable object weight, 110 pounds per linear foot. This wall design best meets the needs of EBC Properties.

Firm Ideas is available to answer any questions you may have concerning this report or the project in general. You can contact Firm Ideas at (512) 400-3333 or by e-mail at *engineers@mail.utexas.edu.*

Sincerely,

Jeffrey Hunt
Project Engineer

FRONT MATTER: LETTER OF TRANSMITTAL

The *letter of transmittal* is a way of conveying the report to the right person: the decision maker who will act upon the work reported in the document. Like any letter, it is directed to a person, not a company or a department. Essentially, the letter gives a brief summary of the executive summary, answering the following questions: What was the scope of the work done, and why was it done? What was found? What conclusions and recommendations can be made on the basis of the work done? If a recommendation is made, how much will it cost to implement it? Sometimes the transmittal letter also addresses concerns that management might have with the results of the work—environmental concerns, especially, or possible legal consequences. The author may not have specifically studied the environmental consequences, for instance, of developing a piece of land, but may have become aware through developing a master plan that neighborhood groups are concerned about runoff to an aquifer.

Our sample letter does not convey any such concerns, but it offers specific conclusions about the cost and performance of the recommended wall design. Comparison with the other wall designs studied is left to the executive summary.

Sometimes, the transmittal letter is clipped onto the front of the report, so that the decision maker for whom it is primarily intended can remove it. That way, the letter remains more private, and the rest of the report can be photocopied and circulated to other readers without including remarks meant just for the primary reader. If your letter does not need to contain any such remarks, then binding the letter into the report makes sense so that they do not become separated.

Remember to sign your letters!

Evaluation of Wall Designs That Control Airborne Sound Transmission

Prepared for

EBC Properties

Prepared by

Jeff Hunt

Firm Ideas

May 1, 2004

FRONT MATTER: TITLE PAGE

This sample report has a title page as well as a cover page. Certainly, in many technical documents, the cover and title page are one and the same. Separating them, however, allows you to create a more visually prominent design for the front cover (see the graphics on this sample's cover) and to include less text on the cover. Again, check with your company or instructor to see which format is expected or required. If the date is not included on the front cover, it must definitely appear on the title page. Providing the date is an important way to establish your document's place in the continuum of work and research on the particular topic or problem.

Company or agency policy determines whether the author's name actually appears anywhere on a technical document. Oftentimes, the company or agency is considered to be the author, because in reality many reports are collaboratively produced. Chances are that as a working engineer, you will not research and write reports all by yourself. But you may be the project lead or have some other such position that requires you to take responsibility for the document.

This title page has a typical format: the client (in this case, only the company, not a person) and the author are listed after "Prepared for" and "Prepared by," respectively. If you look closely, you will notice that the client's name is in a slightly bigger font than the author's. Truly, the individual author's name is probably the least important identity on the title page.

Table of Contents

List of Illustrations i

Executive Summary ii

Introduction 1
- Background 1
- Purpose and Scope 1
- Methodology 2
- Organization 2

Wall Design Alternatives 3
- Basics of Sound Control 3
- Wall Designs 3
- Summary of Wall Designs 6

Solution Criteria 7
- Performance 7
- Marerial Cost 7
- Labor Cost 8
- Mountable-Object Weight 8

Evaluation of Wall Designs 9
- Performance 9
- Marerial Cost 10
- Labor Cost 10
- Mountable-Object Weight 11

Factual Summary and Rating Procedure 12
- Factual Summary 12
- Rating Procedure 14

Conclusions 16

Recommendation 17

List of References 17

Appendix A: Material and Labor Cost Calculations A-1

List of Illustrations

Figures

Figure 1. Comparison of Performance........ 9

Figure 2. Comparison of Material Cost........ 10

Figure 3. Comparison of Labor Cost........ 11

Figure 4. Comparison of Mountable-Object Weight........ 11

Figure 5. Example of Linear Interpolation—Performance........ 14

Figure 6. Breakdown of Total Weighted Points........ 15

Tables

Table 1. Wall Design Performance........ 12

Table 2. Wall Design Material Cost........ 12

Table 3. Wall Design Labor Cost........ 13

Table 4. Wall Design Maximum Mountable-Object Weight........ 13

Table 5. Total Weighted-Points Calculations Using Equation (1)........ 15

Table A-1. Material Cost Calculations........ A-1

Table A-2. Labor Cost Calculations........ A-2

FRONT MATTER: TABLE OF CONTENTS AND LIST OF ILLUSTRATIONS

The *Table of Contents* (TOC) is more important than you might realize. Readers can glance at it and get a pretty good idea of the scope and subjects covered in the document. It is, after all, an outline, and thus it aids both the writer during the writing process and the reader during the reading process. But any TOC is only as good as its design and phrasing are clear and accurate, respectively. The headings must match the actual headings used in your report. But do not feel that you must include all levels of headings; two or three levels are usually sufficient.

Your word processor can help you create clear TOCs, so get to know the capabilities of yours. A program such as Microsoft Word can create an outline view that becomes the TOC, and thus, both the document's headings and the TOC can be updated at the same time. Do not use the space bar to try and insert the spaces in between a heading and the page number. Instead, set tabs or create a table so that the numbers will be lined up properly. The little dots connecting the words and page numbers in the TOC are called a "leader" (found under the Tabs function in Word). The dots do help the reader's eye make the visual connection.

If you have appendices, be sure to give each one a letter (A, B, C, etc.) and an informative title. We want to see at a glance what material is covered in an appendix; it is disconcerting to turn to an appendix and see a bunch of photocopied, handwritten, or printed pages and not have a clue as to what information they contain and how that information relates to your report. The sample report's Appendix A is called "Material and Labor Cost Calculations"; thus, we expect to see detailed cost figures and perhaps some formulas for arriving at the cost data included in the body of the report. Turn to the page following the actual appendix of this book for more information about developing, formatting, and referring to appendices.

A list of illustrations—or a list of figures and tables—is commonly used only if you have more than a total of five numbered graphics in your report.

Generally, neither the table of contents nor the list of illustrations is paginated.

ii

Executive Summary

Firm Ideas was hired by EBC Properties to evaluate wall designs that will control airborne sound transmission in their new apartment complexes. In their existing complexes, EBC Properties has received complaints of transmitted noise through the common wall between two apartments, and the high levels of noise have caused many residents to move out in search of apartments with better sound control. To avoid a similar problem in their new complexes, EBC Properties would like to know which wall design provides adequate sound control through the common wall.

For this report, Firm Ideas evaluated six wall designs that provide airborne sound control:

- Single-Stud Wall with Resilient Channel
- Single-Stud Wall with Sound-Deadening Board
- Staggered-Stud Wall
- Staggered-Stud Wall with Sound-Deadening Board
- Double-Stud Wall
- Double-Stud Wall with Sound-Deadening Board

Firm Ideas used four criteria to evaluate the wall designs. In order of importance, the criteria are performance, material cost, labor cost, and mountable-object weight. Performance was evaluated by obtaining the Sound Transmission Class (STC) values for each wall design. Firm Ideas calculated the material cost and labor cost by using industry-standard cost estimation techniques. The mountable-object weight for each wall design was found in material specifications and on-line sources.

The wall design evaluation by Firm Ideas produced the following findings. All designs met the minimum STC value of 50 required by the *1997 Uniform Building Code—Section 1208*. The Double-Stud Wall with Sound-Deadening Board provided the best performance, with an STC of 58 decibels, and the Staggered-Stud Wall provided the worst performance, with an STC of 50 decibels. The material costs ranged from $10.53 per linear foot for the Staggered-Stud Wall to $16.38 per linear foot for the Double-Stud Wall with Sound-Deadening Board. The labor costs ranged from $6.93 per linear foot for the Staggered-Stud Wall to $13.02 per linear foot for the Double-Stud Wall with Sound-Deadening Board. Finally, the maximum mountable-object weight ranged from 50 pounds per linear foot for the Single-Stud Wall with Sound-Deadening iii

Board, Staggered-Stud Wall with Sound-Deadening Board, and Double-Stud Wall with Sound-Deadening Board to 110 pounds per linear foot for the Double-Stud Wall and Staggered-Stud Wall.

After having evaluated the six wall designs, Firm Ideas recommends that EBC Properties use the Double-Stud Wall as the common wall in their future complexes. This wall design provides moderate performance (STC value of 53 decibels), a relatively low material and labor cost ($11.98 per linear foot and $8.74 per linear foot, respectively), and the highest mountable-object weight. Our study reveals that this wall design is the best value for the money and gives tenants the best combination of benefits: increased sound control and a generous weight allowance for mounting cabinets and fixtures.

BODY OF THE REPORT: *EXECUTIVE SUMMARY*

The ability to summarize, both verbally and in writing, is one of the most important skills a technical professional can master. People want information in as selective and brief a form as possible. Behind every one- to two-page white paper produced for a congressperson headed into an appropriations meeting, there are weeks, if not months, of information gathering, research, and synthesis. Decision makers (whether lawmakers or division heads within an engineering company) must handle many projects and make decisions quickly. Thus, the term "executive summary" was born as a way of labeling for this audience the section of any document that will answer the big questions about time, cost, and resources required for a solution to the problem investigated.

The executive summary must move quickly through the most important information in the report: the reason the project was undertaken, the way in which the project was handled (methodology), the most important results, the conclusions, and any recommendations (along with their costs and benefits). All of this information is presented in much greater detail in the rest of the report, but remember that the executive audience may not read the rest of the report. Remember, too, that rather than photocopying the entire report, meeting organizers may choose to meet and make decisions based only on information in the executive summary, so the information must be accurate, brief, and self-contained. Do not refer to any page, graph, table, citation, or appendix in the rest of the report. The example executive summary here does refer to the *1997 Building Code*, however, because adhering to code is a prerequisite for any feasibility analysis.

Let us look more closely at the five paragraphs of this executive summary. The first paragraph outlines the client's problem and what remedies are being sought (the reason that Firm Ideas undertook the project). The problem with inadequate sound control in the apartment complexes will be described in much more detail in another section.

The second paragraph is a simple listing of the wall designs evaluated by Firm Ideas in their quest to find the most feasible alternative for the client, EBC Properties. How the engineers came up with these six alternatives does not need to be explained here; it will be described later.

The third paragraph sets out the criteria used to evaluate the six alternative wall designs. Since these criteria constitute the heart of this feasibility study, they are presented along with the sources from which information about them was obtained. Since this report happened to be for a technical-communication class, the decision maker (the instructor) wanted to know that the author had consulted reliable sources. In an industry report, such consultation would be assumed and could be listed later in a references section. The four criteria are described much more fully later.

The most important results of the analysis are presented in the fourth paragraph. Instead of giving major results for all six wall designs for all four criteria, the paragraph presents a range of results—enough data to support the conclusion and recommendation given in the final paragraph. Notice that the recommendation in the fifth paragraph is first given as a factual conclusion (the chosen wall design costs $11.98 per linear foot, provides "moderate" performance, etc.) and then stated in terms of its overall benefits. Most recommendations must be justified clearly in terms of cost and benefits.

This summary does not include details about the rating system used to assign points to each wall system. That level of explanation is left to the body of the report. Instead, the summary presents in shortened form a highly selective justification of the recommendation and answers the two questions most on the minds of an executive audience:

- What action is recommended?
- How much will that action cost to implement?

A good benchmark for length of an executive summary is 5–10% of the whole report. Some industry reports actually have very long executive summaries, because the authors are trying to stuff as much information as possible into the section that "executives" read. Be careful of this temptation.

Introduction

1

Background

EBC Properties, an Austin apartment management company, has been receiving complaints about transmitted sound through the common wall in their existing complexes. The high levels of noise have caused some residents to move out of EBC Properties's complexes in order to find apartments that provide better sound control. Meanwhile, EBC Properties is supervising the design of two new apartment complexes for the Austin area. To avoid similar sound control problems in their new complexes, EBC Properties has requested that Firm Ideas evaluate wall designs that provide adequate sound control through the common wall.

Purpose and Scope

To meet the needs of EBC Properties, Firm Ideas has evaluated wall designs that control airborne sound transmission, and we have recommended the best wall design. After EBC Properties reads this report, they plan to use the recommended wall design as the common wall in their new apartment complexes.

The new apartment complexes will be two stories high and constructed with wood framing. Therefore, Firm Ideas only evaluated wall designs framed with wood studs capable of supporting typical dead and live loads of apartments. Keeping these limitations in mind, we have evaluated six wall designs that provide airborne sound control:

- Single-Stud Wall with Resilient Channel
- Single-Stud Wall with Sound-Deadening Board
- Staggered-Stud Wall
- Staggered-Stud Wall with Sound-Deadening Board
- Double-Stud Wall
- Double-Stud Wall with Sound-Deadening Board

These six wall designs differ in their placement of studs (single, staggered, or double row of studs) and their sound-controlling materials (resilient channel and sound-deadening board).

2

Methodology

After choosing the six wall designs for evaluation in this study, Firm Ideas obtained performance data for each wall design. Performance information was obtained from *Construction: Principles, Materials, and Methods*, a building-construction textbook (L. Simmons, 2001). This textbook listed Sound Transmission Class (STC) values (a measure of performance) for each design.

Next, Firm Ideas calculated each design's material cost and labor cost. The costs were calculated using RS Means's *2002 Building Construction Cost Data*, an industry-standard cost estimation book.

Then, Firm Ideas estimated the maximum mountable-object weight for each wall design. These maximum weights were obtained from manufacturer specifications. In addition, we estimated the mountable-object weights for some designs by using recommendations from on-line construction articles.

Finally, we used a point system to determine the best wall design. This point system enabled us to combine the results from our evaluation with the solution criteria to determine the best wall design. A detailed explanation of the point system is given in the Factual Summary and Rating Procedure section.

Organization

The rest of this report is divided into sections that give detailed information on the alternative wall designs, the criteria used to evaluate each wall design, and the evaluation process used to make a recommendation. The first section, Wall Design Alternatives, contains a general discussion on the basics of sound control and includes descriptions of each wall design. The second section, Solution Criteria, describes the solution criteria used to evaluate each wall design, and the third section, Evaluation of Wall Designs, provides an evaluation of each wall design. The evaluations in this section are divided into subsections that compare the wall designs according to the solution criteria. The fourth section, Factual Summary and Rating Procedure, summarizes the comparisons made in the third section and gives the step-by-step procedure that Firm Ideas used to make a recommendation. The Conclusions section presents the findings from the rating procedure, and in the last section, Firm Ideas recommends the best wall design for EBC Properties to use in their new apartment complexes.

BODY OF THE REPORT: *INTRODUCTION*

The introduction to any report must orient the reader enough that he or she can understand quickly the more detailed and technical information contained in the report's central sections. Like the executive summary, the introduction should be written for the greatest possible number of readers—semitechnical and nontechnical as well as technical. How do you orient readers? For client-focused reports (as opposed to journal articles), you make sure to answer these questions:

1. What is the problem or need that necessitated the work reported here?
2. How did the report's author become involved with this problem or need?
3. How did the authors go about attacking the problem, including defining the scope of work?
4. For a recommendation report, how did the author evaluate solutions to the problem?
5. Where can the reader find particular types of information in this report?

The subheadings here reflect the attempt to answer each of these questions. The "Background" subsection answers the first and second questions, "Purpose and Scope" addresses the third question, "Methodology" the fourth, and "Organization" the fifth. Many introductions do not address the fifth question, figuring that a reader can always look at the table of contents. Indeed, if all the "Organization" subsection does is repeat the report's headings, then the section is not necessary. It might be useful, however, to illustrate here what the rating procedure is or explain why the "Evaluation" section is broken down by criteria.

Notice that this section, like all sections in this report, makes liberal use of headings and lists to chunk similar information together and to allow the reader to find particular information quickly.

Wall Design Alternatives

After a brief description of the creation and control of airborne sound transmission, this section describes the wall designs evaluated by Firm Ideas for use in EBC apartment complexes.

Basics of Sound Control

Airborne sound is created from sources such as voices, televisions, and radios (McMullan, 1991). A source sound in one room puts sound energy into the air in the form of sound waves. When these sound waves contact a wall, the waves in the air are converted to vibrations in the wall structure. Then, the vibrations travel through the wall structure and the space in the wall cavity. When the vibrations emerge on the opposite side of the wall, the vibrations are converted back to sound waves, which create a sound in the adjoining room.

Airborne sound transmission through walls can be controlled using three methods:

1. Increasing the weight of the wall: Heavy and dense materials provide good sound control because the sound energy cannot easily vibrate heavy materials.

2. Using discontinuous wall framing: Separating the framing members provides good sound control because the sound vibrations cannot follow a direct path through the wall structure to the other side of the wall.

3. Placing sound-isolating materials in the wall: Adding sound-isolating materials in the wall provides good sound control because the flexible nature and fibrous structure of these materials isolate and absorb sound energy.

Wall Designs

Firm Ideas has evaluated six wall designs (common in apartment complexes) that use all three methods of sound control:

- Single-Stud Wall with Resilient Channel
- Single-Stud Wall with Sound-Deadening Board
- Staggered-Stud Wall
- Staggered-Stud Wall with Sound-Deadening Board
- Double-Stud Wall
- Double-Stud Wall with Sound-Deadening Board

The descriptions of the wall designs are given on the next two pages.

4

Single-Stud Wall with Resilient Channel

This wall design is framed with a single row of 2 × 4 studs spaced 16" apart along a 2 × 4 wood plate. A $\frac{5}{8}$"-thick gypsum board is nailed to one side of the wall, and the gypsum board on the other side is attached to the wood studs by a resilient metal channel. The dense gypsum board provides increased weight to the wall, and the resilient channel acts as a spring to isolate and absorb the vibrations in the studs. The absorption by the channel reduces the vibrations in adjoining gypsum board, thus reducing the sound transmission to the other side of the room. The fiberglass insulation in the wall cavity absorbs additional sound energy (McMullan, 1991; Simmons, 2001).

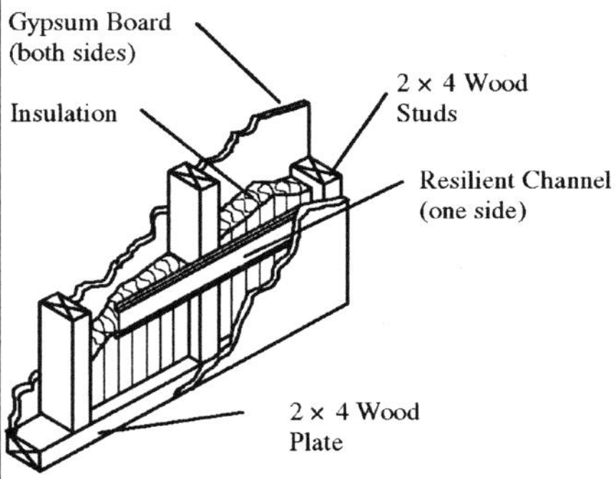

Single-Stud Wall with Sound-Deadening Board

This wall design is framed with a single row of 2 × 4 studs spaced 16" apart along a 2 × 4 wood plate. A ½" -thick sound- deadening board is adhesively applied to each side of the studs. Then, a $\frac{5}{8}$" -thick gypsum board is adhesively applied to the faces of the sound-deadening board. The dense gypsum board provides increased weight to the wall, and the sound-deadening board acts as an isolating material by absorbing vibrations in the studs. This vibration absorption by the sound-deadening board reduces the vibration sent to the opposite gypsum board. The fiberglass insulation in the wall cavity absorbs additional sound energy (McMullan, 1991; Simmons, 2001).

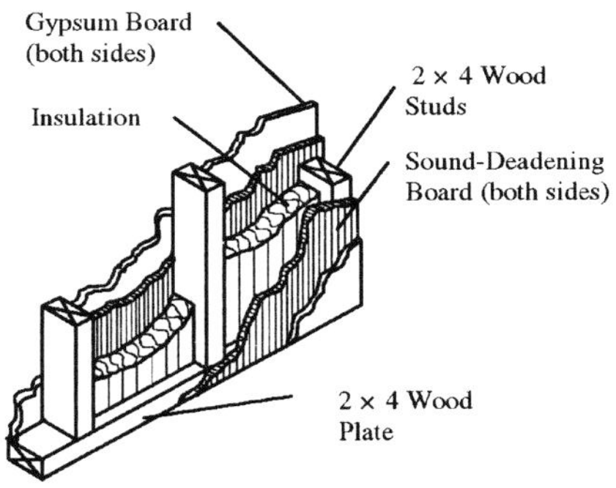

Staggered-Stud Wall

This wall design is framed with a single row of 2 × 4 studs staggered along a 2 × 6 wood plate. The staggered studs are placed every 8", so each $\frac{5}{8}$"-thick gypsum board is nailed to a stud every 16". The dense gypsum board provides increased weight to the wall, and the staggered pattern allows each gypsum board to be nailed to its own set of studs. This discontinuous construction reduces the direct transmission of sound vibration from one side of the wall to the other side of the wall. The fiberglass insulation in the wall cavity absorbs additional sound energy (McMullan, 1991; Simmons, 2001).

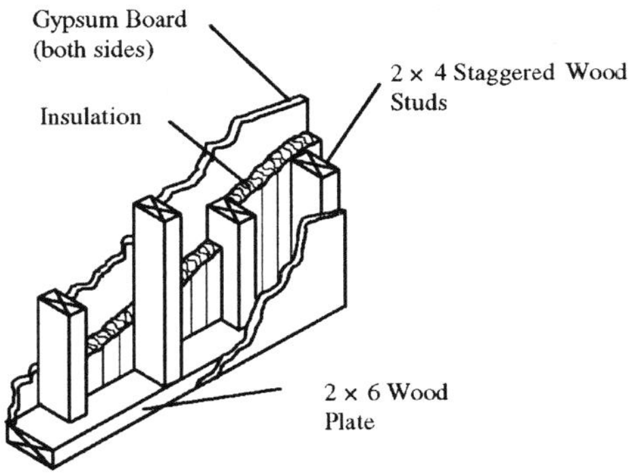

5

Staggered-Stud Wall with Sound-Deadening Board

This wall design is framed exactly the same as the *Staggered-Stud Wall*. However, ½"-thick sound-deadening board is adhesively applied to each side of the studs. Then, $\frac{5}{8}$"-thick gypsum board is adhesively applied to the faces of the sound-deadening board to increase the weight of the wall. The staggered studs provide resistance to sound transmission just as in the *Staggered-Stud Wall*, but the added sound-deadening board absorbs even more sound energy that may travel through the 2 × 6 wood plate or wall cavity. The fiberglass insulation in the wall cavity absorbs additional sound energy (McMullan, 1991; Simmons, 2001).

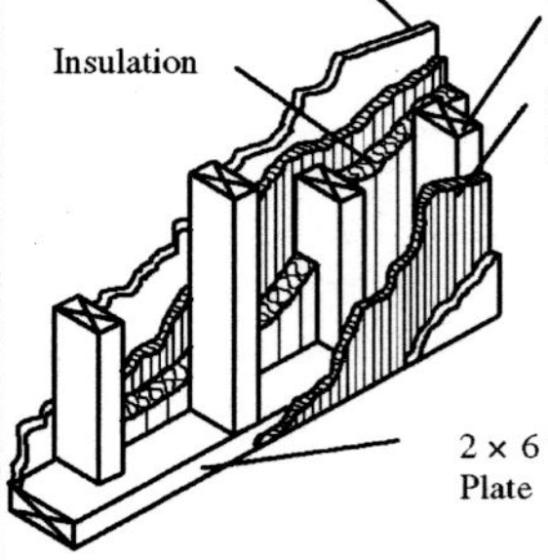

Double-Stud Wall

This wall design is framed with two rows of 2 × 4 wood studs placed on separate 2 × 4 wood plates. The studs in each row are spaced 16" along the plates. A $\frac{5}{8}$"-thick gypsum board is nailed to the outer face of each row of studs to increase the weight of the wall. Because the two rows of studs are completely unattached, the sound waves that vibrate one gypsum board do not directly move through the studs to the opposite gypsum board. This discontinuous construction reduces the direct transmission of sound vibration from one side of the wall to the other side of the wall. The insulation in the wall cavity absorbs additional sound energy (McMullan, 1991; Simmons, 2001).

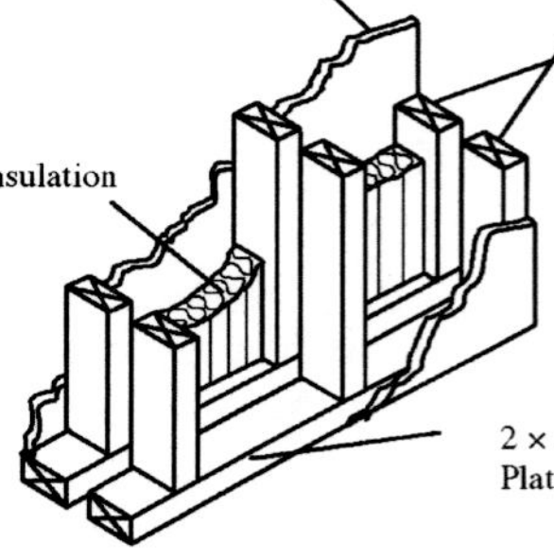

Double-Stud Wall with Sound-Deadening Board

This wall design is framed exactly the same as the *Double-Stud Wall*. However, ½"-thick sound-deadening board is adhesively applied to the outer face of each row of studs. Then, $\frac{5}{8}$"-thick gypsum board is adhesively applied to the faces of the sound-deadening board to increase the weight of the wall. The separated rows of studs provide resistance to sound transmission just as in the *Double-Stud Wall*, but the added sound-deadening board absorbs even more sound energy that may travel through the wall cavity. The insulation in the wall cavity absorbs additional sound energy (McMullan, 1991; Simmons, 2001).

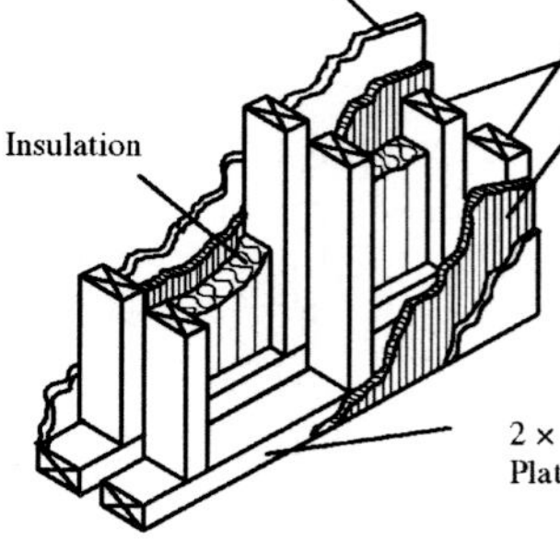

Source for all drawings: Simmons, Leslie. *Construction: Principles, Materials, and Methods*. 2001.

6
Summary of Wall Designs

The similarities and differences among the wall designs are given as follows (grouped by the framing type):

- Single-Stud Walls: Both the Single-Stud Wall with Resilient Channel and the Single-Stud Wall with Sound-Deadening Board are constructed with single-stud framing, gypsum board, and fiberglass insulation. However, the Single-Stud Wall with Resilient Channel uses a resilient metal channel to reduce the vibrations from airborne sound energy, while the Single-Stud Wall with Sound-Deadening Board uses sound-deadening board to reduce the vibrations.
- Staggered-Stud Walls: Both the Staggered-Stud Wall and the Staggered-Stud Wall with Sound-Deadening Board are constructed with staggered-stud framing, gypsum board, and fiberglass insulation. However, the Staggered-Stud Wall uses only the staggered studs to reduce the vibration transmission from airborne sound energy, while the Staggered-Stud Wall with Sound-Deadening Board uses both the staggered studs and sound-deadening board to reduce the vibration transmission.
- Double-Stud Walls: Both the Double-Stud Wall and the Double-Stud Wall with Sound-Deadening Board are constructed with double-stud framing, gypsum board, and fiberglass insulation. However, the Double-Stud Wall uses only a double row of studs to reduce the vibration transmission from airborne sound energy, while the Double-Stud Wall with Sound-Deadening Board uses both the double row of studs and sound-deadening board to reduce the vibration transmission.

BODY OF THE REPORT: THE CENTRAL SECTIONS

Wall Design Alternatives

The "Wall Design Alternatives" section presents and describes the wall systems that Firm Ideas chose as worthy of a more in-depth evaluation of sound-control effectiveness. This section expands upon the simple listing of alternatives given in the Executive Summary and Introduction. Here, the alternative walls are described in words and pictured in sketches. A departure from the norm is the treatment of illustrations as part of the text rather than as figures. The sketches have titles, but no figure numbers. The two pages of sketches look more like part of an instruction manual than a technical report. The decision not to number the figures was made because standard numbering and titling of six different figures would add unnecessary detail to a clear presentation of the alternatives. Before you make a similar decision, check with your boss or instructor. Notice that deciding not to number the sketches does *not* mean that you shouldn't cite their source; see the "Source for all drawings" citation at the bottom of the last sketch.

Like many of the initial sections of a technical report, this section provides some background on the technical concepts involved in the work reported. Airborne sound transmission and the most common methods of controlling it are not subjects that a real-estate development company would necessarily know much about, so the author provides some background. On the other hand, although whole books have been written on the subject, the client does not need to be educated on the full technical details of sound transmission or the nuances of controlling sound. An overview is sufficient here.

Likewise, at the end of this section, the author gives the reader a comparative summary of the wall designs presented separately on previous pages. Such comparative analyses are especially useful in evaluation reports and enable the reader to understand quickly the similarities and differences that will affect the ultimate recommendation. Notice the grouping of wall designs based on framing type, so that six wall designs end up in three bullet points. Chinking information like this helps the reader understand and remember concepts and facts.

Solution Criteria

7

Firm Ideas evaluated each wall design by using four solution criteria. EBC Properties set the performance, material cost, and labor cost criteria, and Firm Ideas added the mountable-object weight criterion. After the solution criteria were set, EBC Properties told Firm Ideas how much each criterion should influence the recommendation. The four solution criteria are given as follows in prioritized order, along with each criterion's influence on the recommendation:

1. **Performance**—constitutes 45% of the recommendation.
   Because the purpose of this project is to provide adequate sound control, EBC Properties stated that the performance of the wall design was the most important criterion. In this evaluation, performance was defined as how well each wall resisted airborne sound transmission. Therefore, Firm Ideas used the wall assembly's Sound Transmission Class (STC) rating to determine its sound transmission resistance. The STC values were obtained from *Construction: Principles, Materials, and Methods*, a building construction textbook (Simmons, 2001). An STC rating reflects the number of decibels a sound loses while passing through a given wall design; a better performing design produces a higher STC rating. All wall designs evaluated in this study have met the minimum STC rating of 50 decibels required by the *1997 Uniform Building Code—Section 1208.*

2. **Material Cost**—constitutes 30% of the recommendation.
   Like most apartment management companies, EBC Properties operates on a limited budget. Consequently, EBC Properties gave significant weight to the material cost of the wall designs. The material cost for each design was determined by the amount and type of the materials in that particular wall. Firm Ideas calculated the material cost by using RS Means's *2002 Building Construction Cost Data*, an industry-standard cost estimation book. This cost was measured in dollars per linear foot for an eight-foot-high wall.

8

3. **Labor Cost**—constitutes 15% of the recommendation.

   The next criterion set by EBC Properties was the labor cost of the wall designs. The labor cost was defined as how much each wall costs to install, based on the amount and type of labor needed to construct the wall in the city of Austin. (The labor cost in the city of Austin is approximately 27% cheaper than the national average labor cost.) Firm Ideas calculated the labor cost by using RS Means's *2002 Building Construction Cost Data.* The labor cost was measured in dollars per linear foot for an eight-foot-high wall.

4. **Mountable-Object Weight**—constitutes 10% of the recommendation.

   After construction is complete, the residents may use the common wall for mounting objects such as cabinets and fixtures; however, these objects affect the sound performance of some wall designs (USG, 2003). The mountable-object weight was defined as the maximum mountable weight each wall design can support without a decrease in sound performance. Firm Ideas obtained the maximum mountable weight from manufacturer specifications and recommendations from on-line construction articles.

BODY OF THE REPORT: THE CENTRAL SECTIONS

Solution Criteria

For a feasibility report, the "Solution Criteria" section constitutes the heart of the evaluation of various solutions to the problem or need being investigated. Only by setting valid, useful criteria can an investigator make technically correct and economically and socially feasible choices. Thus, defining, justifying, and prioritizing these criteria becomes critical in a feasibility investigation. This report not only prioritizes the evaluation criteria, but also provides a scale by which to measure exactly how important each criterion is to the final decision. Notice that the author is careful to point out that these criteria and their relative importance have been agreed to by the client.

If criteria are going to be used to measure the feasibility of various solutions, they must themselves be able to be measured. If cost, for instance, is a criterion for choosing something—a birthday present for your mother, let us say—then you have to know the costs of various possible presents. And those costs must all be in a common currency—U.S. dollars, let us say. In addition, you need to look at possible hidden costs of the candidate items—warranties or maintenance, for example. Notice that for this report on sound-controlling wall systems, cost is separated into material cost and labor cost. Both those costs are measured in dollars per linear foot, but the first is calculated on the basis of the amount and type of material and the second on the time to construct. In addition, performance is specifically defined as the ability to resist airborne sound transmission, measured by the wall's STC rating.

For all criteria, the sources of information that were used for comparative measurement are specifically cited in parentheses at the end of the sentence containing the fact, finding, or quotation from that source. Different engineering journals and professional societies use somewhat different referencing and citation systems. In some systems, a superscript number is used at the end of the sentence. This number refers to an item in a list at the end of the report and signals that the sources are listed in the order they were first cited in the text. In other referencing systems, sources are cited in the text by author and year and are listed at the end of the report in alphabetical order. Since this report is on a civil- or architectural-engineering subject, the author uses the citation and referencing system required by ASCE (American Society of Civil Engineering), which uses author–date citation and an alphabetized list at the end of the report. Each branch of engineering has its own requirements for referencing, which can usually be found on the branch's website. See the ASCE requirements at http://www.pubs.asce.org/authors/index.html#ref.

Evaluation of Wall Designs

Firm Ideas evaluated six wall designs, using the four solution criteria established by EBC Properties and Firm Ideas. The six wall designs are given as follows, along with their abbreviations used in this section:

- Single-Stud Wall with Resilient Channel (SSWw/RC)
- Single-Stud Wall with Sound-Deadening Board (SSWw/SDB)
- Staggered-Stud Wall (StSW)
- Staggered-Stud Wall with Sound-Deadening Board (StSWw/SDB)
- Double-Stud Wall (DSW)
- Double-Stud Wall with Sound-Deadening Board (DSWw/SDB)

The following subsections (organized by the solution criteria) contain the results from our evaluation.

Performance

According to Simmons (2001), the DSWw/SDB has the highest STC value of 58 decibels (dB) and therefore has the best performance. The StSWw/SDB, DSW, and SSWw/SDB provide moderate performance, with STC values of 54, 53, and 52 dB, respectively. Finally, the SSWw/RC and the StSW have the worst performance, with STC values of 51 and 50 dB, respectively. These STC values are shown in Figure 1.

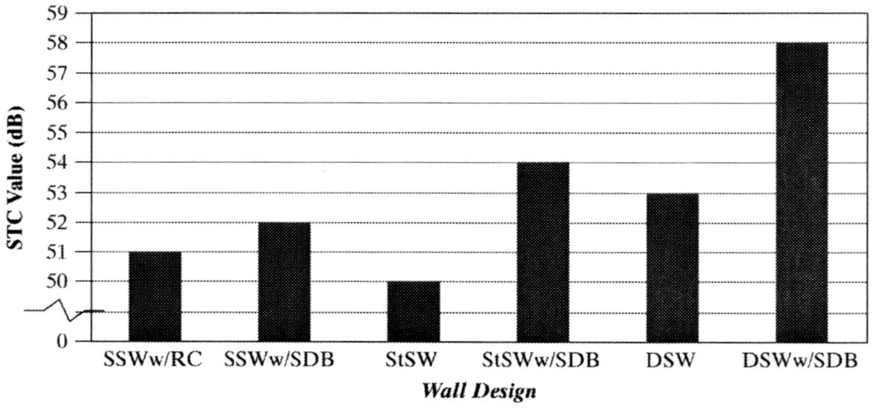

Figure 1. Performance: STC Values for Each Wall Design.
Source of data: Simmons, L., 2001. *Construction: Principles, Materials, and Methods.*

10
Material Cost

According to our calculations, the DSWw/SDB has the highest material cost of $16.38 per linear foot. The StSWw/SDB and the SSWw/SDB have moderate material costs of $14.93 and $13.35 per linear foot, respectively. The SSWw/RC, the DSW, and the StSW have relatively low material costs of $11.72, $11.98, and $10.53 per linear foot, respectively. All material costs are for an eight-foot-high wall. The material cost calculations we developed are shown in Appendix A, and the material costs are summarized in Figure 2.

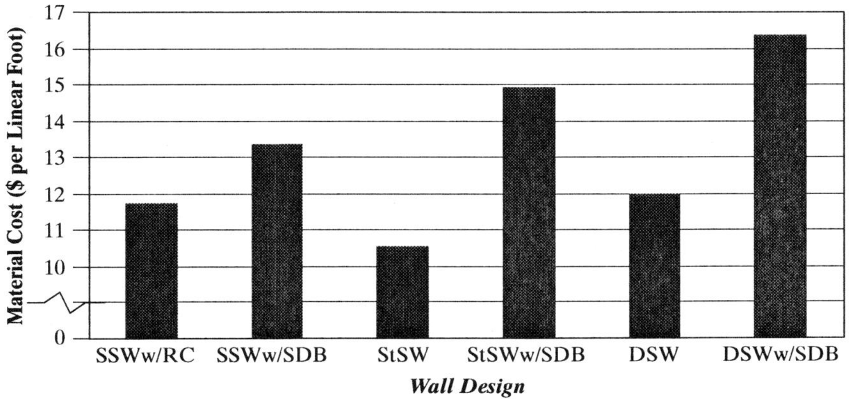

Figure 2. Material Cost per Linear Foot for Each Eight-Foot-High Wall Design.
Cost computed using RS Mean's *2002 Building Construction Cost Data.*

Labor Cost

The labor costs follow a pattern similar to that of the material costs. The DSWw/SDB has the highest labor cost of $13.02 per linear foot. The StSWw/SDB and the SSWw/SDB have moderate labor costs of $11.22 and $10.34 per linear foot, respectively. The SSWw/RC, the DSW, and the StSW have relatively low labor costs of $8.74, $8.58, and $6.93 per linear foot, respectively. All labor costs are for an eight-foot-high wall. The labor cost calculations are given in Appendix A, and the labor costs are summarized in Figure 3.

11

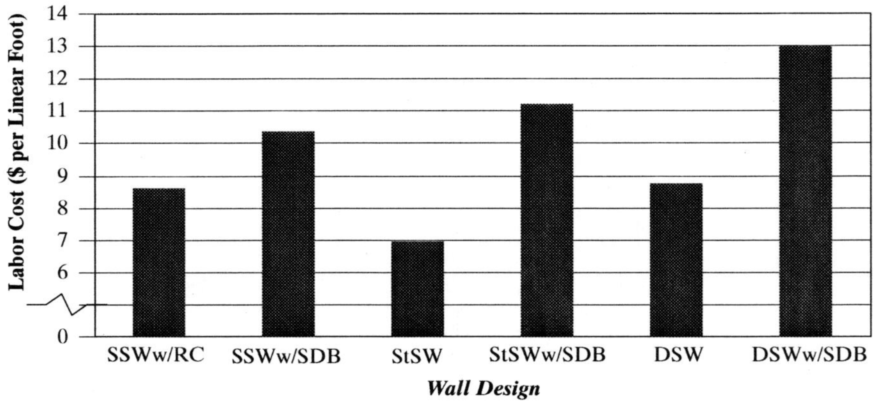

Figure 3. Labor Cost per Linear Foot for Each Eight-Foot-High Wall Design.
Cost computed with RS Means's *2002 Building Construction Cost Data*.

Mountable-Object Weight

According to Simmons (2001), the DSW and StSW have the highest mountable-object weight of 110 pounds per linear foot. The SSWw/RC has the next highest mountable-object weight of 67 pounds per square foot. The mountable-object weight for the SSWw/RC is less than for the DSW and StSW because a large amount of weight prevents the resilient channel from flexing and absorbing wall vibrations (Simmons, 2001). The SSWw/SDB, StSWw/SDB, and DSWw/SDB are the designs with the lowest mountable-object weights of 50 pounds per linear foot. These three designs have the lowest weight because when the sound-deadening board becomes overly compressed, the board loses its ability to flex and absorb wall vibrations (Georgia Pacific, 2003a). The mountable-object weights for each design are shown in Figure 4.

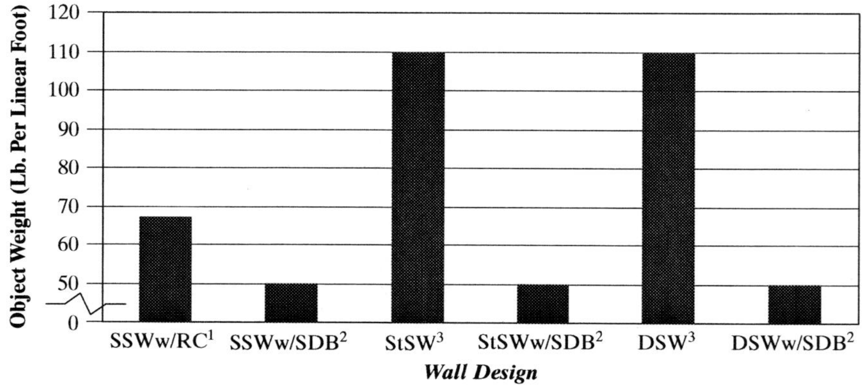

Figure 4. Maximum Mountable-Object Weight: Mountable-Object Weight (Pounds per Linear Foot) Each Wall Design Can Support.
Sources of data: 1. Handyman Club of America, 2003;
2. Georgia Pacific, 2003a;
3. Georgia Pacific, 2003b.

BODY OF THE REPORT: THE CENTRAL SECTIONS

Evaluation of Wall Designs

As you would suspect, the evaluation section contains a number of charts comparing the alternative wall solutions according to various criteria. Describing the findings in words *and* charting the results graphically helps the reader wade through a lot of detail and understand the gist of the comparisons. We can see quickly, for instance, that the wall system presented on the far right of the charts, the Double-Stud Wall with Sound-Deadening Board (DSWw/SDB), is the highest in all criteria except the last, mountable-object weight. Being the highest is not always a good thing, of course, especially when it comes to cost. So, we see in a nutshell that high performance comes at a high cost and that performance as defined in terms of STC rating does not include the capability of having heavy objects mounted on the wall. We do not even have to read the narrative of the findings to suspect that the DSWw/SDB will *not* be the recommended choice.

Notice that the sources of the *data* represented in each figure are listed underneath each figure's title. There is some disagreement on how detailed figure-data references need to be. If you have full publication information included in the list of references at the end of your report, you should not have to worry about providing details in the citation; for ASCE standards, the author and date should suffice. In citing the sources of cost estimates, however, this author felt that the full citation—RS Means's *2002 Building Construction Cost Data*—was a good idea.

Notice the break marks on the ordinate (the y-axis) of Figures 1–4. This mark is a responsible way of showing that the scale has been distorted. The interval from 0 to 50 decibels is a lot more compressed than the intervals between 50 and 51, 51 and 52, etc. Starting both axes at zero is a good idea, if you can possibly do it, but the values here all fall between 51 and 58, so there is no point in taking up a lot of room on the graph to show the scale below that range. Some graphing programs let you create a break mark, and some do not. Even if you cannot make one, it is still a good idea to start both axes at zero.

12

Factual Summary and Rating Procedure

Factual Summary

After evaluating the six wall designs, Firm Ideas reached the factual conclusions shown in Tables 1–4. The designs are ordered from best to worst for each solution criterion. A narrative summary of these conclusions follows Tables 1–4.

Table 1. Wall Design Performance

Wall Design	STC (dB)
Double-Stud Wall with Sound-Deadening Board	58
Staggered-Stud Wall with Sound-Deadening Board	54
Double-Stud Wall	53
Single-Stud Wall with Sound-Deadening Board	52
Single-Stud Wall with Resilient Channel	51
Staggered-Stud Wall	50

Best Design in Criterion

Decreasing Performance

Worst Design in Criterion

Source: Simmons, Leslie. *Construction: Principles, Materials, and Methods*. 2001.

Table 2. Wall Design Material Cost

Wall Design	Material Cost ($ per Linear Ft)
Staggered-Stud Wall	10.53
Single-Stud Wall with Resilient Channel	11.72
Double-Stud Wall	11.98
Single-Stud Wall with Sound-Deadening Board	13.35
Staggered-Stud Wall with Sound-Deadening Board	14.93
Double-Stud Wall with Sound-Deadening Board	16.38

Best Design in Criterion

Increasing Material Cost

Worst Design in Criterion

Cost computed with RS Means's 2002 *Building Construction Cost Data.*

13

Table 3. Wall Design Labor Cost

Wall Design	Labor Cost ($ per Linear Ft)
Staggered-Stud Wall	6.93
Single-Stud Wall with Resilient Channel	8.58
Double-Stud Wall	8.74
Single-Stud Wall with Sound-Deadening Board	10.34
Staggered-Stud Wall with Sound-Deadening Board	11.22
Double-Stud Wall with Sound-Deadening Board	13.02

Best Design in Criterion

Increasing Labor Cost

Worst Design in Criterion

Cost computed with RS Means's 2002 *Building Construction Cost Data.*

Table 4. Wall Design Maximum Mountable-Object Weight

Wall Design	Maximum Mountable-Object Weight (Lb per Linear Ft)
Double-Stud Wall	110
Staggered-Stud Wall	110
Single-Stud Wall with Resilient Channel	67
Single-Stud Wall with Sound-Deadening Board	50
Staggered-Stud Wall with Sound-Deadening Board	50
Double-Stud Wall with Sound-Deadening Board	50

Best Design in Criterion

Decreasing Mountable-Object Weight

Worst Design in Criterion

Sources: Handyman Club of America; Georgia Pacific.

In the performance criterion, the Double-Stud Wall with Sound-Deadening Board was the best design, and the Staggered-Stud Wall was the worst design. Regarding material cost, the Staggered-Stud Wall was the best design, and the Double-Stud Wall with Sound-Deadening Board was the worst design. For labor cost, the Staggered-Stud Wall was the best design, and the Double-Stud Wall with Sound-Deadening Board was the worst design. In the mountable-object weight criterion, the Double-Stud Wall and Staggered-Stud Wall were the best designs, and the Single-Stud Wall with Sound-Deadening Board, Staggered-Stud Wall with Sound-Deadening Board, and Double-Stud Wall with Sound-Deadening Board were the least desirable designs.

14

Rating Procedure

After evaluating all the wall designs, Firm Ideas determined the best wall design, using a rating procedure. This procedure is outlined in the following steps:

1. Firm Ideas entered the information from each solution criterion into Tables 1–4 on the previous two pages. These tables order the designs from best to worst in each criterion.
2. For each solution criterion, we determined the number of points for each design:
    a. The best design received 10 points.
    b. The worst design or designs received 1 point each.
    c. The point values for the remaining designs were determined using linear interpolation. The linear interpolation for the performance criterion is given in Figure 5 as an example:
        i. A line was drawn from point (50,1) to point (58,10). This is the line in Figure 5.
        ii. A line was drawn from each design's STC value to the sloping line (resulting in the vertical arrows in Figure 5).
        iii. A line was drawn from the intersection of each vertical arrow and the sloping line to the *y*-axis (horizontal arrows). The point where the design's horizontal arrows intersect the *y*-axis is the design's point value.

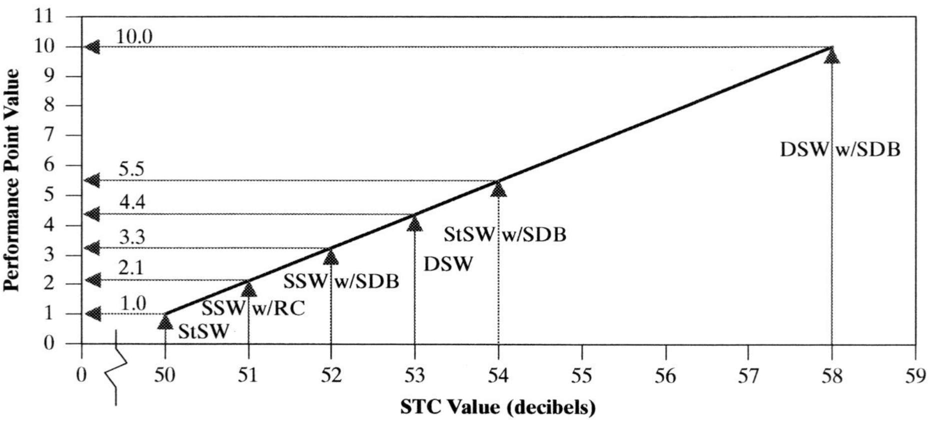

Figure 5. Example of Linear Interpolation—Performance. In this figure, the points are calculated for the performance criterion.

3. Using the point values determined from Step 2 and the solution-criteria weights determined by EBC Properties, Firm Ideas calculated the Total Weighted Points for each wall design as described in the "Conclusions" section.

BODY OF THE REPORT: THE CENTRAL SECTIONS

Factual Summary and Rating Procedure

Since the previous section presented so much numerical data, and because the alternative wall systems are hard to keep straight (whether as acronyms or long tongue-tying phrases), the author has included a "Factual Summary and Rating Procedure" section that gives the highlights of the findings in a visual form that makes misunderstanding impossible. As we noted in the last section, sometimes the "highest" bar in a bar chart means that the corresponding component performed well in that category, while other times it signifies the opposite. By showing how the wall systems are ordered from best to worst, each table presented in this section clarifies the comparisons for even the most inattentive reader.

The last subsection then describes the procedure whereby the wall systems were rated according to the importance of each category, or criterion. Linear interpolation is a common concept for engineers, but not for less technical people, so the interpolation is explained both in words and in Figure 5.

Together, these two subsections prepare us for the final sections, "Conclusions" and "Recommendation." We certainly know which wall "won" in the performance criterion, but the rating procedure prepares us to factor in a lot of other information before coming to conclusions.

Conclusions

For this study on sound-controlling wall systems, the six designs were compared according to four criteria and rated according to the procedure outlined in the previous section of this report. Firm Ideas calculated the Total Weighted Points for each wall design by using Equation (1).

$$\text{Total Weighted Points} = (PP \times 0.45) + (MCP \times 0.30) + (LCP \times 0.15) + (MOP \times 0.10) \quad (1)$$

PP is the Performance Point Value determined from Step 2;
MCP is the Material Cost Point Value determined from Step 2;
LCP is the Labor Cost Point Value determined from Step 2;
MOP is the Mountable-Object Weight Point Value determined from Step 2.

The calculations for the Total Weighted Points are shown in Table 5. Please see the tables in the "Factual Summary and Rating Procedure"for an explanation of the wall-system acronyms.

Table 5. Total-Weighted-Points Calculations Using Equation (1).

Wall Designs	PP	× Weight +	MCP	× Weight +	LCP	× Weight +	MOP	× Weight =	Total Weighted Points
SSWw/RC	1.0	× 0.45 +	8.2	× 0.30 +	7.6	× 0.15 +	3.6	× 0.10 =	**4.9**
SSWw/SDB	2.1	× 0.45 +	5.7	× 0.30 +	5.0	× 0.15 +	1.0	× 0.10 =	**4.0**
StSW	3.3	× 0.45 +	10.0	× 0.30 +	10.0	× 0.15 +	10.0	× 0.10 =	**6.0**
StSWw/SDB	4.4	× 0.45 +	3.2	× 0.30 +	3.7	× 0.15 +	1.0	× 0.10 =	**4.1**
DSW	5.5	× 0.45 +	7.8	× 0.30 +	7.3	× 0.15 +	10.0	× 0.10 =	**6.4**
DSWw/SDB	10.0	× 0.45 +	1.0	× 0.30 +	1.0	× 0.15 +	1.0	× 0.10 =	**5.1**

The Double-Stud Wall (DSW) had the most Total Weighted Points, and the Single-Stud Wall with Sound-Deadening Board (SSWw/SD) had the fewest Total Weighted Points. For a better visual comparison, the information in Table 5 is also given in Figure 6. This figure shows the breakdown of the Total Weighted Points for each wall design.

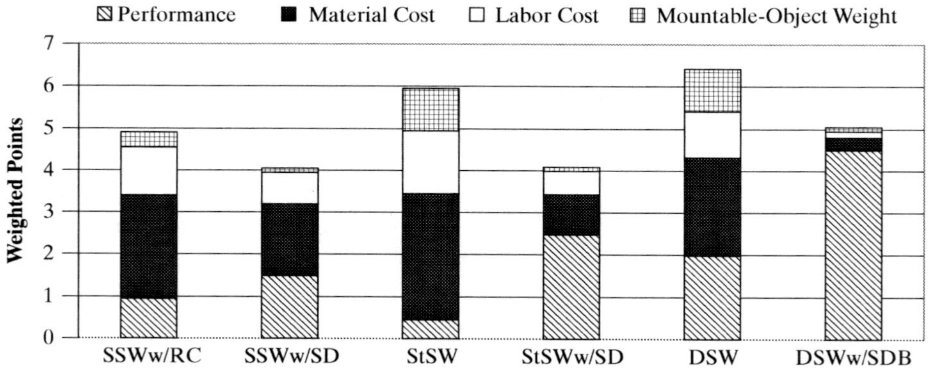

Figure 6. Breakdown of Total Weighted Points. This chart shows the breakdown of the total weighted points by solution criterion for each wall design.

BODY OF THE REPORT: CONCLUSIONS

At first glance, you might think that a conclusions section is unnecessary, given the factual summary presented just previously. Conclusions go further, however, than to simply present factual findings; they interpret and judge those findings. In this report, the interpretation is quantified by a rating system that applied a certain level of importance to each criterion and then assigned points to each wall system on the basis of relative importance. The point system is designed to give a total picture of how each wall rates across the board, including all categories. The interpretation is first given numerically (the equation and Table 5), then in narrative form (*X* is best, and *Y* is worst), and then in visual terms (Figure 6). The equation used to calculate the Total Weighted Points is explained clearly and should present no problem for any high school–educated reader. Figure 6 includes a caption ("This chart shows. . . .") as a way of stressing quickly for all readers what is significant about the information shown on the chart. Like the executive summary and the introduction, conclusions should be written with a wide audience in mind.

Often, the conclusions are given along with the recommendation in a section called "Conclusions and Recommendations." Certainly, in this report, such a combined section would make sense. But separating the recommendation out allows for a greater focus on exactly what action is being recommended and avoids the tendency to weaken the recommendation by hiding it in the middle of a section or by using lots of passive voice.

16

Recommendation

After evaluating the six wall designs and following the recommendation procedure, Firm Ideas recommends that EBC Properties use the Double-Stud Wall as the common wall in their new complexes. This wall design provides moderate performance (STC value of 53 decibels) at a relatively low material cost and labor cost ($11.98 per linear foot and $8.74 per linear foot, respectively, for an eight-foot-high wall). This cost was the third lowest out of the six wall designs investigated. In addition, the Double-Stud Wall can support the highest mountable-object weight: 110 pounds per linear foot. Our evaluation reveals that this wall design is the best value for EBC Properties and gives their tenants the best combination of benefits: increased sound control and a generous weight allowance for mounting cabinets and fixtures. This wall design best meets the needs of EBC Properties.

BODY OF THE REPORT: RECOMMENDATION

The recommendation answers the client's most important questions: What should be done to solve the problem, and how much will it cost to do it? The recommendation should reveal good knowledge of the client's concerns, and the language should be straightforward and firm. If you do not feel quite sure about your recommended action, you should not make a recommendation, but rather say that further study is needed or that none of the solutions is feasible at present. Sometimes, recommendations can be broken down into short-term (sometimes called Band-Aid® solutions) and long-term recommendations. The implementation of long-term solutions often has to wait for adequate funding.

List of References

17

Bergman, P. (2002). "Soundproofing." *Ultra High Fidelity Magazine*. Issue 63. http://www.uhfmag.com/Issue63/soundproofing.html (10 March 2003).

Georgia-Pacific Co. "Georgia-Pacific Tough Rock Sound Deadening Board." http://www.gp.com/gypsum/gypsumboard/submittal/sounddeadening.html (1 April 2003).

Georgia-Pacific Co. "Georgia-Pacific Tough Rock Gypsum Board." http://www.gp.com/gypsum/gypsumboard/submittal/gypsumboard.html (1 April 2003).

Handyman Club of America. "Quiet." http://www.handymanclub.com/document.asp (1 April 2003).

Harris, C. (1991). *Handbook of Acoustical Measurements and Noise Control*. 3d ed. New York: McGraw-Hill.

International Conference of Building Officials. *1997 Uniform Building Code*. Volume 1. Whitter, CA: IBCO.

McMullan, R. (1991). *Noise Control in Buildings*. Boston: BSP Professional.

RS Means. *2002 Building Construction Cost Data*. 60th ed. Kingston, MA: RS Means.

Simmons, L. (2001). *Construction: Principles, Materials, and Methods*. 7th ed. New York: John Wiley & Sons.

Thorburn Associates, Inc. "Resilient Channels: How Metal Absorbs Sound Waves." http://www.ta-inc.com/newsletter/rslchnls.htm (14 March 2003).

USG Manufacturers. "Sound Construction." http://www.usg.com/Design_Solutions/2_3_sound_construct.asp (3 March 2003).

BACK MATTER: LIST OF REFERENCES

The "List of References" section (sometimes called "Works Cited") provides evidence that you thoroughly researched your subject and made use of creditable sources of information. Not all engineering reports contain references. A structural design report, for instance, may justify the design it presents on the basis of the structure's and the site's constraints alone, without referring to other designs published other places. For most university courses and for many feasibility reports, however, you want to show that you have gathered and synthesized good information in the process of evaluating engineering solutions.

Why is this list called "List of References" rather than "Works Cited"? Well, the title "Works Cited" means exactly what it says: All the material included in the list is cited somewhere in the text of the report. Thus, for example, for the eighth item on the list—RS Means's *2002 Building Construction Cost Data*—there are numerous citations in the body of the report, the first one appearing in the "Methodology" subsection of the Introduction. On the other hand, naming the list "List of References" means that it is more of a bibliography; you consulted each and every source on the list, but you may not have cited every one in your text. Thus, the article by P. Bergman, "Soundproofing," in the on-line *Ultra High Fidelity Magazine*, is not cited specifically in the body of this report; apparently, the author read this piece for background information but did not use in his or her report any specific fact, finding, or quotation presented by the article. In fact, three of the sources listed here are not referred to in the text (Bergman, Harris, and Thorburn).

For electronic sources, notice the date in parentheses at the end of the citation. This is the date the report author accessed the web site. If the site does not list a date this access date will be the only date in the citation. Since web sites can came and go, this date may provide the only proof of a particular electronic publication's existence.

The style of referencing in this list follows the ASCE guidelines for authors, found on the ASCE website http://www.pubs.asce.org/authors/index.html. ASCE requirements make sense here, since this is an architectural- and civil-engineering report. Make sure to check with your boss or instructor about which style of referencing is conventionally used in your field, class, or company. Often, larger companies (such as IBM) will have their own style guides for everything from abbreviations to reference lists.

A-1

Appendix A: Material and Labor Cost Calculations

Table A-1. Material Cost Calculations

Wall Design	Material	Cost per Unit of Material (dollars)	×	Amount of Material per Linear Foot	×	Austin Location Factor	=	Material Cost per Linear Foot
Single-Stud Wall with Resilient Channel	$\frac{5}{8}$" gypsum board	0.22 per SF	×	16 SF	×	0.948	=	3.34
	2 × 4 Wood Studs (16" o.c.) + plates	3.20 per LF	×	1 LF	×	0.948	=	3.03
	3.5" Insulation	0.34 per SF	×	8 SF	×	0.948	=	2.58
	Resilient Channel	0.73 per LF	×	4 LF	×	0.948	=	2.77
						Total =		**11.72**
Single-Stud Wall with Sound-Deadening Board	$\frac{5}{8}$" gypsum board	0.22 per SF	×	16 SF	×	0.948	=	3.34
	2 × 4 Wood Studs (16" o.c.) + plates	3.20 per LF	×	1 LF	×	0.948	=	3.03
	3.5" Insulation	0.34 per SF	×	8 SF	×	0.948	=	2.58
	Sound D. Board	0.29 per SF	×	16 SF	×	0.948	=	4.40
						Total =		**13.35**
Staggered-Stud Wall	$\frac{5}{8}$" gypsum board	0.22 per SF	×	16 SF	×	0.948	=	3.34
	2 × 4 Wood Studs (8" o.c.) + plates	4.87 per LF	×	1 LF	×	0.948	=	4.62
	3.5" Insulation	0.34 per SF	×	8 SF	×	0.948	=	2.58
						Total =		**10.53**
Staggered-Stud Wall with Sound-Deadening Board	$\frac{5}{8}$" gypsum board	0.22 per SF	×	16 SF	×	0.948	=	3.34
	2 × 4 Wood Studs (8" o.c.) + plates	4.87 per LF	×	1 LF	×	0.948	=	4.62
	3.5" Insulation	0.34 per SF	×	8 SF	×	0.948	=	2.58
	Sound D. Board	0.29 per SF	×	16 SF	×	0.948	=	4.40
						Total =		**14.93**
Double-Stud Wall	$\frac{5}{8}$" gypsum board	0.22 per SF	×	16 SF	×	0.948	=	3.34
	2 × 4 Wood Studs (16" o.c.) + plates	6.40 per LF	×	1 LF	×	0.948	=	6.07
	3.5" Insulation	0.34 per SF	×	8 SF	×	0.948	=	2.58
						Total =		**11.98**
Double-Stud Wall with Sound-Deadening Board	$\frac{5}{8}$" gypsum board	0.22 per SF	×	16 SF	×	0.948	=	3.34
	2 × 4 Wood Studs (16" o.c.) + plates	6.40 per LF	×	1 LF	×	0.948	=	6.07
	3.5" Insulation	0.34 per SF	×	8 SF	×	0.948	=	2.58
	Sound D. Board	0.29 per SF	×	16 SF	×	0.948	=	4.40
						Total =		**16.38**

A-2

Table A-2. Labor Cost Calculations

Wall Design	Material	Cost per Unit of Material (dollars)	×	Amount of Material per Linear Foot	×	Austin Location Factor	=	Labor Cost per Linear Foot
Single-Stud Wall with Resilient Channel	$\frac{5}{8}$" gypsum board	0.24 per SF	×	16 SF	×	0.67	=	2.57
	2 × 4 Wood Studs (16" o.c.) + plates	4.00 per LF	×	1 LF	×	0.67	=	2.68
	3.5" Insulation	0.15 per SF	×	8 SF	×	0.67	=	0.80
	Resilient Channel	0.94 per LF	×	4 LF	×	0.67	=	2.52
						Total	=	**8.58**
Single-Stud Wall with Sound-Deadening Board	$\frac{5}{8}$" gypsum board	0.24 per SF	×	16 SF	×	0.67	=	2.57
	2 × 4 Wood Studs (16" o.c.) + plates	4.00 per LF	×	1 LF	×	0.67	=	2.68
	3.5" Insulation	0.15 per SF	×	8 SF	×	0.67	=	0.80
	Sound D. Board	0.40 per SF	×	16 SF	×	0.67	=	4.29
						Total	=	**10.34**
Staggered-Stud Wall	$\frac{5}{8}$" gypsum board	0.24 per SF	×	16 SF	×	0.67	=	2.57
	2 × 4 Wood Studs (8" o.c.) + plates	5.30 per LF	×	1 LF	×	0.67	=	3.55
	3.5" Insulation	0.15 per SF	×	8 SF	×	0.67	=	0.80
						Total	=	**6.93**
Staggered-Stud Wall with Sound-Deadening Board	$\frac{5}{8}$" gypsum board	0.24 per SF	×	16 SF	×	0.67	=	2.57
	2 × 4 Wood Studs (8" o.c.) + plates	5.30 per LF	×	1 LF	×	0.67	=	3.55
	3.5" Insulation	0.15 per SF	×	8 SF	×	0.67	=	0.80
	Sound D. Board	0.40 per SF	×	16 SF	×	0.67	=	4.29
						Total	=	**11.22**
Double-Stud Wall	$\frac{5}{8}$" gypsum board	0.24 per SF	×	16 SF	×	0.67	=	2.57
	2 × 4 Wood Studs (16" o.c.) + plates	8.00 per LF	×	1 LF	×	0.67	=	5.36
	3.5" Insulation	0.15 per SF	×	8 SF	×	0.67	=	0.80
						Total	=	**8.74**
Double-Stud Wall with Sound-Deadening Board	$\frac{5}{8}$" gypsum board	0.24 per SF	×	16 SF	×	0.67	=	2.57
	2 × 4 Wood Studs (16" o.c.) + plates	8.00 per LF	×	1 LF	×	0.67	=	5.36
	3.5" Insulation	0.15 per SF	×	8 SF	×	0.67	=	0.80
	Sound D. Board	0.40 per SF	×	16 SF	×	0.67	=	4.29
						Total	=	**13.02**

BACK MATTER: APPENDICES

How do you decide what information to put in an appendix, as opposed to the body of the report? Well, the best way to decide is to ask yourself these two questions:

- Is all of this information critical for the reader to have at this point (in the body of the report)?
- Can I summarize the information in the body of the report and then put the details in an appendix, where they will not interrupt the reader?

If the answer to the first question is "yes," you do not want to relegate the information to an appendix, even if the answer to the second question is "yes." If the answer to the first question is "no," then definitely go ahead and put the information in an appendix. You can almost always summarize information. If you cannot, then go back to the first question; if the answer is still "no," then just refer the reader to the appendix. In any case, always remember to direct the reader to each appendix at the appropriate place in the report.

Typical kinds of information suitable for an appendix include large tables; complex calculations (more than a couple of equations); large or complex schematics, plans, or drawings; large maps; and raw data of any kind. Group similar kinds of data (e.g., calculations, photographs, survey results, rules, and requirements) in one appendix, and give each appendix a letter (A, B, etc.) and a meaningful title that indicates what sort of information it contains. The appendix here has the helpful title "Appendix A: Material and Labor Cost Calculations."

Normally, appendices are not paginated as part of the body of the report; rather, each appendix is paginated separately: A-1, A-2, B-1, B-2, etc.

SUMMARY

- Making the design and format of a document consistent enhances the document's readability. Conversely, inconsistent headings, illogical spacing, and awkward integration of graphics in the text all lead readers to misunderstand or outright ignore your writing.
- Study models of well-presented information, but never copy them slavishly. Notice the type of information presented in each section of the sample document, and then decide whether your information fits that type.

Problems

1. Take the last homework problem you did for any engineering or science class and imagine that you have to explain the problem and its solution in a letter to the Director of Human Resources at an engineering company. The director wants the letter because he wants to see whether you can communicate technical information in a readable style. Use any graphic(s) you want but remember that the director has a business rather than an engineering degree. Use business letter format.
2. Write a proposal to your instructor describing a plan to study improvements to the parking situation for students at your school. Research and ponder some likely alternatives for creating more parking spaces and/or better traffic flow in and around campus. Include these items in a proposal of 2–4 pages:
    a. Clear statement of traffic problem and its causes
    b. Presentation of a couple of possible solutions you will investigate
    c. A plan for completing the research in time to make a recommendation to your instructor by the end of the semester. What specific tasks do you need to accomplish in order to define the problem (observation is an important task!) and research possible solutions. Your *final* report will include suggestions for how best to alleviate the traffic problem, but this proposal presents a plan of study, *not* a recommendation.
    d. A schedule for completing the research and writing tasks. Assume that your instructor will want a final recommendation report.
    e. A short list of your qualifications for doing this investigation.
3. Write a paragraph describing which type and size font you use for writing print documents of more than one page. Explain the reasons for your choices.
4. For your next writing assignment, fill in the Style Sheet Checklist in Figure 4 (ignore any items that do not apply to the assignment). Show the checklist to your instructor.
5. True or false? Full justification of text always looks more professional than a ragged right margin.
6. True or false? Letters and memos are generally single-spaced.
7. A report's letter of transmittal is written primarily to whom?

    The head of the department most affected by the report
    The person who commissioned the study reported in the document
    The CEO of the company who commissioned the report

8. The summary of a document is usually about what percentage of the length of the whole document?

   25%
   10%
   33%

9. Which is the best reason for creating a separate report section for any recommendations?

   to highlight the recommendations
   to highlight the conclusions
   to strengthen the content of the report

10. The style of referencing used in most engineering journals and reports is based on which *one* of the following sources:

   The Modern Language Association
   The American Chemical Society
   The Chicago Manual of Style
   The Society for Technical Communication

FURTHER READING

Chicago Manual of Style, 15th ed., 2003. University of Chicago Press: Chicago.

This is the mothership of style manuals. This big volume contains more than you'll ever want to know about manuscript preparation, documentation, and the publishing process, but it will also have definitive answers for you on all those niggling little questions about style (do you write out the number twelve or use the Arabic numeral?). The sections on "Grammar and Usage" and "Punctuation" are a terrific reference tool for engineers. Most discipline-specific style manuals in engineering are based on the *Chicago Manual*. See the "Documentation" section for a clear, if lengthy, demonstration of the differences between APA and MLA styles. Since you may have been taught the latter in high school, you will want to acquaint yourself with APA, now that you are in a technical field.

Walker, J.R. and Taylor, T. 1998. *The Columbia Guide to Online Style*. Columbia University Press: New York.

This reference book organizes and categorizes the new standards for citing electronic documents. It also includes sections on how to create documents electronically for print publication and how to format documents for online publication.

14

Project Management

This chapter is devoted to the subject of project management. As a student of engineering, you may ask, "Why is this topic so important?" The answer lies in the nature of design problems. Design problems are inherently open ended; the more open ended the task, the greater the need for a management infrastructure to ensure that elements of the project ultimately fit together. Accomplishing this goal requires that all teams and team members work together with a common strategy. Sometimes, managing a project requires only the most basic application of common sense: get things done on time, hold regular meetings to discuss progress, write things down so you don't forget them. At other times, a more formal structure is required; this formal structure often relies on time-tested methods of project management. A good engineer must know how to work within the framework of a team, keep a project on track, maintain good documentation, address legal issues, and work within a well-defined management plan. This chapter introduces several project management skills that are essential elements of engineering design.

SECTIONS

1 Working in Teams
2 Managing Tasks: Keeping the Project on Track
3 Documentation: The Key to Project Success
4 Legal Issues: Intellectual Property, Patents, and Trade Secrets
Key Terms

OBJECTIVES

In this chapter, you will learn about:

- The importance of project management in ensuring design success
- Teamwork as an essential element of engineering design
- How engineers determine the tasks that will lead to a finished product
- Scheduling and managing design tasks
- The role of the project manager
- How members of a design team interact
- Documentation and its vital role in the design process
- Legal issues and their relevance to the design engineer

1 WORKING IN TEAMS

The spirit of rugged individualism is pervasive in society, persisting as a theme in books, movies, and television. The image of a lone hero striving for truth and justice against insurmountable odds appeals to our sense of adventure and daring. Our pioneering spirit dreams of becoming the sole entrepreneur who endures economic hardship to change the face of technology, or perhaps the head of a startup company that single-handedly bests Microsoft or Intel. Yet, in real life, engineers seldom work alone. Most engineering problems are interdisciplinary, and true progress requires cooperative teamwork and the contributions of many individuals. This concept is easy to understand in the context of designing large structures, such as bridges, airports, skyscrapers, or oil refineries. It would be unimaginable to even approach such large tasks without a well-defined hierarchy of managers, job directors, engineers, and craftspeople. Similarly, the great engineering accomplishments in space exploration, such as the Apollo moon landing, the International Space Station, the Hubble space telescope, and the Mars Exploration Rover required hundreds (in some cases thousands) of engineers working with teams of physicists, chemists, astronomers, material scientists, medical specialists, mathematicians, and project managers. Teamwork is also critical to the design of small but complicated devices such as medical prostheses, copy machines, children's mechanical toys, kitchen appliances, and cellular telephones. These devices (not an exhaustive list) cannot be designed by one person alone.

Teamwork is an important skill, and you must master it if you are to be a good engineer. Working in a team, such as the one depicted in Figure 1, requires that you

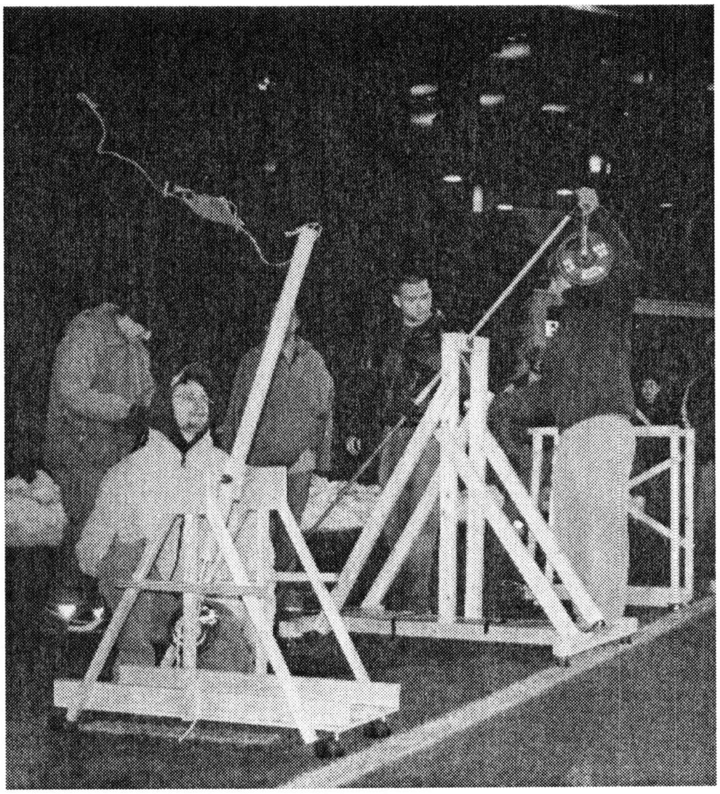

Figure 1. Design team works on the prototype of a Trebuchet catapult. (Photo courtesy of the Daily Free Press.)

speak clearly, write effectively, and have the ability to assimilate another person's point of view. All members of a team must understand how their tasks relate to the responsibilities of the team as a whole.

When addressing a design problem, dealing with the technical issues is only half the story. Determining how people and their time will be managed is just as important. By properly assigning tasks to the right people and ensuring that they work together cooperatively, the success of the project can be ensured. In the sections that follow, we address some of the major issues related to people management.

1.1 Building an Effective Team

An effective team is one that works well together. It functions at its maximum potential and thrives on a symbiosis of the special capabilities of its individual members. One key characteristic of an effective team is a good supportive attitude among fellow teammates. Team morale and a sense of professionalism can be enhanced if team members agree on some basic rules of behavior. The following set of guidelines illustrates one possible approach to building an effective design team.

1. *Agree Upon Goals.* The members of your team should agree upon the goals of the project. This consensus is not as easily achieved as you might think. One teammate may want to solve the problem using a traditional, time-tested approach, while another may want to attempt a far-out, esoteric path to success. Define a realistic set of goals at the outset. If the approach you choose brings surprises, you can always redefine your goals midway through the project.
2. *Define Clear Roles.* Each team member should be assigned a specific function within the team. The responsibilities of each individual should be defined *before* project work begins. Individual roles need not be mutually exclusive, but they should be defined so that all aspects of the design problem fall within the jurisdiction of at least one person. In that way, no task will "fall between the cracks" during the design effort.
3. *Define Procedures.* You and your teammates should agree on a set of procedures for getting things done. Everything should follow a predetermined procedure, including documentation; the ordering of parts; prototype construction; and communicating with professors, clients, and customers. In that way, misunderstandings about conduct can be greatly reduced.
4. *Develop Good Interpersonal Relationships.* You must learn to work with everyone on your team, even with those individuals whom you may personally dislike. In the real world, a client will not care about any conflicts that occur behind the scenes. Engineering professionalism demands that you rise above personality clashes and concentrate on the job at hand. Be nice. Be professional. Forbid name calling, accusations, or assigning fault to team members.
5. *Define Leadership Roles.* Some teams work best when a single person emerges as the chosen leader. Others work better by consensus using distributed leadership or even no leadership at all. Regardless of your team's style, make sure that leadership hierarchy is clearly defined and agreed upon at the start of the project.

PROFESSIONAL SUCCESS: YOU'RE THE TEAM LEADER, ONE TEAMMATE HAS DISAPPEARED, AND YOU'RE DOING ALL THE WORK

It's impossible to get along with all people all the time. When you work closely with other individuals, personal conflicts are inevitable. At times, these disagreements occur because one team member has failed to meet assigned responsibilities. At other times, the conflict arises from fundamental differences in personal outlook or priorities. However complicated your team relationships may become, remember that your customer does not care about them. Your customer is interested in receiving a well-designed product that reflects your best engineering abilities. It's up to you to resolve team conflicts internally. This resolution may mean that some team members will do more work than others, even if they will not be rewarded for their extra efforts. A good leader understands this trade-off and devises a plan to work around an errant teammate. Such situations may seem frustrating and unfair, but they happen in the real world all the time. Learning how to deal with them as a student is part of your engineering education.

PRACTICE!

Define the roles, goals, and procedures that might apply to the following.

1. The design and construction of an eight-lane highway system under a major metropolitan city (*www.bigdig.com*).
2. A team of software engineers developing an auction website (*www.ebay.com*).
3. A team of electrical and mechanical engineers developing a radio-controlled robot (*www.usfirst.org*).
4. A team of biomedical engineers developing an artificial heart for mechanical implantation inside human subjects (*www.abiomed.com*).
5. A software system for scheduling production in a "just in time" textile factory that fulfills customer orders as they arrive (*www.linq.com*).
6. The design and construction of the Mars Exploration Rover (*origin.mars5.jpl.nasa.gov*).
7. An oil pipeline from Prudhoe Bay to Valdez, Alaska (*www.alyeska-pipe.com*).
8. A package tracking system for an air/ground shipping company (*www.fedex.com*).
9. A system for compiling, printing, and distributing a daily national newspaper (*www.usatoday.com*).
10. The design of a mineral extraction plant that produces copper, uranium, gold, and silver (*www.wmc.com*).
11. An oil refinery that must produce a variety of products, including gasoline, diesel, and home heating oil.
12. A machine to selectively apply pesticides to an agricultural area only where parasites have been detected by distributed sensors.
13. A system for monitoring the energy flow and power production in a multi-plant electric power generation system.
14. A system for allowing an expert surgeon to conduct an operation via remote control from a distant location.
15. The design and construction of a portable air conditioning system for roadside tent camping.
16. The design of a command center for communicating with football players wearing individual earpieces.

17. The design of a food court in a shopping mall.
18. The design of a transportation system consisting of autonomous (driver-free) subway trains.
19. The development of an order fulfillment system for a pizza restaurant.
20. The design of an agricultural management system for a large farm collective.

1.2 Organizational Chart

When engineers work on a team-oriented project, they often establish some hierarchy among individuals. It would be nice if an engineering team could always function as a simple group of colleagues, but inevitably some team members will be burdened more than others unless everyone's responsibilities are clearly spelled out. Likewise, it will be easy for some tasks to fall between the cracks if roles and responsibilities are not understood by all. One vehicle for specifying the management structure of a team is called the *organizational chart.* An organizational chart indicates who is responsible for each aspect of an engineering project. It also describes the hierarchy and reporting structure of the team. Figure 2 illustrates a simple organizational chart that might be used by students working on a design project for an engineering class. The objective of this particular project is to design a heart rate monitor for a professor doing research in cardiac defibrillation. Cadence acts as the team leader, and she, in turn, reports to the course teaching assistant for leadership and guidance. In the corporate world, where the structuring of workers and bosses can become complex, organizational charts are essential because each employee must understand the entire responsibility chain beginning with upper management. Note that organizational charts are used by all sorts of organizations, not just engineering firms. Figure 3, for example, provides an abbreviated description of the organization of the National Aeronautics and Space Administration (NASA). The actual NASA organization chart is more complex than the one shown in Figure 3.

1.3 The Job Description

When you work for an engineering company (or any company, for that matter), your responsibilities are spelled out in a document called your *job description*. A job description

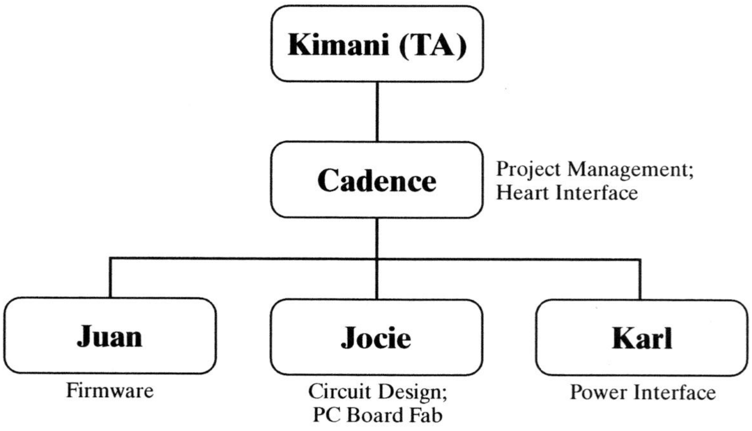

Figure 2. Team organizational chart for heart defibrilator project.

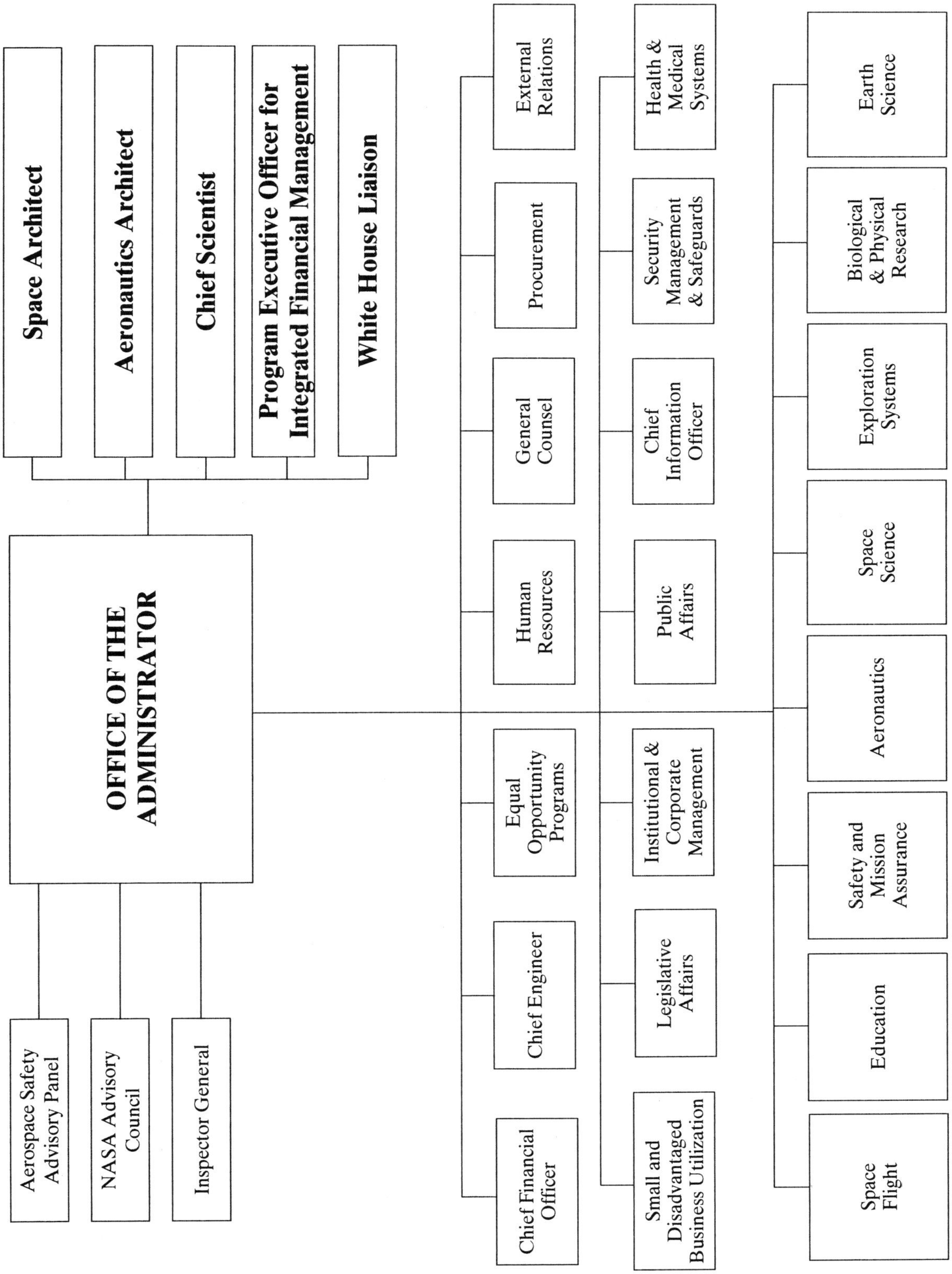

Figure 3. Abbreviated organizational chart for the National Aeronautics and Space Administration (NASA).

can be as informal as a simple letter of employment, or it can be a detailed, multi-page document. In general, the larger the company and the more complex the organizational chart, the more detailed the job description for each person. Smaller companies may have more loosely defined job descriptions because fewer people are available to handle all the tasks required to keep the company running. A job description is tied to the work to be performed and not to any specific individual who may hold the job at any particular time. Job descriptions become most prominent when a job is vacant and is subsequently advertised.

A typical formal job description might contain the following information:

- Job title
- Grade level (defines the salary range of the job)
- Immediate supervisor (specified as another job position, not a named individual)
- List of responsibilities

Note that job descriptions often form the basis of employment postings as well. One example of a specific engineering job description might read as follows:

Photonics Engineer, Photonics Center, SG 74

Tracking Code 1743/G274

Job Description

Develop and manage the Photonics Center's new Integrated Optics Laboratory, consisting of an integration/packaging facility. Coordinate new equipment purchases, installation, and commissioning for the laboratory. Collaborate with Photonics Center staff and partner companies to create processes and manage workflow of the lab. Train interns, Photonics Center staff, and partner companies to use the equipment in the lab, and supervise all equipment use. Recommend future capital equipment investments for the lab. Responsible for the maintenance of the lab: order supplies, establish and enforce policies and procedures, and maintain and calibrate equipment.

Required Skills

Master of science or equivalent in engineering (EE, ME, or Optical) and four to six years of experience with hybrid integration processes and packaging equipment. Excellent interpersonal skills with the ability to work with a variety of individuals.

1.4 Team Contact List

As obvious as it may seem, it's important to maintain a contact list for all personnel working on a particular project. One team member (or a staff person if one exists) should have the responsibility of maintaining the list. It should contain e-mail addresses and telephone numbers for everyone on the team, as well as home or cell phone numbers and instructions for when it's permissible to call outside of working hours. If resources permit, it's a good idea to maintain the list as a web page so that it is easily updated and constrained to a *single*, accurate list. Password protection may be needed in some situations.

1.5 Team Meetings

It's important for an engineering design team to meet regularly to review project status, design issues, and problem resolution. Frequency is entirely up to the team, but it is usually decided by the team leader. Meeting once per week seems to be the norm in many engineering organizations, including college design courses. Others may choose to meet only when deemed necessary (called meeting "on demand"), while other projects may require daily meetings. However often your team meets, it will be important to impose some type of structure so that meetings do not degrade into free-for-all chaos. Although a preset agenda is always helpful, setting the tone for the meeting is the responsibility of the team leader. This task can be particularly difficult in a course situation where the team is composed of otherwise equal peers. Even in company settings, personality differences between employees can get in the way of team cooperation and progress. Learning to deal with these situations in college will help you prepare for the job market. It is an important component of your engineering education.

1.6 Working with Other Teams in the Organization

More often than not, a complex design effort will involve more than one team. Assigning multiple teams to a project is common practice in companies that manufacture products for sale. The classic division of labor, still used in many companies, divides responsibilities into the broad areas of research, development, manufacturing, and marketing. The *research team* conducts fundamental studies that seek to probe new forefronts of knowledge relating to the company's core expertise. The *development team*, often comprised of different individuals, is assigned the task of transforming fundamental discoveries into product concepts that culminate in working prototypes. The *manufacturing team* must take the prototype and develop a method for mass producing the product in a cost effective and profitable way. The *marketing team*, which often contains engineers who understand how the product works, has the job of seeking and securing customers.

Traditional thinking in project management asks these four teams to interact serially, with each one passing its work "over the wall" to the next group down the chain of command. In such a scenario, Research thinks up the ideas, Development reduces them to practice, Manufacturing determines how to make the product in quantity, and Marketing seeks out new customers. Traditionally, the only feedback in the loop might be from Marketing back to Research, wherein the former communicates customers' needs to the latter. For long-term reference, teams typically operate as separate entities, with each holding on to its own set of priorities and rivalries. Often these rivalries bog down the design process and greatly increase the time span from concept to finished product.

More modern project management philosophies recognize the importance of continuous interaction between the various components of a company's organizational chart. The most successful companies are those in which the needs and priorities of each team are considered from the start and given appropriate weight when decisions are made. In such an interactive team environment, for example, Development might not be given sole authority to specify materials and dimensions of a product. Rather, manufacturing will ask that options be limited to pre-existing processes that can be enacted without requiring expensive new tooling. Marketing may insist that the design incorporate ergonomic features that will allow the product to fare well against a competitor's product. Research may seek the assistance of Development and Manufacturing to decide which projects to address, and Development may seek the help of Marketing to assess customer needs. Manufacturing may collaborate with Development to adopt rapid prototyping methods.

In some companies, the classic team divisions described above may not have distinct boundaries at all. Research, Development, and Manufacturing may be the responsibilities of one blended group of individuals. This approach has found success, for example, in the Saturn Corporation, a subsidiary of General Motors. Their team philosophy and work unit structure are summarized in the abridged diagram of Figure 4. The structure of Saturn is based on "work units" and "modules." Work units are responsible for the primary tasks related to automobile production. Modules are groups of work units that perform similar types of labor. Both work units and modules are responsible for all decisions made within the circle; each circle has an elected representative on the next highest order circle. When first introduced, this self-directed, team-integrated ap-

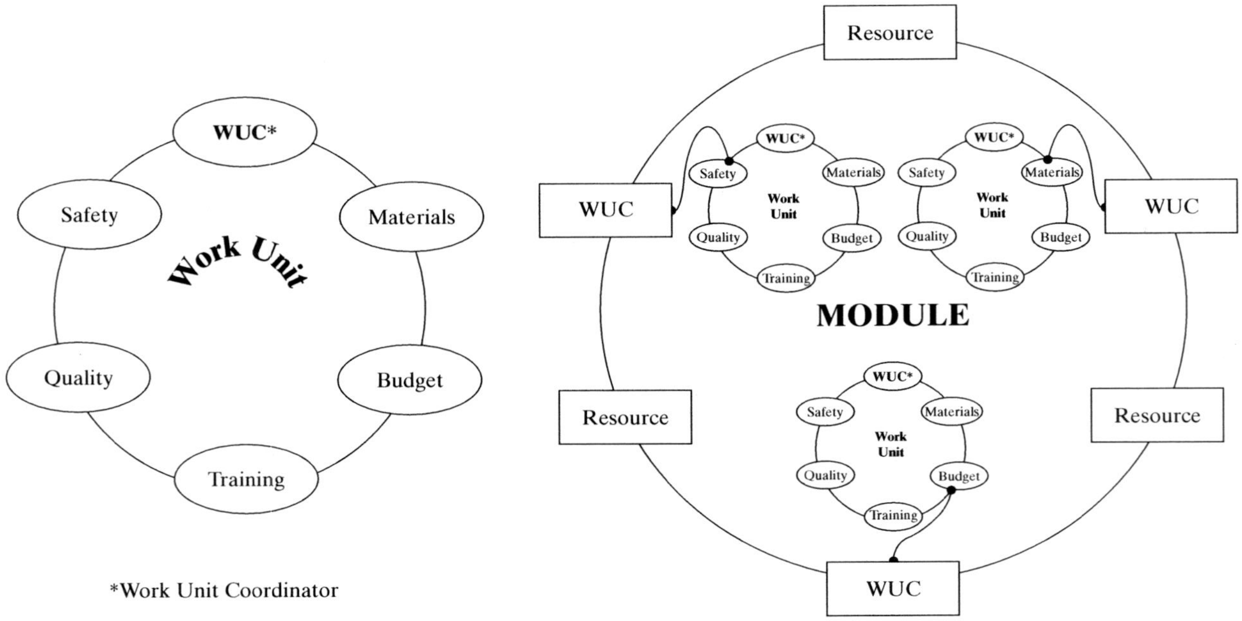

Figure 4. Work unit and module organizational structure for the Saturn automobile manufacturing company.

proach to project management was unique to the U.S. auto industry and still stands out today as a refreshing way to approach large-scale endeavors.

When confronted with the classic, compartmentalized approach to product design, it's important for engineers to avoid rivalries and "turf wars." Remember that ultimate success requires the symbiotic cooperation of all teams involved in the design process. While the various team structures may occupy distinct branches of the company's organizational chart, as in Figure 5, it's the unseen bridges between them that will keep the company strong and lead to rapid and successful product development.

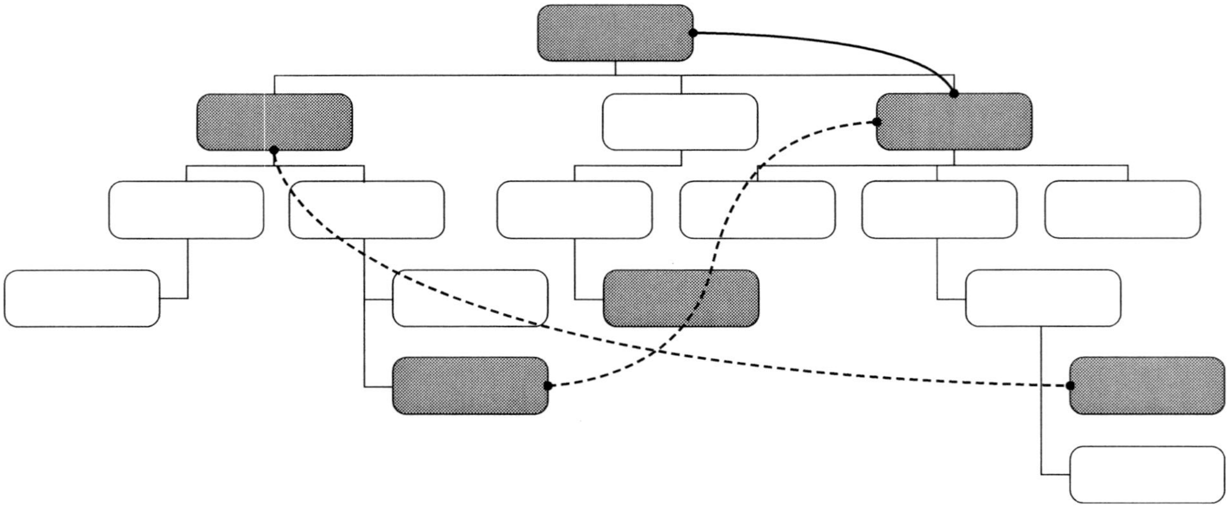

Figure 5. The unseen bridges between branches of an organizational chart keep the company strong and lead to rapid and successful product development.

PRACTICE!

1. Devise an organizational chart and time line for building a special effects apparatus for a drama production at your college or university.
2. A large, nonprofit organization is planning a walk-a-thon in which 3,000 people will walk 60 miles in 3 days and will sleep in tent camps each night. Develop an organizational chart for all persons involved in planning and implementing this event.
3. A student organization is planning to enter the Mini Baja vehicle competition for students sponsored by the Society of Automotive Engineers (*students.sae.org*). Develop an organizational chart that will guide the team through the design and construction of a competition vehicle.
4. Imagine that you work for a company that makes genetically engineered laboratory mice. Define an organizational chart for such a company.
5. Even sports teams have organizational charts. Choose your favorite professional or college sports team and draw its organizational chart.
6. Your college or university has an administrative organizational structure. Learn about this structure and draw an organizational chart, from the President down to the level of either department heads or deans.
7. Draw the organizational chart for a major civil engineering company of your choice.
8. Write the job description for an engineer assigned to a team designing a plastics processing plant.
9. Write the job description for someone who works for a web-based hardware store.
10. Write the job description for an engineer who has just joined a company that makes underwater hydrophones. The company has a total of ten employees.
11. Write the job description for an engineer who works for a large defense contractor that makes military aircraft.
12. Write the job description for an engineer who works for a company that makes pipeline valves.
13. Write the job description for an engineer who works for a company and helps them optimize factory floor operations for a cosmetics manufacturer.
14. Suppose you want to hire an industrial engineer for a company that makes large kitchen appliances. Write an advertisement that you might put on a nationally circulated job site (e.g., *www.monster.com*).
15. Suppose that you want to hire a petroleum engineer to oversee the operations of a refinery construction project. Write the advertisement that you might submit to a national trade magazine.

2 MANAGING TASKS: KEEPING THE PROJECT ON TRACK

Time management is critical to the success of any engineering project. In a perfect world, engineers would have as much time as needed to work on all aspects of a project; in the real world, however, deadlines have a nasty habit of creating pressure to get the product out the door. Demands for demonstrations of progress, prototype tests, something for "sales and marketing to show," and the pressures of corporate life require that an engineer develop a sense of how much time will be needed for each aspect of product development.

Even the simplest design project requires time management. A systematic approach to design tasks is always preferable to a random, hit-or-miss approach. While the subject of time management can (and does) occupy the contents of entire books, several time management tools form a basic set that should be understood by all engineers.

2.1 Checklist

A simple checklist enables the monitoring of engineering tasks related to a specific design project. It differs very little from the "to-do" list that you might compile for a set of weekend chores. The tasks enumerated on an engineering checklist need not be completed in any particular order. An example of a simple checklist that might apply to the design of an entry for an egg-drop contest is shown in Figure 6.

2.2 Time Line

A *time line* is a valuable tool for keeping a project on schedule. A time line is similar to a checklist, but its various tasks are expected to be completed sequentially by specified milestone dates. The time line is very appropriate when the entire team works in unison, with the differentiation of tasks occurring serially in time. If a given task is in danger of not being completed before its designated milestone date, it's the job of the project manager to allocate more time, and overtime if necessary, so that the task can be completed on schedule. A typical time line (in this case, one prepared by students designing a vehicle for a class design project) is shown in Figure 7.

2.3 Gantt Chart

When a project is complex and involves many people, a simple time line may be inadequate for managing all aspects of the project. Similarly, if the project's various tasks are interdependent, so that the completion of one phase depends on the success of several others, the *Gantt Chart* of Fig. 8 may be a more appropriate time-management tool. The Gantt chart is simply a two-dimensional plot in which the horizontal axis reflects time measured in days, weeks, or months, and the vertical axis represents either the tasks to be completed or the individuals responsible for those tasks. Unlike the checklist, which simply enumerates the tasks to be completed, and the one-dimensional time line, which merely displays sequential time allocations for each phase of the project, the Gantt chart shows how much time is allotted to multiple tasks performed in parallel. It also provides for overlapping time periods that help indicate the interdependency between the various

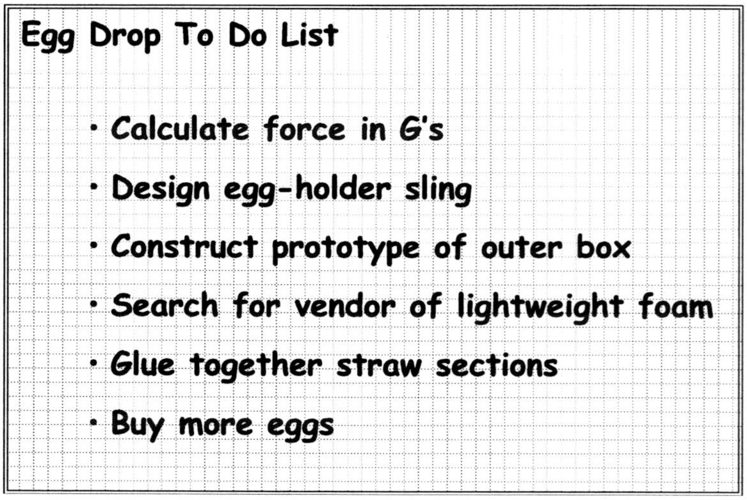

Figure 6. Checklist for the design of an entry for an egg-drop contest.

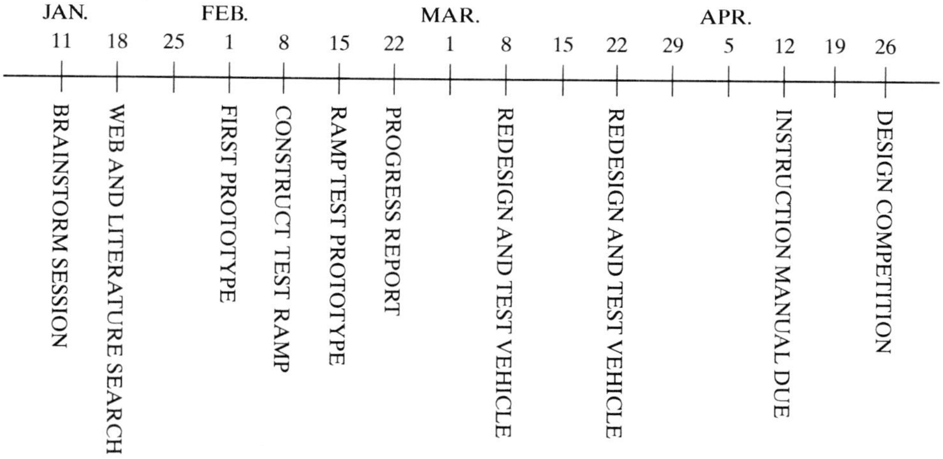

Figure 7. Time line for tasks pertaining to a vehicle design competition.

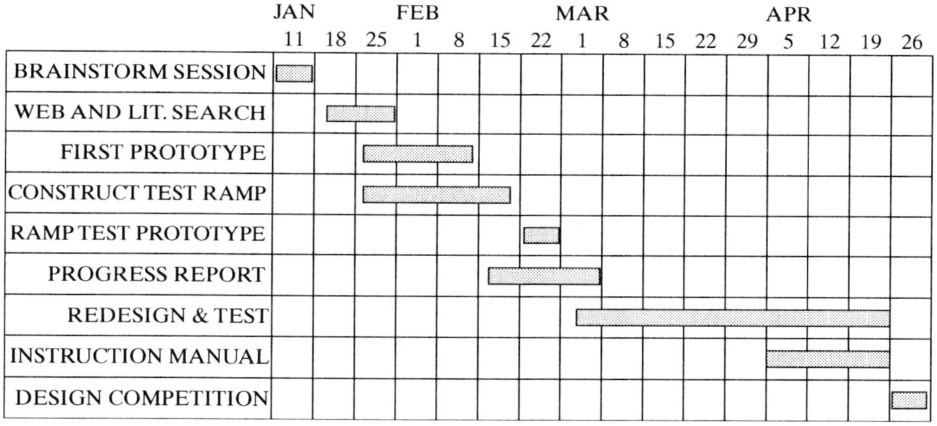

Figure 8. A Gantt chart provides a more comprehensive, two-dimensional method of scheduling the tasks shown in Figure 7.

aspects of the project. When a particular task has been completed, it can be shaded on the Gantt chart, so that the status of the project can be determined at a glance.

2.4 PERT Chart

The Project Evaluation and Review Technique (PERT) was first proposed by the United States Navy and developed by the consulting firm of Booz, Allen, and Hamilton in 1958. Its purpose at the time was to coordinate the activities of over 10,000 separate subcontractors involved in the Polaris missile development program. The PERT method is very similar to the *critical path method* (CPM), and these terms are essentially interchangeable. The PERT technique is fundamentally a method for prioritizing and scheduling complex, interrelated activities. The PERT chart helps to identify the most time-critical events in the design process.

The essence of the technique is embodied in a graphical network called the *PERT chart*, depicted in generic form in Figure 9. The PERT chart consists of numbered mile-

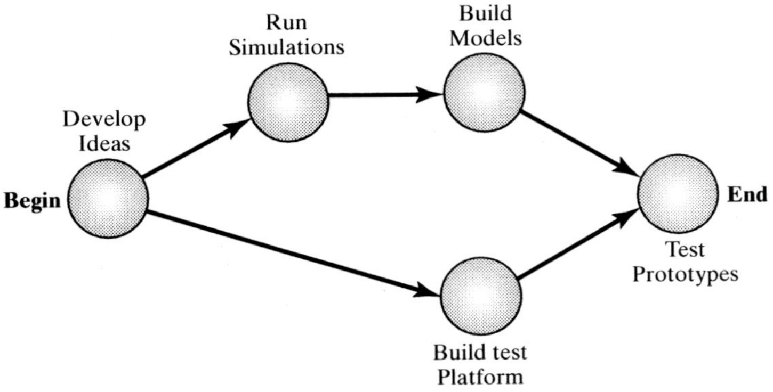

Figure 9. The generic form of a PERT chart. The circles represent milestone points, and each arrow represents a task leading up to a node. Each task is allocated a completion time.

stone circles and pathway branches. Each branch is labeled by a time interval allocated to the completion of the task to which the branch leads. Branches are labeled in appropriate time units (e.g., days, weeks, or months). Like the Gantt chart, the PERT chart summarizes the time allocated for each task and notes task completion milestones. Unlike the Gantt chart, however, the PERT chart also shows the way in which the tasks depend on one another. This interdependency is indicated by the branch lines that connect the task-completion circles of the chart.

A PERT chart must have a starting point, or *node*, and a single ending milestone (the last node to which all pathways must lead). As in the Gantt chart, time progresses from left to right. The time allocated for the pathway between any two milestone nodes will be equal to the sum of the times allocated for each task in the pathway. In Figure 10, for ex-

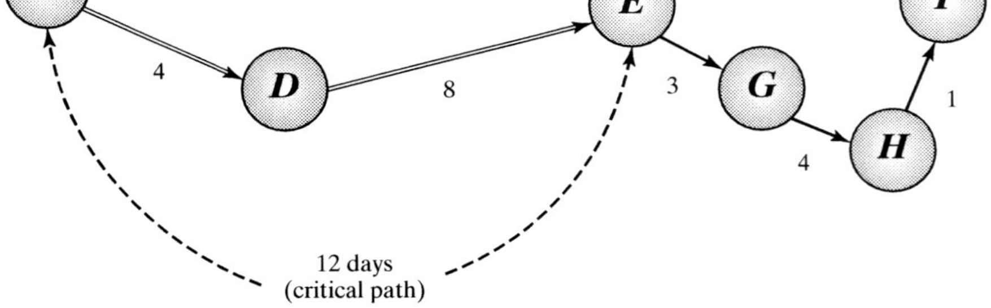

Figure 10. A PERT chart showing milestone nodes and task time allocations in days. On this chart, critical pathways are shown as double lines.

ample, the sequence of tasks leading from milestone A to milestone C is allocated a total of $3 + 5 = 8$ days.

As each task depicted on a PERT chart is completed, the project manager checks it off on the chart. The manager can thus monitor the progress of the entire project and be alerted to any possible path delays. Some project managers prefer the PERT chart over the Gantt chart because it clearly illustrates task dependencies. A PERT chart, however, can be much more difficult to interpret, especially on complex projects. Alternatively, some project managers may choose to use both techniques.

When compiling a PERT chart for project management, it's possible (and often preferable) for the sum of branch times over one pathway to be less than the sum of times over a parallel pathway. The excess, called *slack time*, can be used to compensate for the inevitable delays that will accompany the design process. If the delay experienced along the way does not exceed the slack time of the pathway, the entire project will still be on track. Of most interest, therefore, are so-called *critical pathways*. The latter are those branches containing zero slack time. Any delays in a critical pathway can jeopardize the time flow of the entire project and should be monitored carefully. The pathway from node A to node E (via node D) in Figure 10, for example, is a critical pathway.

EXAMPLE 1: MANAGING A BUILDING CONSTRUCTION PROJECT

The major steps required to erect a framed house (somewhat simplified in this case) are summarized by the PERT chart of Figure 11. Such a chart might typically be used by a general contractor who must hire many different subcontractors to complete various aspects of the project. Node 0 denotes the starting point of construction. Before anything else can transpire, the building site must be excavated so that the foundation can be framed, as well as molded and poured. In Figure 11, the former is allocated 4 days, and the latter 2 days. After the foundation has been poured and cured, the external skeleton or frame of the building can be erected. This frame will form the support structure for all the exterior walls of the building, as well as any interior walls specified by the architect. This task has been allocated a total of 5 days.

After the building frame has been completed, the major tasks continue over parallel pathways. The framing for the roof must be completed, followed by the weather protection (in this case, standard roofing shingles). These tasks have been allocated 2 and 3 days, respectively, leaving considerable slack time compared to the other paths. Similarly, following arrival at node 3, work begins in parallel on the installation of the building's exterior walls. The rough plumbing and electrical tasks, to be described in the next paragraph, require that the interior walls remain open, but not the exterior walls.

In parallel with the finishing of the roof and outside walls, the installation of the rough plumbing is now initiated. Rough plumbing refers to the piping—both supply and waste—that must lie within the walls of the structure. Obviously, this task must be completed before the interior walls themselves can be installed. Yet another parallel pathway begins after node 3, consisting of the rough electrical wiring. The latter must also be completed before the interior walls can go up. The rough plumbing and wiring tasks are allocated 3 and 5 days, respectively.

Following the completion of the rough work, the interior walls can now be installed (2 days). Once the interior walls have been completed, the plumbing and electrical crews can return to install fixtures; afterwards, the interior walls may be painted. Usually, this task is performed after the electrical and plumbing work are finished, because the latter tasks might cause workers to mar the finished paintwork. Some contractors, however,

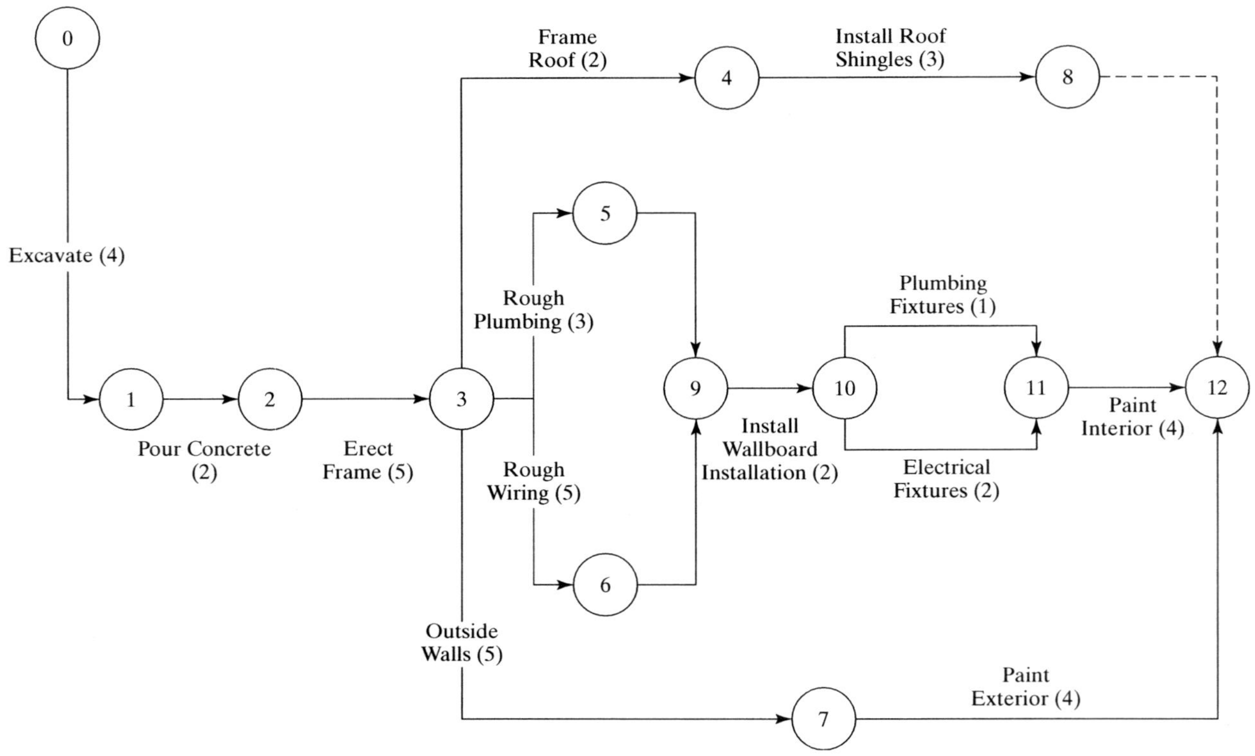

Figure 11. PERT chart that describes the principal steps involved in building a house.

prefer to paint before fixture installation, because the resulting paint job is much cleaner around the edges of the plumbing and electrical fixtures.

The completion of the roof, interior paint, and exterior paint on the PERT chart all arrive at the last node on the diagram (node 12), signifying the completion of the job. The house may now be occupied. The critical pathway in this PERT chart consists of the following tasks, beginning at node 3: rough electrical wiring, wallboard installation, wiring completion, and interior painting. These tasks require a total of $5 + 2 + 2 + 4 = 13$ days. Any delays over this critical pathway will lead to a delay in occupancy of the house.

EXAMPLE 2: MANAGING A SOFTWARE PROJECT

Figure 12 shows a PERT chart used by a development team tasked with the design of a digital hardware device running on firmware (machine-level, binary software). The unit, once manufactured and ready for sale, is to be shipped with help files on a CD ROM. Once the specifications have been set, the PERT chart divides into three distinct pathways—one related to hardware, one to the firmware, and one to the documentation. The critical pathway in this project runs from node 1 to node 2, and then via nodes 3, 5, 6, 8, and 7 to the final node 11, for a total time allocation of $2 + 4 + 2 + 3 + 4 + 1 + 4 = 20$ weeks.

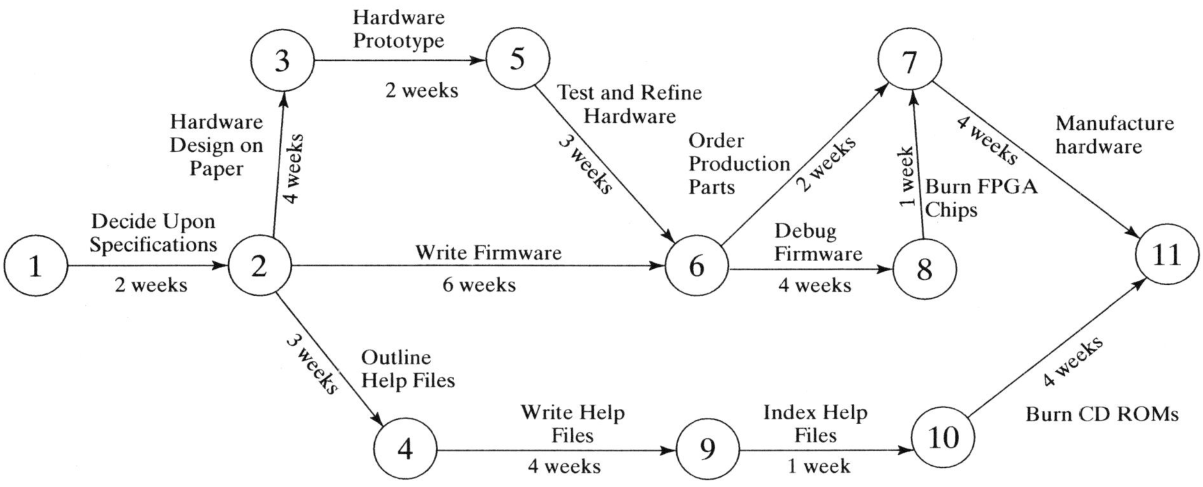

(1) Start project
(2) All specifications finalized
(3) Hardware design on paper
(4) Outline of help files completed
(5) Prototype built
(6) Hardware functional
(7) Ready for fabrication
(8) Firmware ready to integrate with hardware
(9) Help files completed
(10) Help files indexed
(11) Project complete

Figure 12. PERT chart that describes the development of a digital hardware device running firmware and shipped with help files on CD ROM.

PROFESSIONAL SUCCESS: THE LAWS OF TIME ESTIMATION

How long will it take to perform a particular design task? How much time should you allocate to each segment of a PERT chart or Gantt chart? The following three Laws of Time Estimation will help you to determine the time required for a given design task:

1. Everything takes longer than expected.
2. If you've done the same task before on a different project, estimate the amount of time required to finish the task. The actual amount of time required will be about twice as long. Something will always be different in the current project.
3. If you've worked on something *similar* before, estimate the amount of time required to finish the task. The actual amount of time required will be about five times as long.
4. If you've *never* worked on something similar before, estimate the amount of time required to finish the task. The actual amount of time required will be the next highest time unit. For example, something estimated to take an hour will really take a day; something estimated to take a day will take a week, etc.

3 DOCUMENTATION: THE KEY TO PROJECT SUCCESS

Engineering design is never performed in isolation. Even the simplest of projects involves a designer and an end user. More often, a design effort involves considerably more individuals who worry about many facets of the product. Additionally, the uses of the product may involve an entire segment of the population. The design of an automobile, for example, encompasses the work of mechanical, electrical, industrial, materials, and safety engineers, and the entire driving public constitutes the group of end users. A consumer products company involved in the production of cellular telephones will bring together computer, systems, electrical, mechanical, and manufacturing engineers in a multidisciplinary team that may include people from sales and marketing. A public works management corporation, tasked with designing a waste water treatment facility, might solicit the help of civil, chemical, mechanical, and environmental engineers, along with the city planners and systems engineers. Complex engineering projects are successful only if everyone on the design team communicates with everyone else at all phases of the design effort.

One way in which engineers communicate with each other is through careful recordkeeping. Good documentation is essential when you work as a member of a design team. As a practicing engineer, it will be your responsibility to maintain a comprehensive collection of design concepts, sketches, detailed drawings, test results, redesigns, reports, and schematics—whatever records are pertinent to the project. This *documentation trail* will serve as a tool for passing information on to team members who may need to repeat or verify your work. The documentation trail is critical to those who will manufacture your product from a prototype, apply for patents based on your inventions, or even take over your job if you are promoted or move to another company. Written records are also a good way to communicate with yourself. Many an engineer has been unable to reproduce design accomplishments due to sloppy recordkeeping. Indeed, one of the marks of a professional engineer is the discipline needed to keep organized, neat, up-to-date records. Documentation should never be performed as an afterthought. If a project is dropped by one team member, the state of documentation should always be such that another team member can resume the project without delay. As a student of engineering, you should learn the art of recordkeeping and develop good documentation habits early in your career. Most companies, laboratories, and other technical firms require their employees to keep records that document the results of engineering efforts.

3.1 Paper versus Electronic Documentation

Today, just about every piece of engineering documentation, with the exception of the engineer's logbook (described in the next section), is produced on a computer. Examples of documents destined for preservation include text documents, spreadsheets, computer codes, schematics, drawings, design layouts, and simulated test results. Most engineers prefer to store information electronically so that it can be viewed onscreen and printed only as needed. Others prefer to preserve documentation by printing everything on paper and storing the documents in a physical file cabinet. Whichever method you choose, take heed of the following important guidelines:

- *Organize your information.* It's important to store documentation in an organized and logical manner. If the project is small, its documentation should be stored in a single folder (paper or electronic). Larger projects may require a group of folders, each relating to different aspects of the project. The folders should be labeled and dated with informative titles (such as "Propulsion System for XYZ Project") and kept in a place that will be easy to find in the future.
- *Back up your information.* It's equally important to store a duplicate copy of all documentation. This guideline applies to written as well as electronic information. Fire, flood, theft, misplacement, the all-too-common disk crash, and the unfortunate havoc wreaked by malicious computer viruses can lead to the loss of a project's entire documentation trail. Archival storage of records in a different physical location will help to keep a project on track should one of these catastrophes occur.

3.2 The Engineer's Logbook (Notebook)

One important vehicle for recordkeeping is the *engineer's logbook,* sometimes called the *engineer's notebook.* A well-maintained logbook serves as a permanent record that includes all ideas, calculations, innovations, and test results that emerge from the design effort. When engineers work in a team, each member keeps a separate logbook reflecting that individual's assigned tasks. When the project is brought to completion, the logbooks of all team members are placed in an archive and remain the property of the company. An engineering notebook thus serves as an archival record of new ideas and engineering research achievements *whether or not they lead to commercial use.* A complete logbook serves as evidence of inventorship and establishes the date of conception and "reduction to practice" of a new idea. It shows that the inventor (you!) has used diligence in advancing the invention to completion. In this respect, the engineer's logbook is more than just a simple lab notebook. It serves as a valuable document that has legal implications. When you work as an engineer, you have a professional responsibility to your employer, your colleagues, and to the integrity of your job to keep a good logbook.

The notebook shown in Figure 13 is typical of many used in industry, government labs, and research institutions. It has permanently bound and numbered pages, a cardboard cover, and quadrille lines that form a coarse grid pattern. A label fixed to the front cover uniquely identifies the notebook and its contents. The company, laboratory, or project name is printed at the top, and the notebook is assigned a unique number by the user. In some large institutions, a central office may assign notebook numbers to its employees when the notebook is issued.

Figure 13. Cover and label of a typical engineer's logbook.

The techniques for logbook use differ from those used in some science and introductory engineering classes where instructors encourage students to write things down first on loose scratch paper and then recopy relevant items into a neat notebook. This procedure is bad practice for the design engineer. Although notebooks prepared in this way are easier for instructors to grade, the finished notebook seldom resembles a running record of what actually occurred in the laboratory; it is not especially useful for engineering design projects. Design is as much a *process* as it is a final product, and the act of recording ideas as they emerge and events as they happen helps engineers to think and be creative. Also, keeping a record of what did *not* work is just as important as recording what did work; this practice ensures that mistakes will not be repeated in the future.

3.3 Logbook Format

An engineering logbook should be used as a design tool. Enter everything into your logbook, no matter how seemingly irrelevant. Write down ideas as you think of them, even if you have no immediate plans to pursue them. Keep an ongoing record of successes and failures. Record the results of every test—mechanical, structural, electrical, system, flight, flow, pressure, performance—even if the results may not be used in the final design. Stop to write things down. This habit will require discipline but will always be worth the effort. Important information, including some you might otherwise have forgotten, will be in your logbook and will be preserved for a time when you may need it.

Any logbook format that meets your needs and those of your team is suitable, as long as it forms a permanent record of your contributions to the design project. Ban loose paper from the laboratory. It is easily lost, misfiled, or damaged. Resist the temptation to reach for loose paper when you need to do a calculation. Instead of grabbing that pad to draw a sketch or discuss an idea, take the time to open your logbook. You'll be glad you did when those numbers and sketches you need are readily available. Unbound paper used for anything other than doodling has no place in an engineering laboratory.

3.4 Using Your Engineer's Logbook

As chief author of your logbook, you have the freedom to set your own objectives for its use. The following guidelines, however, are typical of those used by many engineers and design teams:

1. Each person working on a project should keep a separate logbook specifically for that project. All relevant data should be entered. When the logbook is full, it should be stored in a safe place specifically designated for logbook storage. In that way, everyone will know where to find the logbook when it's needed.
2. All ideas, calculations, experiments, tests, mechanical sketches, flowcharts, circuit diagrams, etc., related to the project should be entered into the logbook. Entries should be dated and written in ink. Pencil has a nasty habit of smudging when pages rub against each other. Any computer-generated plots, graphics, schematics, or photos printed on loose paper should be pasted or taped onto bound logbook pages. This procedure will help prevent loss of important data.
3. Logbook entries should outline the problem addressed, tests performed, calculations made, and so forth, but subjective conclusions about the success of the tests (e.g., "I believe . . . ") should be avoided. The facts should speak for themselves.
4. The voice of the logbook should speak to a third-party reader. Assume that your logbook will be read by teammates, your boss, or perhaps someone from marketing.

Witnessed and understood
Marigold White 10/27/05
Eli Whitney 10/27/05

5. In settings where intellectual property is at stake, the concluding page of each session should be dated and, where appropriate, signed. This practice eliminates all ambiguity with regard to dates of invention and disclosure. Important entries that signify key events in the design process should be periodically and routinely witnessed by at least one other person, and preferably two. Witnesses should endorse and date the relevant pages with the words, "witnessed and understood."
6. Logbook pages should not be left blank. If a portion of a page would otherwise be left blank, a vertical or slanted line should be drawn through it. Pages should be numbered consecutively and not be torn out. These measures are necessary should your logbook ever become part of a legal proceeding where the integrity of the information comes into question. Do not make changes using correction fluid. Cross out instead. This precaution will prevent you from creating obscure or questionable entries should your logbook be entered as legal evidence in patent or liability actions. Although this precaution probably won't be relevant to logbooks you keep for college design courses, it's a good idea to begin following it now so that the procedure becomes a career habit.

EXAMPLE 3: AN ENGINEER'S LOGBOOK

The following example illustrates proper use of an engineering logbook. Imagine that the logbook pages shown describe your team's design for a telescoping flag for a motorized wheelchair. The intent is to increase visibility when the user operates the wheelchair in city streets. The first page, Figure 14, shows your preliminary sketch of a basic concept based on the use of a commercially available telescoping car radio antenna. The second page, Figure 15, contains some calculations that estimate the battery drain as the flag is raised and lowered. The entries on the third page, Figure 16, show a list of parts and materials that your team decides to purchase. These parts will enable you to build a prototype and test your system's ability to raise and lower the flag on demand.

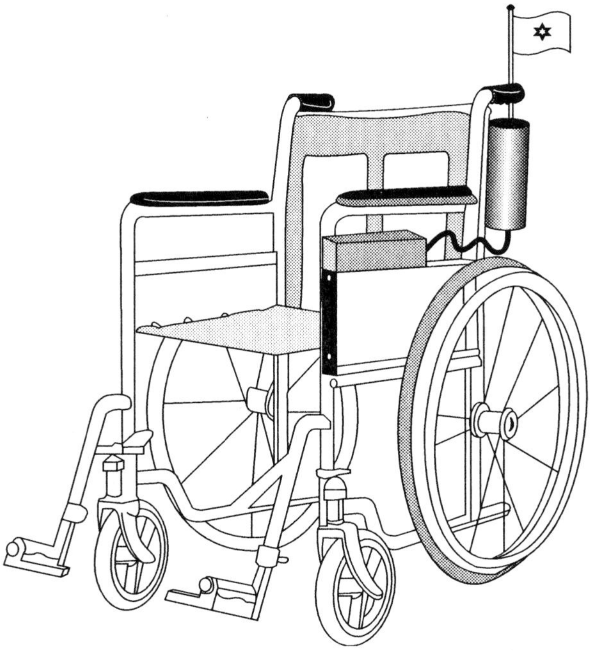

Figure 14. Preliminary logbook sketch for a telescoping wheelchair flag.

Battery Energy Requirements

Measured motor current: $0.45\ A \approx \frac{1}{2} A$

Rod Speed: (estimated) abt. 10 cm/sec

→ *Motor* ON *time* = Dist/Rate = $\frac{121\ cm}{10\ cm/sec} = 12.1$ sec

Here 12 seconds = $\frac{1}{5}$ *minute* = $\frac{1}{300}$ *hour*

. . . so each extension requires about. . .

0.45 A × 1 hour $\approx 1.5 \times 10^{-3}$ *A-H*

A 0.3 amp-hour battery can provide about

0.3 A-H ≈ 200 extensions of flag.

Note to Mike: Good for a start

Figure 15. Calculations to show estimates of battery drain.

10/2/05

* Wheelchair Flag Project

* Parts List (other than auto car-antenna thingy)

- Nuts and bolts (#8 × ½" with washers)
- Machine screws (#6 × ¾" long)
- Ubolts (4)
- Flexible pipe clamps—3" (4)
- Epoxy glue (1 tube with hardener)
- Electrical tape (1 roll)
- Solder (Can we borrow a small soldering iron? Don't need to use it much.)
- Switch (from Radio Shack)
- Wire (2 conductor; stranded)

Figure 16. List of parts and materials recorded in logbook.

3.5 Technical Reports and Memoranda

Logbooks represent only one of the methods that a design team uses to keep a good documentation trail. Engineers also communicate by writing technical reports at significant milestones of a design project. A technical report describes a particular accomplishment and perhaps provides some project history or background material as well. The report may contain theory, data, test results, calculations, design parameters, or fabrication dimensions. Technical reports form the backbone of a company's technical database. Reports are typically stored in archival format, each with its own title and catalog number. Information for technical reports is easily gathered from logbooks that are accurate and up to date. When the time comes to write a patent disclosure, journal paper, or product application note, the technical report becomes an indispensable reference tool. A technical report is also an appropriate way to explain why a particular idea did not work or was not attempted. Taking the time to write a technical report about a negative result or design failure can save considerable time should a design concept be revisited by engineers who were not present when the original project was undertaken.

Yet another way in which engineers communicate their ideas to other engineers and scientists is through the writing of formal journal articles. A journal article is appropriate when the work represents new knowledge unknown to those working in the field. While most often used to report experimental and theoretical findings, journal articles are also very frequently used to report design innovations. The structure of a typical journal paper and its relevant components are discussed in more detail in other areas of this book.

Each of the various professional organizations representing engineers worldwide publishes one or more peer-reviewed journals to which anyone can submit a manuscript for consideration. It's also very common for a paper to be authored by multiple authors, especially when the work is the result of a team effort. In addition to the journals published by professional societies, numerous other technical journals are produced by publishing companies.

Peer review is the process by which a paper is evaluated and critiqued, then either accepted or rejected for publication. The editor of the journal will typically send a submitted paper to one or more knowledgeable reviewers who will return a list of comments and questions for the author. It is common for a submitted paper to be returned at least one time to the author for revision. It's most unusual for a paper to be accepted unchanged upon its first submission. Conversely, some papers may be rejected as unsuitable for publication. This peer-review method ensures that papers appearing in the best journals are accurate, relevant, and up to date.

3.6 Schematics and Drawings

Documentation does not always appear in the form of text. Graphical records, such as drawings, circuit schematics, photographs, and plots, also become part of the documentation trail. These items typically are created with the help of software tools. If the documents are stored electronically, then all files related to a particular project should be organized in a logical hierarchy. Some engineers choose to keep all files for a project in a single file folder on the computer. Others prefer to sort files by the applications that produce them (e.g., CAD drawings in one folder, spreadsheet files in another). Other engineers like to transfer all the computer files related to a given project to a single removable disk or flash memory (USB) drive that can be stored in a physically remote, lockable cabinet. If paper is the storage medium, then graphical output should be printed on paper and kept in a folder along with other written records. Regardless of which storage method is chosen, the information related to a particular project should be carefully preserved in a format that will prevent loss. Backup in a remote physical location provides important insurance for mission-critical projects.

3.7 Software Documentation and the Role of the Engineering Notebook

Of all design endeavors, software writing is the one most prone to poor documentation. The revision loop of a software design cycle can be extremely rapid, because the typical software development tool enables a programmer to make small changes and test their effects immediately. This rapid-fire method of development is good for prototyping but invites poor documentation habits. Seldom does the software engineer find a good time to stop and document the flow of a program, because most pauses are short and change is frequent. As a result, the documentation for many software programs is added after the fact, if at all.

If you find yourself writing software, get into the habit of including documentation in your program as you go along. All software development tools provide a means for embedding comments in the program code. Add them frequently to explain why you've taken a certain approach or written a particular section of the code. Explain the meaning of object names and program variables. Your program documentation should enable other engineers on your team to completely understand and take over your sections of the code simply by reading the comment lines. Good documentation will also be invaluable should you need to modify your code at a later time. It's amazing how quickly a programmer can forget the internal logic of a program after setting it aside for only a short time.

If your program is destined for commercial sale, then good internal documentation and truly helpful "help" files are essential. Documentation included inside the program on a regular basis will easily translate into help files and an online instruction manual when the need arises. One trick used by top-notch software developers is to write the help and instruction files as the program code is developed, rather than as an afterthought. Changes to the help files can be made at the same time that changes are made in the program code. The abundance of commercial software packages with pathetic or poorly written help files or instruction manuals is testimony to generations of software engineers who have perpetuated a tendency toward poor documentation habits. If you master the skill of documenting software, your software products will be better utilized and more successful than those having poor documentation.

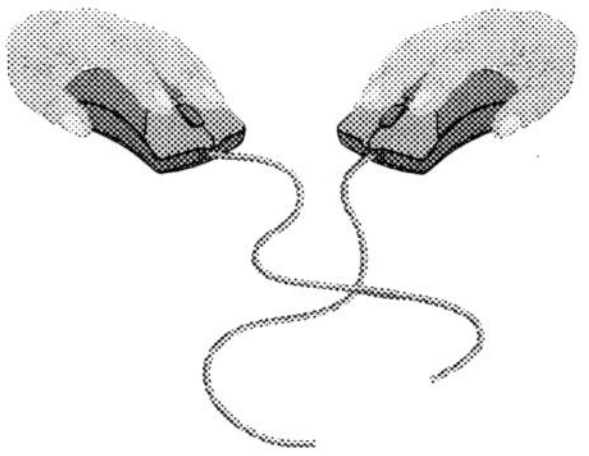

Although keeping engineering logbooks is less relevant to software development than to other types of engineering, logbooks can still play a special role. On the pages of your notebook, you can outline the overall structure of the program and the interconnections between its various modules. You can enter sketches of graphical user interfaces without writing actual computer code. You can similarly draw block diagrams of relational databases and make lists of the variables to be used in the software.

PROFESSIONAL SUCCESS: HOW TO KEEP GOOD RECORDS ALL THE TIME

If you want to keep a good documentation trail, get into the habit of carrying your logbook with you wherever you go. In that way, it will be available whenever you have a thought or idea that needs recording. Buy a medium-size notebook that can fit easily into your backpack. Clip a pen right inside the front cover. Be sure to write your name and contact information on the cover in case you misplace your logbook! A tiny, 3″ × 5″ bound notebook will do nicely. Although writing space will be limited because of the smaller size, you'll be more likely to carry it if it's not overly large.

PRACTICE!

1. Refer to the logbook calculations in Figure 15. Revise the estimate of the number of extensions if the motor requires 2.1 A.
2. Refer to the logbook calculations in Figure 15. Revise the estimate of the number of extensions if the motor requires 300 mA, and a 200 A-H battery (similar to a lead-acid car battery) is used. Rewrite the logbook page.
3. Do some Internet research and identify sources and vendors for the telescoping antenna shown in Figure 14.
4. For the flag system outlined in Figure 14, what are the advantages and disadvantages of using a separate battery for the flagpole (as opposed to tapping the primary storage battery that powers the wheelchair.) Write a logbook page that records your ideas.
5. On a simulated logbook page, outline some ideas for marketing your telescoping flag invention. Include cost estimates for its manufacture.
6. Imagine that you are one of the Wright brothers developing the first airplane. Sketch a typical logbook page from the project.
7. Research the history of the famous 20th-century American inventor Lee DeForest. Sketch one page of a logbook that he might have kept during the time of his invention.
8. Sketch a logbook page that might have been recorded by Josephine Cochran, the inventor of the modern dishwasher.
9. Imagine what it was like to be Philo T. Farnsworth, the original inventor of the television. Sketch several logbook pages that summarize how the system works.
10. Keep a brainstorming logbook for one week. Record all ideas that come to mind regarding methods to produce a personalized air transportation device.

3.8 The Importance of Logbooks: A Case Study

This case involves a small, Massachusetts-based biomedical engineering company called Abiomed (pronounced ab-ee-oh-med). Among its other heart-related projects, Abiomed was a leading contender for the development of a self-contained, artificial human heart designed to be a permanent replacement for individuals who could otherwise stay alive only with a human heart transplant. As noted on the company website (*www.abiomed.com*), heart disease is the leading cause of death in developed countries. In the United States alone, hundreds of thousands of people could be saved each year if transplants were available, yet only about 2,000 transplantable hearts become available in any given year. Hence, the need for a permanent artificial heart is real and widespread. Doctors at Jewish Hospital in Louisville, Kentucky, installed the first Abiocor artificial heart in a human patient in July 2001.

Abiomed began its total artificial heart project in earnest in the early 1990s. The major components of the heart system, depicted in Figure 17, include the central implanted pump that connects to the subject's principal blood vessels, and a system for transferring electrical power to run the pump through the skin from a battery pack worn outside the body. One key feature of the power transfer process, the lack of wires piercing the skin, is essential for the long-term efficacy of the artificial heart, because skin perforations are prime entry points for infection and require constant medical supervision in a skilled-care facility. Rather than requiring the patients to be tethered to a console, the fully self-contained Abiomed device is designed to provide the patient some semblance of a normal, mobile, home life. The system for transferring

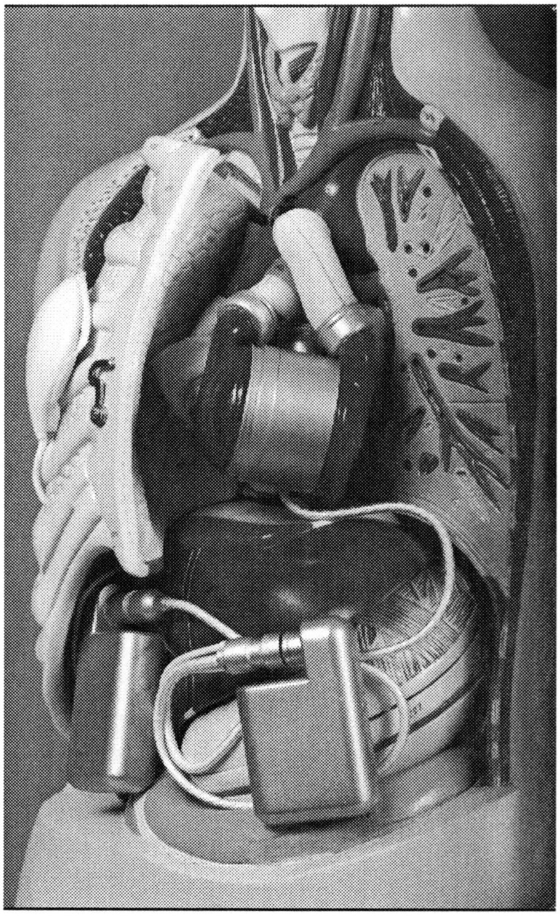

Figure 17. Abiomed's total artificial heart system.

electrical power, called the transcutaneous energy transfer device, or "TET," is essentially a pair of concentric, high-frequency magnetic coils. One is implanted under the skin, and the other is worn outside the skin. (Transcutaneous means "through the skin.") Abiomed wished to have its engineers fully focus on the daunting task of developing the heart pump itself, so it hired another biomedical engineering company, World Heart Corporation of Ottawa, Canada, to design the energy transfer module.

After about four years of effort, World Heart was still unable to produce a TET that met Abiomed's stringent technical specifications. Although World Heart claimed to be converging on a solution, Abiomed was not convinced that a satisfactory TET device was imminent. Faced with an impending critical animal test that would determine future funding of its entire heart project from the National Institutes of Health, Abiomed decided to sever its reliance on the World Heart TET. The Abiomed CEO instructed one of his engineers, Dr. Z, to develop a home-grown TET device as quickly as possible. Dr. Z was a very capable fellow, and after only four months of effort, he succeeded in designing and testing a fully working version of a TET device that met all the requirements of Abiomed's impending heart-pump test.

This development led to a lawsuit by World Heart, who claimed that Abiomed could not have developed its own TET device in the mere time span of four months without having stolen secrets and technology from World Heart. After all, World Heart claimed, its engineers had worked on the project for four years and were only just beginning to converge on a possible solution. How could the Abiomed engineer have designed a superior TET in only four months? Abiomed countered with a claim that its short path to success was due solely to the high competency of its engineer, and that, in fact, it had taken special precautions to ensure that nothing would be stolen from the previous unsuccessful World Heart design effort. The suit went to court and, in the end, Abiomed prevailed. The jury recognized that Abiomed's TET design, while performing the same basic transcutaneous energy-transfer function as the World Heart device, was completely different with regard to all details of implementation. The Abiomed device, the jury concluded, used different circuits, materials, magnetic construction, and semiconductor components.

During the trial, a key component of Abiomed's defense was the logbook kept by the engineer who designed the TET. His logbook was used to prove that Abiomed had designed its own independent version of the TET. Following the company's customary logbook policy, as well as sound engineering practice, Dr. Z had kept careful records of his TET design, having entered every design concept, schematic, circuit layout, and test result that emerged during the design process. He had even noted the various circuit configurations that had gone up in smoke on the test bench before his first working model emerged. Each page of his logbook had been dated, and each lab session involving successful tests had been signed and countersigned by another Abiomed engineer. During trial, entire pages of Dr. Z's logbook were reproduced and projected on a large screen for the jury to see. Dr. Z's logbook played a crucial role in the success of Abiomed's legal defense.

The presentation of Dr. Z's logbook in court, while critical to the outcome of the trial, did not proceed flawlessly, however. Several pages of the logbook involving work done in January 1996 had been incorrectly dated with the year 1995. Following a common mistake that many individuals make when the year changes, Dr. Z had, without thinking, hastily written the year from the previous month of December. The lawyers for World Heart were quick to seize upon this error as prima facie evidence that something was "fishy." They claimed Dr. Z had forged portions of his logbook in an attempt to present a false picture of his accomplishments. In the end, the jury was not convinced and realized that Dr. Z's error was nothing more than a common calendar mistake. Nevertheless, this seemingly small lack of attention to detail in logbook procedure put the case against Abiomed in jeopardy for a time.

PROFESSIONAL SUCCESS: DEVELOPING GOOD LOGBOOK HABITS

Most of us follow routines without thinking. When we wake up in the morning, we brush our teeth. When we eat, we instinctively reach for a clean plate. When we get into a car, we (hopefully) buckle our seatbelts automatically. An as engineer, the urge to write things down in a logbook should become as instinctive as these other common tasks. In contrast to these personal procedures for our own well being, however, we engineers are not trained from childhood to record our experiences in a notebook. Developing this instinct requires practice, but it can become part of your routine over time. When personal computers and the Internet first came into being, most people did not think very much about the novelty of e-mail. Now checking one's mail has become a daily routine for most. You developed this unnatural skill by practicing it over time. It should be the same with your engineer's logbook. Force yourself to get into the habit of using your logbook whenever you practice design. Over time, it will become as natural as brushing your teeth.

4 LEGAL ISSUES: INTELLECTUAL PROPERTY, PATENTS, AND TRADE SECRETS

It would be nice if engineers could focus solely on technical issues. After all, engineering students spend an enormous amount of time learning about fundamentals and applications, plus many technical skills. While most curricula in engineering include elements that address the social impact of engineering, including ethical, economic, and social issues, arguably the most important of these is the role that legal issues play in the design process.

The juxtaposition between legal issues and design practice falls squarely with the domain of project management. Hence, we conclude this chapter with a discussion of patent lore. The worldwide system of patents and licensing, part of the broader world known as *intellectual property*, has as its noble goal the protection of new ideas and inventions. The patent system originated as a means of encouraging innovation by preventing large companies with lots of money from gobbling up the ideas of smaller inventors and dominating the market. While well intentioned in concept, the patent system has evolved into a maze of regulations and practices that can be most perplexing to engineers uninitiated in the world of legal procedures. For this reason, most large companies employ attorneys whose job it is to navigate the company through the often bewildering world of intellectual property.

Another important arena in which legal issues intersect engineering is in the area of product liability. The western world, particularly the United States, has become a litigious society in which companies can be sued for all sorts of real or perceived product defects leading to injury, loss of property, or loss of life. The more dramatic stories, particularly those related to automobile accidents, are revealed to the general public via the media. But countless other cases that have filled civil courtrooms everywhere are not as widely disseminated. Understanding how the management of a design project impacts the issue of product liability is a critical engineering skill.

This section introduces some of the basic legal concepts that underscore the relationship between engineering and the law. While the material presented here merely skims the surface of intellectual property law, it will introduce you to the language of the legal profession and build your awareness of legal issues as they relate to engineering design.

4.1 Patents

A patent can be issued for any invention provided that certain criteria are met. The invention must introduce a new concept or way of doing something. The product must be something that can actually be produced, and it must have been *reduced to practice* (i.e., produced in at least prototype form) prior to the initial patent application. Also, the subject of the patent must not simply be an obvious synthesis of preexisting ideas. The first page of a typical patent document is shown in Figure 18.

A patent is valid for seventeen years (in the United States) during which time the inventor has sole rights to produce or license the invention. A patent *license* is a legal contract binding the inventor, who may not have the resources to actually produce the invention, with a company that has the resources and desires to manufacture and market the invention. In the United States, patents are awarded, or *issued* by the United States Patent and Trademark Office (USPTO, or simply "Patent Office"). This organization maintains a comprehensive public website (*www.uspto.gov*) from which anyone may download patents dating back as far as 1976. Earlier patents may be ordered from the patent archive as hardcopy for a small fee.

4.2 Patent Jargon

Lawyers have their own vocabulary for describing the patent process. The concept of *prior art* refers to the body of knowledge that existed at the time of the patented invention.

(12) **United States Patent**
Bifano et al.

(10) **Patent No.: US 6,529,311 B1**
(45) **Date of Patent: Mar. 4, 2003**

(54) **MEMS-BASED SPATIAL-LIGHT MODULATOR WITH INTEGRATED ELECTRONICS**

(75) Inventors: **Thomas Bifano,** Mansfield, MA (US); **Mark Horenstein,** West Roxbury, MA (US)

(73) Assignee: **The Trustees of Boston University,** Boston, MA (US)

(*) Notice: Subject to any disclaimer, the term of this patent is extended or adjusted under 35 U.S.C. 154(b) by 0 days.

(21) Appl. No.: **09/702,054**

(22) Filed: **Oct. 30, 2000**

Related U.S. Application Data

(60) Provisional application No. 60/161,939 filed on Oct. 28, 1999.

(51) **Int. Cl.**7 .... **G02B 26/00**

(52) **U.S. Cl.** **359/291;** 359/290; 359/223; 359/224; 359/298

(58) **Field of Search** 359/290, 291, 359/223, 224, 298, 248, 900

(56) **References Cited**

U.S PATENT DOCUMENTS

4,954,789 A	*	9/1990	Sampsell	330/4.3
5,312,513 A	*	5/1994	Florence et al.	156/643
5,535,047 A	*	7/1996	Hornbeck	359/295
5,867,302 A		2/1999	Fleming	359/291
6,028,689 A		2/2000	Michalicek et al.	359/224
6,123,985 A		9/2000	Robinson et al.	428/162
6,181,460 B1		1/2001	Tran et al.	359/291
6,329,738 B1	*	12/2001	Hung et al.	310/309
6,396,619 B1		5/2002	Huibers et al.	359/291

* cited by examiner

Primary Examiner—Hung Xuan Dang
Assistant Examiner—Tuyen Tra
(74) *Attorney, Agent, or Firm*—Weingarten, Schurgin, Gagnebin & Lebovici LLP

(57) **ABSTRACT**

Method and apparatus for forming an array of reflective elements for spatial light modulation. The array includes a substrate supporting electronically addressable actuators, each associated with a corresponding reflective element, a coupling attaching each actuator to the corresponding reflective element to place each reflective element in a substantially planar surface. Each electronically addressable actuator responds to predetermined addressing from a processing circuit to reposition the corresponding reflective element out of the planar surface a predetermined distance identified in the predetermined electronic addressing.

9 Claims, 7 Drawing Sheets

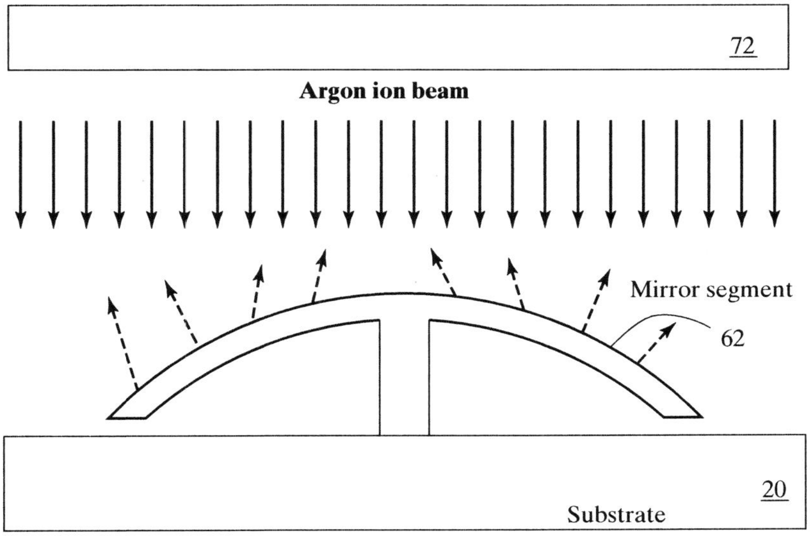

Figure 18. Top header of a patent issued by the U.S. Patent and Trademark Office.

A device cannot be patented if it is described anywhere in the prior art. The invention must be reproducible, and the language of the patent must be sufficiently clear that it can be understood by an individual *skilled in the art*. The latter phrase is used to describe someone who, at the time of the invention, would have been familiar with the field of endeavor to which the invention applies and would have had the requisite education typical of those working in the general area of technology. To be patentable, a new invention must pass the tests of *obviousness* and *anticipation*. In legal terms, an idea is "obvious" if elements of prior art can be combined to produce the invention, *and* if, at the time of the patent application, the general literature provided suggestions for doing so. An invention is "anticipated" if a single piece of prior art describes the invention in its entirety. These tests are applied as part of the application process, known simply as the patent's *prosecution history*. If an invention is deemed obvious or anticipated in the prior art by the USPTO, then it is not patentable.

4.3 The Process of Obtaining a Patent

When an individual or company seeks to obtain patent protection for an invention, several well-defined steps must be taken to ensure success.* The time span from invention to patent issue is typically a year or more and usually requires the assistance of an attorney. At each step in the process, the patent may be rejected by the Patent Office. Conversely, being issued a patent for a particular invention is an exciting event often marked by celebration at the receiving company.

Disclosure—The first step for the inventor is to file a *patent disclosure*. A disclosure stakes claim to the invention and establishes the date of origin of the invention. Compared to a formal patent document, which utilizes a very specific language structure and can be quite lengthy, a disclosure is a much simpler document that describes the invention in plain language. Filing a disclosure does not guarantee that the invention will be patentable, but it contains enough information to uniquely identify both the inventor and the invention, and it firmly establishes the date of invention. An example of a patent disclosure, filed by Boston University on behalf of the author of this text, is shown in Figure 19. This disclosure describes a unique circuit used to control micro-electromechanical (MEMS) devices for optical applications.

Patent Prosecution—Following the filing of a disclosure, the organization seeking the patent has the responsibility of convincing the Patent Office that the invention is unique and worthy of patent protection. This process is formally known as "patent prosecution." Prosecution begins when the inventor, or, more typically, the attorney acting on behalf of the inventor, submits a draft of the formal patent document for review by the Patent Office. Also submitted are all the documents known to describe the prior art at the time of the invention. Patent prosecution is an iterative process during which the Patent Office may question some of the claims made in the patent in light of the prior art, or introduce other pieces of prior art not known to the inventor. An exchange of arguments for and against often transpires between the Patent Office and the inventor's attorney until an agreement is reached as to what exact features of the invention will be covered by the patent. These features are jointly referred to as the patent's *claims*. Patent claims may be *literal*, describing a physical apparatus, or they may be *method claims* that describe a manufacturing technology or way of using the new invention.

---

*The steps outlined in this section describe the United States patent process. Procedures in other countries may differ.

BOSTON UNIVERSITY TECHNOLOGY DISCLOSURE

1. Title of Invention:
Method for Linearizing Deflection of a MEMS Device Using Binary Electrodes and Voltage Modulation

2. Inventor(s) Information:

Name(s)	Position Title(s)	School/Department or Collaborating Organization	Telephone Number
Mark N. Horenstein	Professor	Dept. of Elec. and Comp. Eng.	617-353-2800

3. Events:

	Event Description:	Date:	Location:	References & Comments:
A.	First description of complete invention, oral or written (*conception*)	2/27/2003	Precision Engineering Laboratory	Discussions with T. Bifano and P. Bierden
B.	Invention development records, notes, drawings (Evidence of *diligence*)	3/1/2003	Rm. PHO 527 8 Saint Mary's St.	Analysis and Development
C.	First successful demonstration, if any (first actual *reduction to practice*)	3/4/2003	Rm. PHO 527 8 Saint Mary's St.	First Successful Simulation
D.	First publication containing full description of invention			
E.	External disclosures (in the past, or expected in the future, with date): None			

I (We) hereby agree to assign all right, title and interest to this invention to Boston University and agree to execute all documents as requested, assigning to Boston University our right in any patent application filed on this invention, and to cooperate with the Boston University Office of Technology Transfer in the protection of this invention. Boston University will share any royalty income derived from the invention with the inventor(s) according to its standard policies.

Mark N. Horenstein — 3/20/2003
Inventor's Signature — Date

44 Cummington St., Boston, MA 02215
Home Address

000-00-0000 — United States
Social Security Number (required) — Country of Citizenship

4. Please attach the following (both hard and electronic copies):
- A detailed description of the invention;
- A two or three sentence, non-confidential description of its usefulness;
- A one page non-confidential description of its usefulness;
- A list of potential licensees.

5. Technology disclosed to and understood by:

Elicia Keyes — Elicia Keyes
Name of Non-Inventor Witness — **Signature of Non-Inventor Witness**

Figure 19. Example of a patent disclosure.

4.4 The Anatomy of a Patent

Much like a technical journal article, a patent has a predefined structure that helps the reader understand the key elements of the described invention as well as the specific claims of the patent. The major elements of a patent are illustrated in block diagram form in Figure 20. Most of these patent components, described in detail below, can be

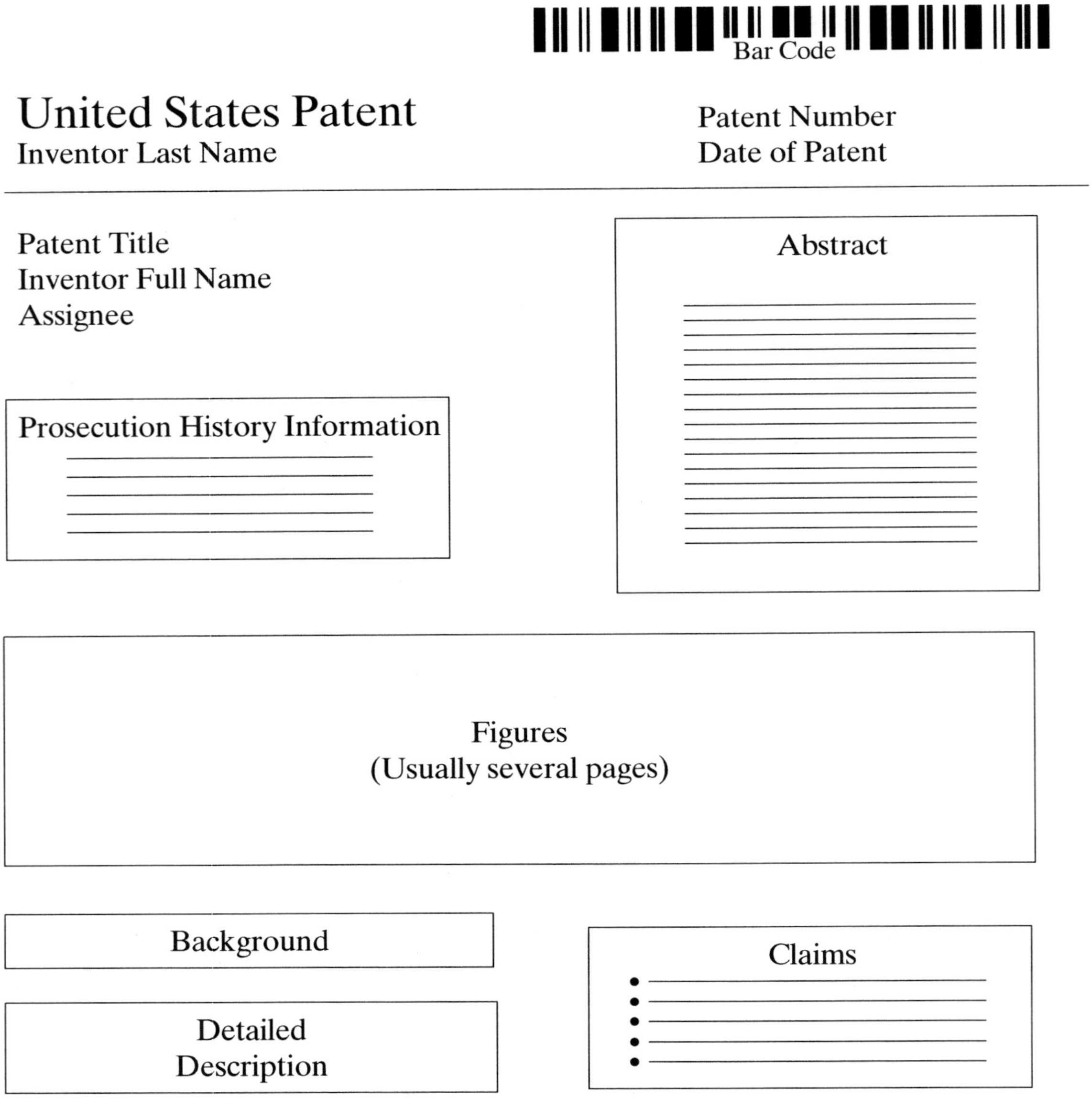

Figure 20. Structure of a patent shown in block diagram form.

identified on the top page of the patent shown in Figure 18. The next-to-last page of the patent, showing the last few paragraphs of the detailed description, and the first of the numbered claims, is provided in Figure 21.

Patent Number—Every patent issued by the USPTO is given a unique number assigned according to the order of issue. Currently, this number is well into the six millions.

Patent Title—As part of the application process, the inventor must choose a title that describes the key focus of the patent.

US 6,529,311 B1

3

an oxygen plasma. The polymer photoresist layer **50** is exposed and developed to leave apertures **52** in FIGS. **8**A and B for the deposition of a metalization or metal layer **54** for the anchors and for the diaphragm and its flexures described above. The metal deposition **54** as shown in FIGS. **9**A and B can be sputtered material such as a chromium-aluminum composite, aluminum, gold, or nickel.

The metal layer is patterned and etched in FIGS. **10**A and B to leave the flexure **30** supported diaphragm **20**. A polymer photoresist and reactive ion etch, such as in a chlorine atmosphere, may be used to create and separate the flexures **30** and diaphragm **20**.

At this point, the photoresist may be released by wet solvent procedures as shown in FIGS. **11**A and B or the mirror structure may be begun as shown in FIGS. **12**A and B using a second layer of polymer resist **56**. The resist is patterned to leave upon being developed an aperture **58** for the formation in the steps of FIGS. **13**A and **B** of a metalized or metal layer **60** for the mirror. The procedures are similar to those in forming the metal layer **54** for the diaphragms **20**. The metal layer **60** is patterned and etched as before in FIGS. **14**A and B to separate the mirror structures **62** and their struts or posts **64**. Finally the whole structure is subjected to a wet solvent procedure to remove all polymer. At some point in the procedure, such as at FIGS. **13**A and B, before polymer release, the device may be given a surface polishing to remove surface roughness and improve the quality of reflection.

In some cases and in reference to FIGS. **16–19**, the mirror **62** will exhibit a stress induced curvature resulting from the stresses built into the metal layer **60** during formation and release of supporting polymer at the conclusion of fabrication. These stresses **64**, as shown in FIG. **16**, vary over the depth of the mirror **62**, and in fact change polarity. Thus, the stresses can be balanced giving a planar mirror surface by removal of portions of the mirror element **62** until a point is reached where the stresses combine to keep the mirror surface flat. This point can be reached in the process of removal of surface layers as shown in FIG. **19**. The procedure for removal may utilize an ion beam **70** in an argon atmosphere to cut back the mirror surface. The process can be monitored by an interferometer **72** to detect the point of maximum flatness.

The circuitry **26** shown in FIG. **3** can be of several forms as illustrated in FIGS. **20** and **21**. In FIG. **20** a CPU **80**, off chip, applies instructions including addressing information designating, in a repeating sequence over the whole array **12**, each of mirrors **16** to be moved and data typically in the form of a voltage indicating the amount and polarity of displacement of that mirror. This information is fed to the circuit **26** at each mirror assembly to an address decoder **82** and voltage decoder **84** where the voltage is stored in a capacitive memory **86**. A driver gate **88** is activated when the corresponding mirror is addressed to apply that voltage through it to a capacitive plate **90** which in turn applies an electrostatic force to the actuator diaphragm **20**. Sufficient motion can be achieved with a low voltage of, for example a few volts compatible with CMOS circuitry, to achieve the 360 degree change in light phase on the mirrors.

An alternative CMOS circuit is illustrated in FIG. **21** where an on or off chip processor **100** applies via a data bus **102** to respective decoders **104** addressing and displacement information. In a typical application of eight bit data, a 256×256 mirror array can be addressed and a data byte of eight bits used to achieve a resolution of 256 displacement positions. In this example, the decoder determines from the addressing when its corresponding mirror is being addressed and then uses each of the eight bits to apply a low voltage to corresponding capacitive plates in an array **108**. The plates are sequentially sized, typically each plate being twice the size of its neighbor. Each data bit applies or does not apply a fixed voltage to the corresponding plate based on the bit being of one state or the other, achieving a combined force proportional to the area of activated plates and a resolution of 256 positions.

4

The invention can be broadly scaled to different size arrays and mirror areas. A total mirror displacement of half a micron can be provided to achieve the desired phase change in the optical spectrum. The spacing of the diaphragms **20** and CMOS circuits **26** is a function of the voltage available and the total desired displacement, response time and other factors within the grasp of those skilled in the art.

The invention is not intended to be limited by any of the above description and is to be interpreted on the scope of the following claims.

What is claimed is:

1. A method for fabrication of an array of individually positioned reflectors for spatial light modulation comprising the steps of:

providing plural addressable circuits in an array on a substrate;

forming a flexure supported platform associated with each said circuit, said platform being flexure supported from said substrate; and

attaching a reflecting element to said platform on a side opposite from said substrate thereby providing a plurality o f reflecting elements of said array aligned in a surface;

said circuit having electronics operative in response to a signal addressed thereto and operative in association with said platform to apply a force to move said platform and said reflecting element together to a predetermined position out of alignment with the array surface;

wherein said circuits providing step includes the step of providing a conductive plate on said substrate facing said platform, an addressable gate and a voltage memory, said gate operative to apply a voltage in said memory to said plate in response to said signal;

said signal having a voltage component corresponding to the predetermined position to which said reflective element is to be moved and an address component for activating said gate of an appropriately addressed circuit.

2. A method for fabrication of an array of individually positioned reflectors for spatial light modulation comprising the steps of:

providing plural addressable circuits in an array on a substrate;

forming a flexure supported platform associated with each said circuit, said platform being flexure supported from said substrate; and

attaching a reflecting element to said platform on a side opposite from said substrate thereby providing a plurality of reflecting elements of said array aligned in a surface;

said circuit having electronics operative in response to a signal addressed thereto and operative in association with said platform to apply a force to move said platform and said reflecting element togetber to a predetermined position out of alignment with the array surface;

Figure 21. Next-to-last page of the patent of Figure 18. The last paragraphs of the detailed description and the beginning of the numbered claims can be found on this page.

Inventor—The inventor of a patent must always be a person (or persons). The inventor is the individual first responsible for bringing the object of the invention from concept to practice.

Assignee—The assignee of a patent can be an individual, set of individuals, or a company. The assignee has the legal and commercial rights to the patent and also the right to license the patent to others for a privately negotiated fee. When an individual has arrived at an invention while on the payroll of an employer, the assignee is usually the latter.

Abstract—The abstract provides a stand-alone summary of the invention. It allows the casual reader to understand the context and impact of the patent without reading the entire document.

Illustrations—Traditionally, the figures that accompany a patent appear before the written sections. Each element of a figure that will be addressed in the text is labeled with a unique number that is repeated every time that element appears in the patent. This construct may appear cumbersome to the average person, and it can be tedious to read, but it is a time-tested method that reduces ambiguities in the patent document.

Background of the Invention—This section describes the area of technology to which the patent applies and the problem solved by the invention. It also describes the prior art leading up to the time of the invention.

Description of the Invention—Usually the longest portion of the patent document, this section provides a detailed outline of the invention. The goal of this section is to enable an individual "skilled in the art" to completely reconstruct and produce the patented invention.

Claims—In this final section of the patent, the specific elements of the invention claimed by the assignee are outlined in detail. It is the set of claims, not the preceding description of the invention, that forms the legal basis for the patent and defines the specific legal protection granted to the inventor. The first claim listed is called the *primary* claim. Further claims that rely on the primary claim for validity are called *dependent claims*. Lawyers use the phrase *claim limitation* to describe the individual pieces of intellectual property protected by the claims.

Although the description of patents provided above is simplified, it captures the essence of patent structure and format. A more comprehensive look at patents and the patent system can be acquired by perusing the website of the United States Patent and Trademark Office.

KEY TERMS

Documentation
Gantt chart
Organizational chart
Patent
PERT chart
Project management
Teamwork
Time line

Problems

1. Develop a simple checklist for your homework assignments over the coming week.
2. Prepare a checklist of tasks required to tune a bicycle for optimum performance.
3. Prepare a checklist that will help guide you in the design of a child's safety seat.

4. Imagine that you work for a company that is designing a hybrid gas–electric car for commercial sale. Create an organizational chart for the company and a Gantt chart for designing the vehicle's drive train.
5. Choose an engineering company with which you are familiar or in which you have an interest. Develop an organizational chart for the company. Information about a company's structure and personnel usually can be found on the company's website.
6. Imagine that you wish to start your own company to write software tools for doing business on the Internet. Create an organizational chart that outlines the positions you'll need to fill in order to get the company started.
7. Develop a time line for the completion of the prototype of an automobile powered from fuel cells rather than an internal combustion engine.
8. Develop a time line for completing your course requirements over the time span of 4 academic years.
9. Suppose that you've been given the assignment to develop a human-powered airplane. Develop a time line for completing this ambitious assignment.
10. Create a Gantt chart for an entry into the national solar-powered vehicle design competition, First Solar (*www.firstsolar.com*).
11. Prepare a Gantt chart for the design of a retrofit fire escape for a 10-story city building.
12. Prepare your own version of a Gantt chart that might have been used for the construction of the Golden Gate Bridge in San Francisco, California.
13. Develop a Gantt chart for housing an educational conference on engineering design. Consider all needed arrangements, including food, transportation, lodging, and meeting facilities.
14. Develop a Gantt chart for the design of a cell phone tower.
15. Consider the case of four engineers who are designing an entry into a material handling competition (see *www.mhiq.org*). Develop a Gantt chart for the design of a system based on a complete list of ideas that you develop. The vertical axis of your Gantt chart should reflect members of the design team, not specific project tasks.
16. Construct a "fuzzy" Gantt chart depicting the tasks required to design a software system for tracking inventory at a hardware store. Each time allocation should include an error estimate—an increment of fuzzy time—that allows for possible early or late completion of every task on the chart. Based on your fuzzy estimates, what are the longest and shortest possible time durations of the design effort? The vertical axis of your chart should indicate software developers, not individual design tasks.
17. Consider the fuzzy Gantt chart scenario of the previous problem. Develop one that might apply to the design of a pilotless model surveillance airplane.
18. Can you create a Gantt chart for the creation of the Earth? Your chart can be based on either a religious, evolutionary, or cosmological perspective, as you prefer.

19. Do some research into the steps required to build a two-story, commercial-grade building. Then construct a PERT chart applicable to the design and construction of a public library.

20. Construct a PERT chart that you imagine to be applicable to the design of the International Space Station.

21. Imagine that you lead a team of engineers designing a breakwater for a coastal sea port. Develop a PERT chart for the completion of this project. Consider all elements, from basic research and data analysis to final construction. Identify any critical pathways in your chart.

22. Consider the case of a car design competition in which the object is to be the first to drop a small beanbag (a "Hacky Sack®") into a 10-cm square hole cut into a 3-m square tabletop. Develop a PERT chart for the competition of a successful design effort. Identify any critical pathways in your chart and then adjust time allocations in the branches so that the project minimizes the former but results in a project that can be completed within one academic semester.

23. Make a list of conditions under which you might choose a Gantt chart over a PERT chart for time management. Then make a similar list for the opposite case.

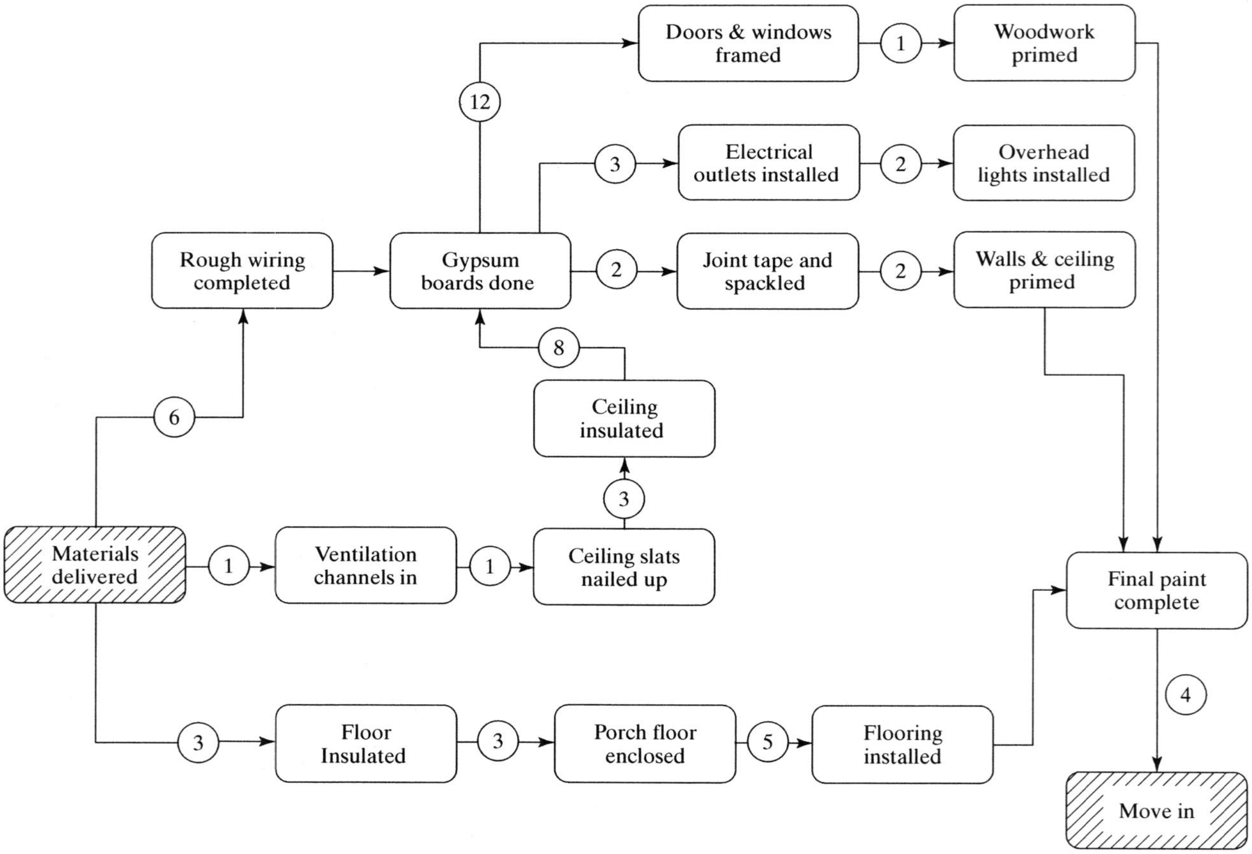

Figure 22. PERT chart for Problems 24–26.

24. Examine the PERT chart of Figure 22. Make a list of all critical pathways as well as the location and duration of all slack time intervals.

25. For the PERT chart of Figure 22, what is the total time allocated to the completion of the entire design task?

26. For the PERT chart of Figure 22, identify all the different pathways and their total allocated path intervals.

27. Develop a PERT chart for the design of a tunnel linking Beijing to New York City.

28. Outline a PERT chart for the completion of the first self-sustaining lunar exploration base.

29. Develop a PERT chart for university commencement exercises.

30. Begin to keep a logbook of your class activities. Enter sketches and records of design assignments, inventions, and ideas.

31. Sketch a logbook page that describes a design concept for a recumbent bicycle.

32. Prepare a logbook page that describes the inner workings of a common DVD player.

33. Develop a logbook entry describing a machine that shells and stores peanuts.

34. Develop a logbook page for the design of the cooling system for an indoor ice-hockey rink.

35. Pretend that you are Alexander Graham Bell, the inventor of the telephone. Prepare several logbook pages that describe your invention.

36. Pretend that you are Marie Curie, the person who discovered the radioactive element radium. Prepare several logbook pages that describe the activities leading to your discovery.

37. Pretend that you are Dr. Zephram Cockrane, the inventor of plasma warp drive on the television and movie series *Star Trek*. Prepare several logbook pages that describe your invention.

38. Imagine that you are Elias Howe, the first inventor to perfect the sewing machine by putting the eye of the needle in its tip. This innovation made possible the bobbin system still in use in sewing machines today. Prepare several logbook pages that describe your invention and its initial tests.

39. Reconstruct logbook pages as they might have been written by Johan Vaaler, the Norwegian inventor of the common paperclip. (Write your entries in English).

40. Imagine that you are Dr. Z, the engineer involved in the Abiomed case of Section 3.8. Sketch the logbook pages that outline the basic operating concept of your transcutaneous energy transfer device.

41. The invention of the incandescent light bulb is largely attributed to the famous American inventor, Thomas Edison. Reconstruct the logbook pages that Edison may have kept describing his classic design efforts.

42. The cotton gin was developed by an American inventor, Eli Whitney, around 1800. This invention had a profound effect on the economic history of the early United States. Reconstruct logbook pages in which Whitney outlines the basic features and development of his invention. Note that patent law in America was in its infancy around the time that Whitney did his work on the cotton gin.

43. Samuel F.B. Morse, inventor of the telegraph in the 1830s and the pioneer who launched the world's first "information age," was actually an artist by profession when he developed his classic invention. During a long voyage home from study in France, he passed his time thinking about conversations he had heard concerning ongoing experiments in Europe on electricity and magnetism. He developed his ideas for the telegraph while returning to the United States. Sketch the logbook pages that Morse may have kept during his long sea voyage.

44. The first pocket calculator was designed by Jack Kilby, an engineer for Texas Instruments. Look up the history of this inventor, and see if you can reconstruct the probable appearance of one or more pages from his logbook.

15

Introduction to Engineering Ethics

On August 10, 1978, a Ford Pinto was hit from behind on a highway in Indiana. The impact of the collision caused the Pinto's fuel tank to rupture and burst into flames, leading to the deaths of three teenage girls riding in the car. This was not the first time that a Pinto had caught on fire as the result of a rear-end collision. In the seven years since the introduction of the Pinto, there had been some 50 lawsuits related to rear-end collisions. However, this time Ford was charged in a criminal court for the deaths of the passengers.

This case was a significant departure from the norm and had important implications for the Ford engineers and managers. A civil lawsuit could only result in Ford being required to pay damages to the victim's estates. A criminal proceeding, on the other hand, would indicate that Ford was grossly negligent in the deaths of the passengers and could result in jail terms for the Ford engineers or managers who worked on the Pinto.

The case against Ford hinged on charges that it was known that the gas-tank design was flawed and was not in line with accepted engineering standards, even though it did meet applicable federal safety standards at the time. During the trial, it was determined that Ford engineers were aware of the dangers of this design, but management,

SECTIONS

1.1 Background Ideas
1.2 Why Study Engineering Ethics?
1.3 Engineering Is Managing the Unknown
1.4 Personal vs. Business Ethics
1.5 The Origins of Ethical Thought
1.6 Ethics and the Law
1.7 Ethics Problems Are Like Design Problems
1.8 Case Studies

OBJECTIVES

After reading this chapter, you will be able to

- Know why it is important to study engineering ethics
- Understand the distinction between business and personal ethics, and
- See how ethical problem solving and engineering design are similar.

concerned with getting the Pinto to market rapidly at a price competitive with subcompact cars already introduced or planned by other manufacturers, had constrained the engineers to use this design.

The dilemma faced by the design engineers who worked on the Pinto was to balance the safety of the people who would be riding in the car against the need to produce the Pinto at a price that would be competitive in the market. They had to attempt to balance their duty to the public against their duty to their employer. Ultimately, the attempt by Ford to save a few dollars in manufacturing costs led to the expenditure of millions of dollars in defending lawsuits and payments to victims. Of course, there were also uncountable costs in lost sales due to bad publicity and a public perception that Ford did not engineer its products to be safe.

1.1 BACKGROUND IDEAS

The Pinto case is just one example of the ethical problems faced by engineers in the course of their professional practice. Ethical cases can go far beyond issues of public safety and may involve bribery, fraud, environmental protection, fairness, honesty in research and testing, and conflicts of interest. During their undergraduate education, engineers receive training in basic and engineering sciences, problem-solving methodology, and engineering design, but generally receive little training in business practices, safety, and ethics.

This problem has been partially corrected, as many engineering education programs now have courses in what's called engineering ethics. Indeed, the Accreditation Board for Engineering and Technology (ABET), the body responsible for accrediting undergraduate engineering programs in the United States, has mandated that ethics topics be incorporated into undergraduate engineering curricula. The purpose of this book is to provide a text and a resource for the study of engineering ethics and to help future engineers be prepared for confronting and resolving ethical dilemmas, such as the design of an unsafe product like the Pinto, that they might encounter during their professional careers.

A good place to start a discussion of ethics in engineering is with definitions of ethics and engineering ethics. Ethics is the study of the characteristics of morals. Ethics also deals with the moral choices that are made by each person in his or her relationship with other persons. As engineers, we are concerned with ethics because these definitions apply to all of the choices an individual makes in life, including those made while practicing engineering.

For our purposes, the definition of ethics can be narrowed a little. Engineering ethics is the rules and standards governing the conduct of engineers in their role as professionals. Engineering ethics encompasses the more general definition of ethics, but applies it more specifically to situations involving engineers in their professional lives. Thus, engineering ethics is a body of philosophy indicating the ways that engineers should conduct themselves in their professional capacity.

1.2 WHY STUDY ENGINEERING ETHICS?

Why is it important for engineering students to study engineering ethics? Several notorious cases that have received a great deal of media attention in the past few years have led engineers to gain an increased sense of their professional responsibilities. These cases have led to an awareness of the importance of ethics within the engineering profession as engineers realize how their technical work has far-reaching impacts on society. The work of engineers can affect public health and safety and can influence business practices and even politics.

One result of this increase in awareness is that nearly every major corporation now has an ethics office that has the responsibility to ensure that employees have the ability to express their concerns about issues such as safety and corporate business practices in a way that will yield results and won't result in retaliation against the employee. Ethics offices also try to foster an ethical culture that will help to head off ethical problems in a corporation before they start.

The goal of this book and courses in engineering ethics is to sensitize you to important ethical issues before you have to confront them. You will study important cases from the past so that you will know what situations other engineers have faced and will know what to do when similar situations arise in your professional career. Finally, you will learn techniques for analyzing and resolving ethical problems when they arise.

Our goal is frequently summed up using the term "moral autonomy." Moral autonomy is the ability to think critically and independently about moral issues and to apply this moral thinking to situations that arise in the course of professional engineering practice. The goal of this book, then, is to foster the moral autonomy of future engineers.

The question asked at the beginning of this section can also be asked in a slightly different way. Why should a future engineer bother studying ethics at all? After all, at this point in your life, you're already either a good person or a bad person. Good people already know the right thing to do, and bad people aren't going to do the right thing no matter how much ethical training they receive. The answer to this question lies in the nature of the ethical problems that are often encountered by an engineer. In most situations, the correct response is very obvious. For example, it is clear that to knowingly equip the Pinto with wheel lugs made from substandard, weak steel that is susceptible to breaking is unethical and wrong. This action could lead to the loss of a wheel while driving and could cause numerous accidents and put many lives at risk. Of course, such a design decision would also be a commercial disaster for Ford.

However, many times, the ethical problems encountered in engineering practice are very complex and involve conflicting ethical principles. For example, the engineers working on the Pinto were presented with a very clear dilemma. Trade-offs were made so that the Pinto could be successfully marketed at a reasonable price. One of these trade-offs involved the placement of the gas tank, which led to the accident in Indiana. (This case will be presented in more depth in a later section of this book on safety.) So, for the Ford engineers and managers, the question became the following: Where does an engineering team strike the balance between safety and affordability and, simultaneously, the ability of the company to sell the car and make a profit.

These are the types of situations that we will discuss in this book. The goal, then, is not to train you to do the right thing when the ethical choice is obvious and you already know the right thing to do. Rather, the goal is to train you to analyze complex problems and learn to resolve these problems in the most ethical manner.

1.3 ENGINEERING IS MANAGING THE UNKNOWN

One source of the ethical issues encountered in the course of engineering practice is a lack of knowledge. This is by no means an unusual situation in engineering. Engineers often encounter situations in which they don't have all of the information that is needed. By its nature, engineering design is about creating new devices and products. When something is new, many questions need to be answered. How well does it work? How will it affect people? What changes will this lead to in society? How well will this work under all of the conditions that it will be exposed to? Is it safe? If there are some safety concerns, how bad are they? What are the effects of doing nothing?

So, to a large extent, an engineer's job is about managing the unknown. How does an engineer accomplish this? Really, as an engineer you can never be absolutely certain that your design will never harm anyone or cause detrimental changes to society. But you must test your design as thoroughly as time and resources permit to ensure that it operates safely and as planned. Also, you must use your creativity to attempt to foresee the possible consequences of your work.

1.4 PERSONAL VS. BUSINESS ETHICS

In discussing engineering ethics, it is important to make a distinction between personal ethics and professional, or business, ethics, although there isn't always a clear boundary between the two. Personal ethics deals with how we treat others in our day-to-day lives. Many of these principles are applicable to ethical situations that occur in business and engineering. However, professional ethics often involves choices on an organizational level rather than a personal level. Many of the problems will seem different because they involve relationships between two corporations, between a corporation and the government, or between corporations and groups of individuals. Frequently, these types of relationships pose problems that are not encountered in personal ethics.

1.5 THE ORIGINS OF ETHICAL THOUGHT

Before proceeding, it is important to acknowledge in a general way the origins of the ethical philosophies that we will be discussing in this book. The Western ethical thought that is discussed here originated in the philosophy of the ancient Greeks and their predecessors. It has been developed through subsequent centuries by many thinkers in the Judeo–Christian tradition. Interestingly, non-Western cultures have independently developed similar ethical principles.

Although for many individuals, personal ethics are rooted in religious beliefs, this is not true for everyone. Certainly, there are many ethical people who are not religious, and there are numerous examples of nominally religious people who are not ethical. So while the ethical principles that we will discuss come to us filtered through a religious tradition, these principles are now cultural norms in the West, and as such, they are widely accepted regardless of their origin. We won't need to refer explicitly to religion in order to discuss ethics in the engineering profession.

1.6 ETHICS AND THE LAW

We should also mention the role of law in engineering ethics. The practice of engineerin and business is governed by many laws on the international, federal, state, and local levels. Many of these laws are based on ethical principles, although many are purely of a practical, rather than a philosophical, nature. There is also a distinction between what is legal and what is ethical. Many things that are legal could be considered unethical. For example, designing a process that releases a known toxic, but unregulated, substance into the environment is probably unethical, although it is legal.

Conversely, just because something is illegal doesn't mean that it is unethical. For example, there might be substances that were once thought to be harmful, but have now been shown to be safe, that you wish to incorporate into a product. If the law has not caught up with the latest scientific findings, it might be illegal to release these substances into the environment, even though there is no ethical problem with doing so.

As an engineer, you are always minimally safe if you follow the requirements of the applicable laws. But in engineering ethics, we seek to go beyond the dictates of the law. Our interest is in areas where ethical principles conflict *and* there is no legal guidance for how to resolve the conflict.

1.7 ETHICS PROBLEMS ARE LIKE DESIGN PROBLEMS

At first, many engineering students find the types of problems and discussions that take place in an engineering ethics class a little alien. The problems are more open ended and are not as susceptible to formulaic answers as are problems typically assigned in other engineering classes. Ethics problems rarely have a correct answer that will be arrived at by everyone in the class. Surprisingly, however, the types of problem-solving techniques that we will use in this book and the nature of the answers that result bear a striking resemblance to the most fundamental engineering activity: engineering design.

The essence of engineering practice is the design of products, structures, and processes. The design problem is stated in terms of specifications: A device must be designed that meets criteria for performance, aesthetics, and price. Within the limits of these specifications, there are many correct solutions. There will, of course, be some solutions that are better than others in terms of higher performance or lower cost. Frequently, there will be two (or more) designs that are very different, yet perform identically. For example, competing automobile manufacturers may design a car to meet the same market niche, yet each manufacturer's solution to the problem will be somewhat different. In fact, we will see later that although the Pinto was susceptible to explosion after rear-end impact, other similar subcompact automobiles were not. In engineering design, there is no unique correct answer!

Ethical problem solving shares these attributes with engineering design. Although there will be no unique correct solution to most of the problems we will examine, there will be a range of solutions that are clearly right, some of which are better than others. There will also be a range of solutions that are clearly wrong. There are other similarities between engineering ethics and engineering design. Both apply a large body of knowledge to the solution of a problem, and both involve the use of analytical skills. So although the nature of the solutions to the problems in ethics will be different from those in most engineering classes, approaches to the problems and the ultimate solution will be very similar to those in engineering practice.

1.8 CASE STUDIES

Before starting to learn the theoretical ideas regarding engineering ethics and before looking at some interesting real-life cases that will illustrate these ideas, let's begin by looking at a very well-known engineering ethics case: the space shuttle *Challenger* accident. This case is presented in depth following this chapter, but at this point we will look at a brief synopsis of the case to further illustrate the types of ethical issues and questions that arise in the course of engineering practice.

Many readers are already familiar with some aspects of this case. The space shuttle *Challenger* was launched in extremely cold weather. During the launch, an O-ring on one of the solid-propellant boosters, made more brittle by the cold, failed. This failure led to the explosion during liftoff. Engineers who had designed this booster had concerns about launching under these cold conditions and recommended that the launch be delayed, but they were overruled by their management (some of whom were trained as engineers), who didn't feel that there was enough data to support a delay in the launch. The shuttle was launched, resulting in the well-documented accident.

On the surface, there appear to be no engineering ethical issues here to discuss. Rather, it seems to simply be an accident. The engineers properly recommended that there be no launch, but they were overruled by management. In the strictest sense this can be considered an accident—no one wanted the *Challenger* to explode—but there are still many interesting questions that should be asked. When there are safety concerns, what is the engineer's responsibility before the launch decision is made? After the launch decision is made, but before the actual launch, what duty does the engineer have? If the decision doesn't go the engineer's way, should she complain to upper management? Or should she bring the problem to the attention of the press? After the accident has occurred, what are the duties and responsibilities of the engineers? If the launch were successful, but the *post mortem* showed that the O-ring had failed and an accident had very nearly occurred, what would be the engineer's responsibility? Even if an engineer moves into management, should he separate engineering from management decisions?

These types of questions will be the subject of this book. In subsequent chapters, ideas about the nature of the engineering profession, ethical theories, and the application of these theories to situations that are likely to occur in professional practice will be presented. Many other real-life cases taken from newspaper accounts and books will be discussed to examine what engineers should do when confronted with ethically troubling situations. Most of these cases will be *post mortem* examinations of disasters, but two will involve analysis of situations in which disaster was averted when many of the individuals involved made ethically sound choices and cooperated to solve a problem.

A word of warning is necessary before these cases are studied. The cliché "Hindsight is 20/20" will seem very true as we examine many of these cases. When studying a case several years after the fact and knowing the ultimate outcome, it is easy to see what the right decision should have been. Obviously, had NASA owned a crystal ball and been able to predict the future, the *Challenger* would never have been launched. Had Ford known the number of people who would be killed as a result of gas-tank failures in the Pinto and the subsequent financial losses in lawsuits and criminal cases, it would have found a better solution to the problem of gas-tank placement. However, we rarely have such clear predictive abilities and must base decisions on our best guess of what the outcome will be. It will be important in studying the cases presented here to try to look at them from the point of view of the individuals who were involved at the time, using their best judgment about how to proceed, and not to judge the cases solely based on the outcome.

APPLICATION: THE SPACE SHUTTLE *CHALLENGER* ACCIDENT

The explosion of the space shuttle *Challenger* is perhaps the most widely written about case in engineering ethics because of the extensive media coverage at the time of the accident and also because of the many available government reports and transcripts of congressional hearings regarding the explosion. The case illustrates many important ethical issues that engineers face: What is the proper role of the engineer when safety issues are a concern? Who should have the ultimate decision-making authority to order a launch? Should the ordering of a launch be an engineering or a managerial decision? This case has already been presented briefly, and we will now take a more in-depth look.

BACKGROUND

The space shuttle was designed to be a reusable launch vehicle. The vehicle consists of an orbiter, which looks much like a medium-sized airliner (minus the engines!), two solid-propellant boosters, and a single liquid-propellant booster. At takeoff, all of the boosters are ignited and lift the orbiter out of the earth's atmosphere. The solid rocket boosters are only used early in the flight and are jettisoned soon after takeoff, parachute back to earth, and are recovered from the ocean. They are subsequently repacked with fuel and are reused. The liquid-propellant booster is used to finish lifting the shuttle into orbit, at which point the booster is jettisoned and burns up during reentry. The liquid booster is the only part of the shuttle vehicle that is not reusable. After completion of the mission, the orbiter uses its limited thrust capabilities to reenter the atmosphere and glides to a landing.

The accident on January 28, 1986 was blamed on a failure of one of the solid rocket boosters. Solid rocket boosters have the advantage that they deliver far more thrust per pound of fuel than do their liquid-fueled counterparts, but have the disadvantage that once the fuel is lit, there is no way to turn the booster off or even to control the amount of thrust produced. In contrast, a liquid-fuel rocket can be controlled by throttling the supply of fuel to the combustion chamber or can be shut off by stopping the flow of fuel entirely.

In 1974, the National Aeronautics and Space Administration (NASA) awarded the contract to design and build the solid rocket boosters for the shuttle to Morton Thiokol. The design that was submitted by Thiokol was a scaled-up version of the Titan missile, which had been used successfully for many years to launch satellites. This design was accepted by NASA in 1976. The solid rocket consists of several cylindrical pieces that are filled with solid propellant and stacked one on top of the other to form the completed booster. The assembly of the propellant-filled cylinders was performed at Thiokol's plant in Utah. The cylinders were then shipped to the Kennedy Space Center in Florida for assembly into a completed booster.

A key aspect of the booster design are the joints where the individual cylinders come together, known as the field joints, illustrated schematically in Figure 1.1a. These are tang and clevis joints, fastened with 177 clevis pins. The joints are sealed by two O-rings, a primary and a secondary. The O-rings are designed to prevent hot gases from the combustion of the solid propellant from escaping. The O-rings are made from a type of synthetic rubber and so are not particularly heat resistant. To prevent the hot gases from damaging the O-rings, a heat-resistant putty is placed in the joint. The Titan booster had only one O-ring in the field joint. The second O-ring was added to the booster for the shuttle to provide an extra margin of safety since, unlike the Titan, this booster would be used for a manned space craft.

EARLY PROBLEMS WITH THE SOLID ROCKET BOOSTERS

Problems with the field-joint design had been recognized long before the launch of the *Challenger*. When the rocket is ignited, the internal pressure causes the booster wall to expand outward, putting pressure on the field joint. This pressure causes the joint to open slightly, a process called "joint rotation," illustrated in Figure 1.1b. The joint was designed so that the internal pressure pushes on the putty, displacing the primary O-ring into this gap, helping to seal it. During testing of the boosters in 1977, Thiokol became aware that this joint-rotation problem was more severe than on the Titan and discussed it with NASA. Design changes were made, including an increase in the thickness of the O-ring, to try to control this problem.

Further testing revealed problems with the secondary seal, and more changes were initiated to correct that problem. In November of 1981, after the second shuttle flight, a postlaunch examination of the booster field joints indicated that the O-rings were being eroded by hot gases during the launch. Although there was no failure of the joint, there was some concern about this situation, and Thiokol looked into the use of different types of putty and alternative methods for applying it to solve the problem. Despite these efforts, approximately half of the shuttle flights before the *Challenger* accident had experienced some degree of O-ring erosion. Of course, this type of testing and redesign is not unusual in engineering. Seldom do things work correctly the first time, and modifications to the original design are often required.

It should be pointed out that erosion of the O-rings is not necessarily a bad thing. Since the solid rocket boosters are only used for the first few minutes of the flight, it might be perfectly acceptable to design a joint in which O-rings erode in a controlled manner. As long as the O-rings don't completely burn through before the solid boosters run out of fuel and are jettisoned, this design should be fine. However, this was not the way the space shuttle was designed, and O-ring erosion was one of the problems that the Thiokol engineers were addressing.

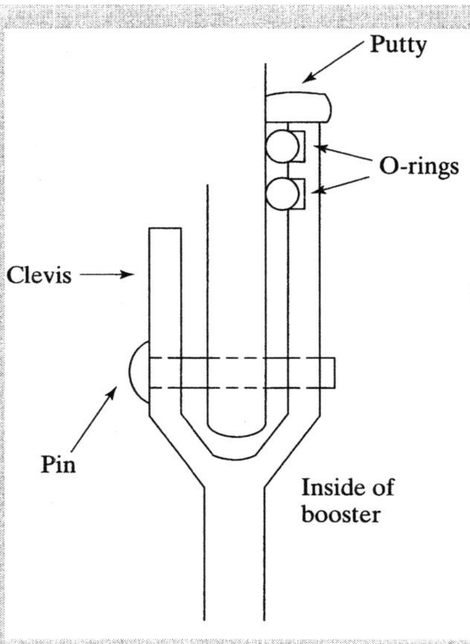

Figure 1.1a A schematic drawing of a tang and clevis joint like the one on the *Challenger* solid rocket boosters.

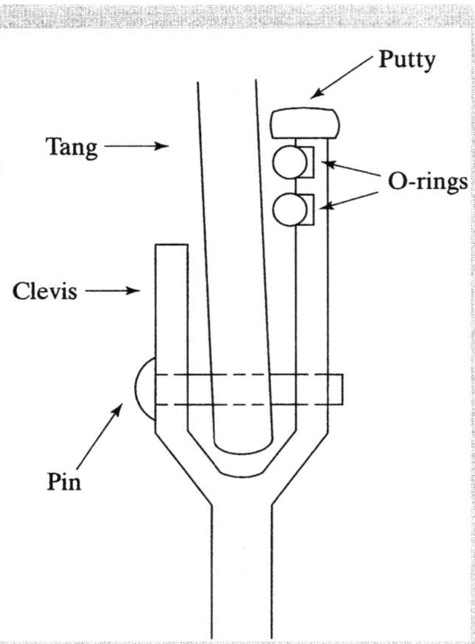

Figure 1.1b The same joint as in Figure 1.1a, but with the effects of joint rotation exaggerated. Note that the O-rings no longer seal the joint.

The first documented joint failure came after the launch on January 24, 1985, which occurred during very cold weather. The postflight examination of the boosters revealed black soot and grease on the outside of the booster, which indicated that hot gases from the booster had blown by the O-ring seals. This observation gave rise to concern about the resiliency of the O-ring materials at reduced temperatures. Thiokol performed tests of the ability of the O-rings to compress to fill the joints and found that they were inadequate. In July of 1985, Thiokol engineers redesigned the field joints without O-rings. Instead, they used steel billets, which should have been better able to withstand the hot gases. Unfortunately, the new design was not ready in time for the *Challenger* flight in early 1986 [Elliot, 1991].

THE POLITICAL CLIMATE

To fully understand and analyze the decision making that took place leading to the fatal launch, it is important also to discuss the political environment under which NASA was operating at the time. NASA's budget was determined by Congress, which was becoming increasingly unhappy with delays in the shuttle project and shuttle performance, which wasn't meeting initial promises. NASA had billed the shuttle as a reliable, inexpensive launch vehicle for a variety of scientific and commercial purposes, including the launching of commercial and military satellites. It had been promised that the shuttle would be capable of frequent flights (several per year) and quick turnarounds and would be competitively priced with more traditional nonreusable launch vehicles. NASA was feeling some urgency in the program because the European Space Agency was developing what seemed to be a cheaper alternative to the shuttle, which could potentially put the shuttle out of business.

These pressures led NASA to schedule a record number of missions for 1986 to prove to Congress that the program was on track. Launching a mission was especially important in January 1986, since the previous mission had been delayed numerous times by both weather and mechanical failures. NASA also felt pressure to get the *Challenger* launched on time so that the next shuttle launch, which was to carry a probe to examine Halley's comet, would be launched before a Russian probe designed to do the same thing. There was additional political pressure to launch the Challenger before the upcoming state-of-the-union address, in which President Reagan hoped to mention the shuttle and a special astronaut—the first teacher in space, Christa McAuliffe—in the context of his comments on education.

THE DAYS BEFORE THE LAUNCH

Even before the accident, the *Challenger* launch didn't go off without a hitch, as NASA had hoped. The first launch date had to be abandoned due to a cold front

TABLE 1-1 Space Shuttle *Challenger* Accident: Who's Who

	ORGANIZATIONS
NASA	The National Aeronautics and Space Administration, responsible for space exploration. The space shuttle is one of NASA's programs.
Marshall Space Flight Center	A NASA facility that was in charge of the solid rocket booster development for the shuttle.
Morton Thiokol	A private company that won the contract from NASA for building the solid rocket boosters for the shuttle.
	PEOPLE
NASA	
Larry Mulloy	Solid Rocket Booster Project manager at Marshall
Morton Thiokol	
Roger Boisjoly, Arnie Johnson	Engineers who worked on the Solid Rocket Booster Development Program.
Joe Kilminster	Engineering manager on the Solid Rocket Booster Development Program.
Alan McDonald	Director of the Solid Rocket Booster Project.
Bob Lund	Vice president for engineering.
Jerald Mason	General manager.

expected to move through the area. The front stalled, and the launch could have taken place on schedule. But the launch had already been postponed in deference to Vice President George Bush, who was to attend. NASA didn't want to antagonize Bush, a strong NASA supporter, by postponing the launch due to inclement weather after he had arrived. Launch of the shuttle was further delayed by a defective microswitch in the hatch-locking mechanism. When this problem was resolved, the front had changed course and was now moving through the area. The front was expected to bring extremely cold weather to the launch site, with temperatures predicted to be in the low 20's (°F) by the new launch time.

Given the expected cold temperatures, NASA checked with all of the shuttle contractors to determine if they foresaw any problems with launching the shuttle in cold temperatures. Alan McDonald, the director of Thiokol's Solid Rocket Motor Project, was concerned about the cold weather problems that had been experienced with the solid rocket boosters. The evening before the rescheduled launch, a teleconference was arranged between engineers and management from the Kennedy Space Center, NASA's Marshall Space Flight Center in Huntsville, Alabama, and Thiokol in Utah to discuss the possible effects of cold temperatures on the performance of the solid rocket boosters. During this teleconference, Roger Boisjoly and Arnie Thompson, two Thiokol engineers who had worked on the solid-propellant booster design, gave an hour-long presentation on how the cold weather would increase the problems of joint rotation and sealing of the joint by the O-rings.

Their point was that the lowest temperature at which the shuttle had previously been launched was 53°F, on January 24, 1985, when there was blow-by of the O-rings. The O-ring temperature at *Challenger's* expected launch time the following morning was predicted to be 29°F, far below the temperature at which NASA had previous experience. After the engineer's presentation, Bob Lund, the vice president for engineering at Morton Thiokol, presented his recommendations. He reasoned that since there had previously been severe O-ring erosion at 53°F and the launch would take place at significantly below this temperature where no data and no experience were available, NASA should delay the launch until the O-ring temperature could be at least 53°F. Interestingly, in the original design, it was specified that the booster should operate properly down to an outside temperature of 31°F.

Larry Mulloy, the Solid Rocket Booster project manager at Marshall and a NASA employee, correctly pointed out that the data were inconclusive and disagreed with the Thiokol engineers. After some discussion, Mulloy asked Joe Kilminster, an engineering manager working on the project, for his opinion. Kilminster backed up the recommendation of his fellow engineers. Others from Marshall expressed their disagreement with the Thiokol engineer's recommendation, which prompted Kilminster to ask to take the discussion off line for a few minutes. Boisjoly and other engineers reiterated to their management that the original decision not to launch was the correct one.

A key fact that ultimately swayed the decision was that in the available data, there seemed to be no correlation between temperature and the degree to which blow-by gasses had eroded the O-rings in previous launches. Thus, it could be concluded that there was really no trend in the data indicating that launch at the expected temperature would necessarily be unsafe. After much discussion, Jerald Mason, a senior manager with Thiokol, turned to Lund and said, "Take off your engineering hat and put on your management hat," a phrase that has become famous in engineering ethics discussions. Lund reversed his previous decision and recommended that the launch proceed. The new recommendation included an indication that there was a safety concern due to the cold weather, but that the data were inconclusive and the launch was recommended. McDonald, who was in Florida, was surprised by this recommendation and attempted to convince NASA to delay the launch, but to no avail.

THE LAUNCH

Contrary to the weather predictions, the overnight temperature was 8°F, colder than the shuttle had ever experienced before. In fact, there was a significant accumulation of ice on the launchpad from safety showers and fire hoses that had been left on to prevent the pipes from freezing. It has been estimated that the aft field joint of the right-hand booster was at 28°F.

NASA routinely documents as many aspects of launches as possible. One part of this monitoring is the extensive use of cameras focused on critical areas of the launch vehicle. One of these cameras, looking at the right booster, recorded puffs of smoke coming from the aft field joint immediately after the boosters were ignited. This smoke is thought to have been caused by the steel cylinder of this segment of the booster expanding outward and causing the field joint to rotate. But, due to the extremely cold temperature, the O-ring didn't seat properly. The heat-resistant putty was also so cold that it didn't protect the O-rings, and hot gases burned past both O-rings. It was later determined that this blow-by occurred over 70° of arc around the O-rings.

Very quickly, the field joint was sealed again by byproducts of the solid rocket-propellant combustion, which formed a glassy oxide on the joint. This oxide formation might have averted the disaster had it not been for a very strong wind shear that the shuttle encountered almost one minute into the flight. The oxides that were temporarily sealing the field joint were shattered by the stresses caused by the wind shear. The joint was now opened again, and hot gases escaped from the solid booster. Since the booster was attached to the large liquid-fuel booster, the flames from the solid-fuel booster blow-by quickly burned through the external tank. The liquid propellant was ignited and the shuttle exploded.

THE AFTERMATH

As a result of the explosion, the shuttle program was grounded as a thorough review of shuttle safety was conducted. Thiokol formed a failure-investigation team on January 31, 1986 which included Roger Boisjoly. There were also many investigations into the cause of the accident, both by the contractors involved (including Thiokol) and by various government bodies. As part of the governmental investigation, President Reagan appointed a blue-ribbon commission, known as the Rogers commission, after its chair. The commission consisted of distinguished scientists and engineers who were asked to look into the cause of the accident and to recommend changes in the shuttle program.

One of the commission members was Richard Feynman, a Nobel prize winner in physics, who ably demonstrated to the country what had gone wrong. In a demonstration that was repeatedly shown on national news programs, he demonstrated the problem with the O-rings by taking a sample of the O-ring material and bending it. The flexibility of the material at room temperature was evident. He then immersed it in ice water. When Feynman again bent the O-ring, it was very clear that the resiliency of the material was severely reduced, a very clear demonstration of what happened to the O-rings on the cold launch date in Florida.

As part of the commission hearings, Boisjoly and other Thiokol engineers were asked to testify. Boisjoly handed over to the commission copies of internal Thiokol memos and reports detailing the design process and the problems that had already been encountered. Naturally, Thiokol was trying to put the best possible spin on the situation, and Boisjoly's actions hurt this effort. According to Boisjoly, after this action he was isolated within the company, his responsibilities for the redesign of the joint were taken away, and he was subtly harassed by Thiokol management [Boisjoly, 1991, and Boisjoly, Curtis, and Mellicam, 1989].

Explosion of the space shuttle *Challenger* soon after lift-off in January 1986. NASA/Johnson Space Center

Eventually, the atmosphere became intolerable for Boisjoly, and he took extended sick leave from his position at Thiokol. The joint was redesigned, and the shuttle has since flown numerous successful missions. However, the ambitious launch schedule originally intended by NASA has never been met. It was reported in 2001 that NASA has spent $5 million to study the possibility of installing some type of escape system to protect the shuttle crew in the event of an accident. No decision has yet been made. Possibilities include ejection seats or an escape capsule that would work during the first three minutes of flight. These features were incorporated into earlier manned space vehicles and in fact were in place on the shuttle until 1982. Whether such a system would have saved the astronauts aboard the *Challenger* is unknown.

SUMMARY

Engineering ethics is the study of moral decisions that must be made by engineers in the course of engineering practice. It is important for engineering students to study ethics so that they will be prepared to respond appropriately to ethical challenges during their careers. Often, the correct answer to an ethical problem will not be obvious and will require some analysis using ethical theories. The types of problems that we will encounter in studying engineering ethics are very similar to the design problems that engineers work on every day. As in design, there will not be a single correct answer. Rather, engineering ethics problems will have multiple correct solutions, with some solutions being better than others.

REFERENCES

ROGER BOISJOLY, "The Challenger Disaster: Moral Responsibility and the Working Engineer," in Deborah G. Johnson, *Ethical Issues in Engineering*, Prentice Hall, Upper Saddle River, NJ, 1991, pp. 6–14.

NORBERT ELLIOT, ERIC KATZ, AND ROBERT LYNCH, "The Challenger Tragedy: A Case Study in Organizational Communication and Professional Ethics," *Business and Professional Ethics Journal*, vol. 12, 1990, pp. 91–108.

JOSEPH R. HERKERT, "Management's Hat Trick: Misuse of 'Engineering Judgment' in the Challenger Incident," *Journal of Business Ethics*, vol. 10, 1991, pp. 617–620.

PATRICIA H. WERHANE, "Engineers and Management: The Challenge of the Challenger Incident," *Journal of Business Ethics*, vol. 10, 1991, pp. 605–16.

RUSSELL BOISJOLY, Ellen Foster Curtis, and Eugene Mellican, "Roger Boisjoly and the Challenger Disaster: The Ethical Dimensions," *Journal of Business Ethics*, vol. 8, 1989, pp. 217–230.

Problems

1. How different are personal ethics and business ethics? Is this difference true for you personally?
2. What are the roots of your personal ethics?
3. Engineering design generally involves five steps: developing a statement of the problem and/or a set of specifications, gathering information pertinent to the problem, designing several alternatives that meet the specifications, analyzing the alternatives and selecting the best one, and testing and implementing the best design. How is ethical problem solving like this?

SPACE SHUTTLE CHALLENGER

4. The astronauts on the *Challenger* mission were aware of the dangerous nature of riding a complex machine such as the space shuttle, so they can be thought of as having given informed consent to participating in a dangerous enterprise. What role did informed consent play in this case? Do you think that the astronauts had enough information to give informed consent to launch the shuttle that day?
5. Can an engineer who has become a manager truly ever take off her engineer's hat? Should she?
6. Some say that the shuttle was really designed by Congress rather than NASA. What does this statement mean? What are the ramifications if this is true?
7. Aboard the shuttle for this flight was the first teacher in space. Should civilians be allowed on what is basically an experimental launch vehicle? At the time, many felt that the placement of a teacher on the shuttle was for purely political purposes. President Reagan was widely seen as doing nothing while the American educational system decayed. Cynics felt that the teacher-in-space idea was cooked up as a method of diverting attention from this problem and was to be seen as Reagan's doing something for education while he really wasn't doing anything. What are the ethical implications if this scenario is true?
8. Should a launch have been allowed when there was no test data for the expected conditions? Keep in mind that it is probably impossible to test for all possible operating conditions. More generally, should a product be released for use even

when it hasn't been tested over all expected operational conditions? When the data is inconclusive, which way should the decision go?

9. During the aftermath of the accident, Thiokol and NASA investigated possible causes of the explosion. Boisjoly accused Thiokol and NASA of intentionally downplaying the problems with the O-rings while looking for other causes of the accident. If true, what are the ethical implications of this type of investigation?
10. It might be assumed that the management decision to launch was prompted in part by concerns for the health of the company and the space program as a whole. Given the political climate at the time of the launch, if problems and delays continued, ultimately Thiokol might have lost NASA contracts, or NASA budgets might have been severely reduced. Clearly, this scenario could have led to the loss of many jobs at Thiokol and NASA. How might these considerations ethically be factored into the decision?
11. Engineering codes of ethics require engineers to protect the safety and health of the public in the course of their duties. Do the astronauts count as "the public" in this context?
12. What should NASA management have done differently? What should Thiokol management have done differently?
13. What else could Boisjoly and the other engineers at Thiokol have done to prevent the launch from occurring?

16

Professionalism and Codes of Ethics

SECTIONS

2.1 Introduction
2.2 Is Engineering a Profession?
2.3 Codes of Ethics

OBJECTIVES

After reading this chapter, you will be able to

- Determine whether engineering is a profession
- Understand what codes of ethics are, and
- Examine some codes of ethics of professional engineering societies.

Late in 1994, reports began to appear in the news media that the latest generation of Pentium® microprocessors, the heart and soul of personal computers, was flawed. These reports appeared not only in trade journals and magazines aimed at computer specialists, but also in *The New York Times* and other daily newspapers. The stories reported that computers equipped with these chips were unable to correctly perform some relatively simple multiplication and division operations.

At first, Intel, the manufacturer of the Pentium microprocessor, denied that there was a problem. Later, it argued that although there was a problem, the error would be significant only in sophisticated applications, and most people wouldn't even notice that an error had occurred. It was also reported that Intel had been aware of the problem and already was working to fix it. As a result of this publicity, many people who had purchased Pentium-based computers asked to have the defective chip replaced. Until the public outcry had reached huge proportions, Intel refused to replace the chips. Finally, when it was clear that this situation was a public relations disaster for them, Intel agreed to replace the defective chips when customers requested.

Did Intel do anything unethical? To answer this question, we will need to develop a framework for understanding ethical problems. One part of this framework will be the codes of ethics that have been established by professional engineering organizations. These codes help guide engineers in the course of their professional duties and give them insight into ethical problems such as the one just described. The engineering codes of ethics hold that engineers should not make false claims or represent a product to be something that it is not. In some ways, the Pentium case might seem to simply be a public-relations problem. But, looking at the problem with a code of ethics will indicate that there is more to this situation than simple PR, especially since the chip did not operate in the way that Intel claimed it did.

In this chapter, the nature of professions will be examined with the goal of determining whether engineering is a profession. Two representative engineering codes of ethics will be looked at in detail. At the end of this chapter, the Pentium case is presented in more detail along with two other cases, and codes of ethics are applied to analyze what the engineers in these cases should have done.

2.1 INTRODUCTION

When confronted by an ethical problem, what resources are available to an engineer to help find a solution? One of the hallmarks of modern professions are codes of ethics promulgated by various professional societies. These codes serve to guide practitioners of the profession in making decisions about how to conduct themselves and how to resolve ethical issues that might confront them. Are codes of ethics applicable to engineering? To answer this question, we must first consider what professions are and how they function and then decide if this definition applies to engineering. Then we will examine codes of ethics in general and look specifically at some of the codes of engineering professional societies.

2.2 IS ENGINEERING A PROFESSION?

In order to determine whether engineering is a profession, the nature of professions must first be examined. As a starting point, it will be valuable to distinguish the word "profession" from other words that are sometimes used synonymously with "profession": "job" and "occupation." Any work for hire can be considered a job, regardless of the skill level involved and the responsibility granted. Engineering is certainly a job—engineers are paid for their services—but the skills and responsibilities involved in engineering make it more than just a job.

Similarly, the word "occupation" implies employment through which someone makes a living. Engineering, then, is also an occupation. How do the words "job" and "occupation" differ from "profession?"

The words "profession" and "professional" have many uses in modern society that go beyond the definition of a job or occupation. One often hears about "professional athletes" or someone referring to himself as a "professional carpenter," for example. In the first case, the word "professional" is being used to distinguish the practitioner from an unpaid amateur. In the second case, it is used to indicate some degree of skill acquired through many years of experience, with an implication that this practitioner will provide quality services.

Neither of these senses of the word "professional" is applicable to engineers. There are no amateur engineers who perform engineering work without being paid while they train to become professional, paid engineers. Likewise, the length of time one works at an engineering-related job, such as an engineering aide or engineering technician, does not confer professional status no matter how skilled a technician one might become. To see what is meant by the term "professional engineer", we will first examine the nature of professions.

2.2.1 What Is a Profession?

What are the attributes of a profession? There have been many studies of this question, and some consensus as to the nature of *professions* has been achieved. Attributes of a profession include

1. Work that requires sophisticated skills, the use of judgment, and the exercise of discretion. Also, the work is not routine and is not capable of being mechanized;
2. Membership in the profession requires extensive formal education, not simply practical training or apprenticeship;
3. The public allows special societies or organizations that are controlled by members of the profession to set standards for admission to the profession, to set standards of conduct for members, and to enforce these standards; and
4. Significant public good results from the practice of the profession [Martin and Schinzinger, 2000].

The terms "judgment" and "discretion" used in the first part of this definition require a little amplification. Many occupations require judgment every day. A secretary must decide what work to tackle first. An auto mechanic must decide if a part is sufficiently worn to require complete replacement, or if rebuilding will do. This is not the type of judgment implied in this definition. In a profession, "judgment" refers to making significant decisions based on formal training and experience. In general, the decisions will have serious impacts on people's lives and will often have important implications regarding the spending of large amounts of money.

"Discretion" can have two different meanings. The first definition involves being discrete in the performance of one's duties by keeping information about customers, clients, and patients confidential. This confidentiality is essential for engendering a trusting relationship and is a hallmark of professions. While many jobs might involve some discretion, this definition implies a high level of significance to the information that must be kept private by a professional. The other definition of discretion involves the ability to make decisions autonomously. When making a decision, one is often told, "Use your discretion." This definition is similar in many ways to that of the term "judgment" described previously. Many people are allowed to use their discretion in making choices while performing their jobs. However, the significance of the decision marks the difference between a job and a profession.

One thing not mentioned in the definition of a profession is the compensation received by a professional for his services. Although most professionals tend to be relatively well compensated, high pay is not a sufficient condition for professional status. Entertainers and athletes are among the most highly paid members of our society, and yet few would describe them as professionals in the sense described previously. Although professional status often helps one to get better pay and better working conditions, these are more often determined by economic forces.

Earlier, reference was made to "professional" athletes and carpenters. Let's examine these occupations in light of the foregoing definition of professions and see if athletics and carpentry qualify as professions. An athlete who is paid for her appearances is referred to as a professional athlete. Clearly, being a paid athlete does involve sophisticated skills that most people do not possess, and these skills are not capable of mechanization. However, substantial judgment and discretion are not called for on the part of athletes in their "professional" lives, so athletics fails the first part of the definition of "professional." Interestingly, though, professional athletes are frequently viewed as role models and are often disciplined for a lack of discretion in their personal lives.

Athletics requires extensive training, not of a formal nature, but more of a practical nature acquired through practice and coaching. No special societies (as opposed to unions, which will be discussed in more detail later) are required by athletes, and athletics does not meet an important public need; although entertainment is a public need, it certainly doesn't rank highly compared to the needs met by professions such as medicine. So, although they are highly trained and very well compensated, athletes are not professionals.

Similarly, carpenters require special skills to perform their jobs, but many aspects of their work can be mechanized, and little judgment or discretion is required. Training in carpentry is not formal, but rather is practical by way of apprenticeships. No organizations or societies are required. However, carpentry certainly does meet an aspect of the public good—providing shelter is fundamental to society—although perhaps not to the same extent as do professions such as medicine. So, carpentry also doesn't meet the basic requirements to be a profession. We can see, then, that many jobs or occupations whose practitioners might be referred to as professionals don't really meet the basic definition of a profession. Although they may be highly paid or important jobs, they are not professions.

Before continuing with an examination of whether engineering is a profession, let's look at two occupations that are definitely regarded by society as professions: medicine and law. Medicine certainly fits the definition of a profession given previously. It requires very sophisticated skills that can't be mechanized, it requires judgment as to appropriate treatment plans for individual patients, and it requires discretion (physicians have even been granted physician–patient privilege, the duty not to divulge information given in confidence by the patient to the physician). Although medicine requires extensive practical training learned through an apprenticeship called a residency, it also requires much formal training (four years of undergraduate school, three to four years of medical school, and extensive hands-on practice in patient care). Medicine has a special society, the American Medical Association (AMA), to which a large fraction of practicing physicians belong and that participates in the regulation of medical schools, sets standards for practice of the profession, and enforces codes of ethical behavior for its members. Finally, healing the sick and helping to prevent disease clearly involve the public good. By the definition presented previously, medicine clearly qualifies as a profession.

Similarly, law is a profession. It involves sophisticated skills acquired through extensive formal training; has a professional society, the American Bar Association (ABA); and serves an important aspect of the public good. (Although this last point is increasingly becoming a point of debate within American society!) The difference between athletics and carpentry on one hand and law and medicine on the other is clear. The first two really cannot be considered professions, and the latter two most certainly are.

2.2.2 Engineering as a Profession

Using medicine and law as our examples of professions, it is now time to consider whether engineering is a profession. Certainly, engineering requires extensive and sophisticated skills. Otherwise, why spend four years in college just to get a start in engineering? The essence of engineering design is judgment: how to use the available materials, components, devices, etc. to reach a specified objective. Discretion is required in engineering: Engineers are required to keep their employers' or clients' intellectual-property and business information confidential. Also, a primary concern of any engineer is the safety of the public that will use the products and devices he designs. There is always a trade-off between safety and other engineering issues in a design, requiring discretion on the part of the engineer to ensure that the design serves its purpose and fills its market niche safely.

The point about mechanization needs to be addressed a little more carefully with respect to engineering. Certainly, once a design has been performed, it can easily be replicated without the intervention of an engineer. However, each new situation that requires a new design or a modification of an existing design requires an engineer. Industry commonly uses many computer-based tools for generating designs, such as computer-aided design (CAD) software. This shouldn't be mistaken for mechanization of engineering. CAD is simply a tool used by engineers, not a replacement for the skills of an actual engineer. A wrench can't fix an automobile without a mechanic. Likewise, a computer with CAD software can't design an antilock braking system for an automobile without an engineer.

Engineering requires extensive formal training. Four years of undergraduate training leading to a bachelor's degree in an engineering program is essential, followed by work under the supervision of an experienced engineer. Many engineering jobs even require advanced degrees beyond the bachelor's degree. The work of engineers serves the public good by providing communication systems, transportation, energy resources, and medical diagnostic and treatment equipment, to name only a few.

Before passing final judgment on the professional status of engineering, the nature of engineering societies requires a little consideration. Each discipline within engineering has a professional society, such as the Institute of Electrical and Electronics Engineers (IEEE) for electrical engineers and the American Society of Mechanical Engineers (ASME) for mechanical engineers. These societies serve to set professional standards and frequently work with schools of engineering to set standards for admission and curricula. However, these societies differ significantly from the AMA and the ABA. Unlike law and medicine, each specialty of engineering has its own society. There is no overall engineering society that most engineers identify with, although the National Society of Professional Engineers (NSPE) tries to function in this way. In addition, relatively few practicing engineers belong to their professional societies. Thus, the engineering societies are weak compared to the AMA and the ABA.

It is clear that engineering meets all of the definitions of a profession. In addition, it is clear that engineering practice has much in common with medicine and law. Interestingly, although they are professionals, engineers do not yet hold the same status within society that physicians and lawyers do.

2.2.3 Differences Between Engineering and Other Professions

Although we have determined that engineering is a profession, it should be noted that there are significant differences between how engineering is practiced and how law and medicine are practiced. Lawyers are typically self-employed in private practice, essentially an independent business, or in larger group practices with other lawyers. Relatively few are employed by large organizations such as corporations. Until recently, this was also the case for most physicians, although with the accelerating trend towards managed care and HMOs in the past decade, many more physicians work for large corporations rather than in private practice. However, even physicians who are employed by large HMOs are members of organizations in which they retain much of the decision-making power—often, the head of an HMO is a physician—and are a substantial fraction of the total number of employees.

In contrast, engineers generally practice their profession very differently from physicians and lawyers. Most engineers are not self-employed, but more often are a small part of larger companies involving many different occupations, including accountants, marketing specialists, and extensive numbers of less skilled manufacturing employees. The exception to this rule is civil engineers, who generally practice as independent consultants either on their own or in engineering firms similar in many ways to law firms. When employed by large corporations, engineers are rarely in significant managerial positions, except with regard to managing other engineers. Although engineers are paid well compared to the rest of society, they are generally less well compensated than physicians and lawyers.

Training for engineers is different than for physicians and lawyers. One can be employed as an engineer after four years of undergraduate education, unlike law and medicine, for which training in the profession doesn't begin until after the undergraduate program has been completed. As mentioned previously, the engineering societies are not as powerful as the AMA and the ABA, perhaps because of the number of different professional engineering societies. Also, both law and medicine require licenses granted by the state in order to practice. Many engineers, especially those employed by large industrial companies, do not have engineering licenses. It can be debated whether someone who is unlicensed is truly an engineer or whether he is practicing engineering illegally, but the reality is that many of those who are employed as engineers are not licensed. Finally, engineering doesn't have the social stature that law and medicine have (a fact that is reflected in the lower pay that engineers receive as compared to that of lawyers and doctors). Despite these differences, on balance, engineering is still clearly a profession, albeit one that is not as mature as medicine and law and that should be striving to emulate some of the aspects of these professions.

2.2.4 Other Aspects of Professional Societies

We should briefly note that professional societies also serve other, perhaps less noble, purposes than those mentioned previously. Sociologists who study the nature of professional societies describe two different models of professions, sometimes referred to as the social-contract and the business models. The social-contract model views professional societies as being set up primarily to further the public good, as described in the definition of a profession given previously. There is an implicit social contract involved with professions, according to this model. Society grants the professions' perks such as high pay, a high status in society, and the ability to self-regulate. In return for these perks, society gets the services provided by the profession.

A perhaps more cynical view of professions is provided by the business model. According to this model, professions function as a means for furthering the economic advantage of the members. Put another way, professional organizations are labor unions for the elite, strictly limiting the number of practitioners of the profession, controlling the working conditions for professionals, and artificially inflating the salaries of its members. An analysis of both models in terms of law and medicine would show that there are ways in which these professions exhibit aspects of both of these models.

Where does engineering fit into this picture? Engineering is certainly a service-oriented profession and thus fits into the social-contract model quite nicely. Although some engineers might wish to see engineering professional societies function more according to the business model, they currently don't function that way. The engineering societies have virtually no clout with major engineering employers to set wages and working conditions or to help engineers resolve ethical disputes with their employers.

Moreover, there is very little prospect that the engineering societies will function this way in the near future.

2.2.5 If Engineering Were Practiced More Like Medicine

It is perhaps instructive to speculate a little on how engineering might change in the future if our model of the engineering profession were closer to that of law or medicine. One major change would be in the way engineers are educated. Rather than the current system, in which students study engineering as undergraduates and then pursue advanced degrees as appropriate, prospective engineers would probably get a four-year "preengineering" degree in mathematics, physics, chemistry, computer science, or some combination of these fields. After the four-year undergraduate program, students would enter a three- or four-year engineering professional program culminating in a "doctor of engineering" degree (or other appropriately named degree). This program would include extensive study of engineering fundamentals, specialization in a field of study, and perhaps "clinical" training under a practicing engineer.

How would such engineers be employed? The pattern of employment would certainly be different. Engineers in all fields might work for engineering firms similar to the way in which civil engineers work now, consulting on projects for government agencies or large corporations. The corporate employers who now have numerous engineers on their staff would probably have far fewer engineers on the payroll, opting instead for a few professional engineers who would supervise the work of several less highly trained "engineering technicians." Adoption of this model would probably reduce the number of engineers in the work force, leading to higher earnings for those who remain. Those relegated to the ranks of engineering technicians would probably earn less than those currently employed as engineers.

2.3 CODES OF ETHICS

An aspect of professional societies that has not been mentioned yet is the codes of ethics that engineering societies have adopted. These codes express the rights, duties, and obligations of the members of the profession. In this section, we will examine the codes of ethics of professional engineering societies.

It should be noted that although most of the discussion thus far has focused on professionalism and professional societies, codes of ethics are not limited to professional organizations. They can also be found, for example, in corporations and universities as well. We start with some general ideas about what codes of ethics are and what purpose they serve and then examine two professional engineering codes in more detail.

2.3.1 What Is a Code of Ethics?

Primarily, a code of ethics provides a framework for ethical judgment for a professional. The key word here is "framework." No code can be totally comprehensive and cover all possible ethical situations that a professional engineer is likely to encounter. Rather, codes serve as a starting point for ethical decision making. A code can also express the commitment to ethical conduct shared by members of a profession. It is important to note that ethical codes do not establish new ethical principles. They simply reiterate principles and standards that are already accepted as responsible engineering practice. A code expresses these principles in a coherent, comprehensive, and accessible manner. Finally, a code defines the roles and responsibilities of professionals [Harris, Pritchard, and Rabins, 2000].

It is important also to look at what a code of ethics is not. It is not a recipe for ethical behavior; as previously stated, it is only a framework for arriving at good ethical choices. A code of ethics is never a substitute for sound judgment. A code of ethics is not a legal document. One can't be arrested for violating its provisions, although expulsion from the professional society might result from code violations. As mentioned in the previous section, with the current state of engineering societies, expulsion from an engineering society generally will not result in an inability to practice engineering, so there are not necessarily any direct consequences of violating engineering ethical codes. Finally, a code of ethics doesn't create new moral or ethical principles. As described in the previous chapter, these principles are well established in society, and foundations of our ethical and moral principles go back many centuries. Rather, a code of ethics spells out the ways in which moral and ethical principles apply to professional practice. Put another way, a code helps the engineer to apply moral principles to the unique situations encountered in professional practice.

How does a code of ethics achieve these goals? First, a code of ethics helps create an environment within a profession where ethical behavior is the norm. It also serves as a guide or reminder of how to act in specific situations. A code of ethics can also be used to bolster an individual's position with regard to a certain activity: The code provides a little backup for an individual who is being pressured by a superior to behave unethically. A code of ethics can also bolster the individual's position by indicating that there is a collective sense of correct behavior; there is strength in numbers. Finally, a code of ethics can indicate to others that the profession is seriously concerned about responsible, professional conduct [Harris, Pritchard, and Rabins, 2000]. A code of ethics, however, should not be used as "window dressing," an attempt by an organization to appear to be committed to ethical behavior when it really is not.

2.3.2 Objections to Codes

Although codes of ethics are widely used by many organizations, including engineering societies, there are many objections to codes of ethics, specifically as they apply to engineering practice. First, as mentioned previously, relatively few practicing engineers are members of professional societies and so don't necessarily feel compelled to abide by their codes. Many engineers who are members of professional societies are not aware of the existence of the society's code, or if they are aware of it, they have never read it. Even among engineers who know about their society's code, consultation of the code is rare. There are also objections that the engineering codes often have internal conflicts, but don't give a method for resolving the conflict. Finally, codes can be coercive: They foster ethical behavior with a stick rather than with a carrot [Harris, Pritchard, and Rabins, 2000]. Despite these objections, codes are in very widespread use today and are generally thought to serve a useful function.

2.3.3 Codes of the Engineering Societies

Before examining two representative professional codes from Appendix A in more detail, it might be instructive to look briefly at the history of the engineering codes of ethics. Professional engineering societies in the United States began to be organized in the late 19th century, with new societies created as new engineering fields have developed in this century. As these societies matured, many of them created codes of ethics to guide practicing engineers.

Early in the current century, these codes were mostly concerned with issues of how to conduct business. For example, many early codes had clauses forbidding advertising of services or prohibiting competitive bidding by engineers for design projects. Codes also spelled out the duties that engineers had toward their employers. Relatively less emphasis than today was given to issues of service to the public and safety. This imbalance changed greatly in recent decades as public perceptions and concerns about the safety of engineered products and devices have changed. Now, most codes emphasize commitments to safety, public health, and even environmental protection as the most important duties of the engineer.

2.3.4 A Closer Look at Two Codes of Ethics

Having looked at some ideas about what codes of ethics are and how they function, let's look more closely at two of the codes of ethics reproduced in Appendix A: the codes of the IEEE and the NSPE. Although these codes have some common content, the structures of the codes are very different.

The IEEE code is short and deals in generalities, whereas the NSPE code is much longer and more detailed. An explanation of these differences is rooted in the philosophy of the authors of these codes. A short code that is lacking in detail is more likely to be read by members of the society than is a longer code. A short code is also more understandable. It articulates general principles and truly functions as a framework for ethical decision making, as described previously.

A longer code, such as the NSPE code, has the advantage of being more explicit and able to cover more ground. It leaves less to the imagination of the individual and therefore is more useful for application to specific cases. The length of the code, however, makes it less likely to be read and thoroughly understood by most engineers.

There are some specifics of these two codes that are worth noting here. The IEEE code doesn't mention a duty to one's employer. However, the IEEE code does mention a duty to protect the environment, a clause added relatively recently, which is somewhat unique among engineering codes. The NSPE code has a preamble that succinctly presents the duties of the engineer before going on to the more explicit discussions of the rest of the code. Like most codes of ethics, the NSPE code does mention the engineer's duty to his or her employer in Section I.4, where it states that engineers shall "[a]ct . . . for each employer . . . as faithful agents or trustees."

2.3.5 Resolving Internal Conflicts in Codes

One objection to codes of ethics is the internal conflicts that can exist within them, with no instructions on how to resolve these conflicts. An example of this problem would be a situation in which an employer asks or even orders an engineer to implement a design that the engineer feels will be unsafe. It is made clear that the engineer's job is at stake if he doesn't do as instructed. What does the NSPE code tell us about this situation?

In clause I.4, the NSPE code indicates that engineers have a duty to their employers, which implies that the engineer should go ahead with the unsafe design favored by his employer. However, clause I.1 and the preamble make it clear that the safety of the public is also an important concern of an engineer. In fact, it says that the safety of the public is paramount. How can this conflict be resolved?

There is no implication in this or any other code that all clauses are equally important. Rather, there is a hierarchy within the code. Some clauses take precedence over others, although there is generally no explicit indication in the code of what the hierarchy is. The preceding dilemma is easily resolved within the context of this hierarchy. The duty to protect the safety of the public is paramount and takes precedence over the duty to the employer. In this case, the code provides very clear support to the engineer, who must convince his supervisor that the product can't be designed as requested. Unfortunately, not all internal conflicts in codes of ethics are so easily resolved.

2.3.6 Can Codes and Professional Societies Protect Employees?

One important area where professional societies can and should function is as protectors of the rights of employees who are being pressured by their employer to do something unethical or who are accusing their employers or the government of unethical conduct. The codes of the professional societies are of some use in this since they can be used by employees as ammunition against an employer who is sanctioning them for pointing out unethical behavior or who are being asked to engage in unethical acts.

An example of this situation is the action of the IEEE on behalf of three electrical engineers who were fired from their jobs at the Bay Area Rapid Transit (BART) organization when they pointed out deficiencies in the way the control systems for the BART trains were being designed and tested. After being fired, the engineers sued BART, citing the IEEE code of ethics which impelled them to hold as their primary concern the safety of the public who would be using the BART system. The IEEE intervened on their behalf in court, although ultimately the engineers lost the case.

If the codes of ethics of professional societies are to have any meaning, this type of intervention is essential when ethical violations are pointed out. However, since not all engineers are members of professional societies and the engineering societies are relatively weak, the pressure that can be exerted by these organizations is limited.

2.3.7 Other Types of Codes of Ethics

Professional societies aren't the only organizations that have codified their ethical standards. Many other organizations have also developed codes of ethics for various purposes similar to those of the professional engineering organizations. For example, codes for the ethical use of computers have been developed, and student organizations in universities have framed student codes of ethics. In this section, we will examine how codes of ethics function in corporations.

Many of the important ethical questions faced by engineers come up in the context of their work for corporations. Since most practicing engineers are not members of professional organizations, it seems that for many engineers, there is little ethical guidance in the course of their daily work. This problem has led to the adoption of codes of ethics by many corporations.

Even if the professional codes were widely adopted and recognized by practicing engineers, there would still be some value to the corporate codes, since a corporation can tailor its code to the individual circumstances and unique mission of the company. As such, these codes tend to be relatively long and very detailed, incorporating many rules specific to the practices of the company. For example, corporate codes frequently spell out in detail the company policies on business practices, relationships with suppliers, relationships with government agencies, compliance with government regulations, health and safety issues, issues related to environmental protection, equal employment opportunity and affirmative action, sexual harassment, and diversity and racial/ethnic tolerance. Since corporate codes are coercive in nature—your continued employment by the company depends on your compliance with the company code—these codes tend to be longer and more detailed in order to provide very clear and specific guidelines to the employees.

Codes of professional societies, by their nature, can't be this explicit, since there is no means for a professional society to reasonably enforce its code. Due to the typically long lengths of these codes, no example of a corporate *code of ethics* can be included in Appendix A with the other types of codes. However, codes for companies can sometimes be found via the Internet at corporate websites.

Some of the heightened awareness of ethics in corporations stems from the increasing public scrutiny that has accompanied well-publicized disasters, such as the cases presented in this book, as well as from cases of fraud and cost overruns, particularly in the defense industry, that have been exposed in the media. Many large corporations have developed corporate codes of ethics in response to these problems, to help heighten employee's awareness of ethical issues, and to help establish a strong corporate ethics culture. These codes give employees ready access to guidelines and policies of the corporations. But, as with professional codes, it is important to remember that these codes cannot cover all possible situations that an employee might encounter; there is no substitute for good judgment. A code also doesn't substitute for good lines of communications between employees and upper management and for workable methods for fixing ethical problems when they occur.

APPLICATION: CASES

Codes of ethics can be used as a tool for analyzing cases and for gaining some insight into the proper course of action. Before reading these cases, it would be helpful to read a couple of the codes in Appendix A, especially the code most closely related to your field of study, to become familiar with the types of issues that codes deal with. Then, put yourself in the position of an engineer working for these companies—Intel, Paradyne computers, and 3Bs construction—to see what you would have done in each case.

THE INTEL PENTIUM® CHIP

In late 1994, the media began to report that there was a flaw in the new Pentium microprocessor produced by Intel. The microprocessor is the heart of a personal computer and controls all of the operations and calculations that take place. A flaw in the Pentium was especially significant, since it was the microprocessor used in 80% of the personal computers produced in the world at that time.

Apparently, flaws in a complicated integrated circuit such as the Pentium, which at the time contained over one million transistors, are common. However, most of the flaws are undetectable by the user and don't affect the operation of the computer. Many of these flaws are easily compensated for through software. The flaw that came to light in 1994 was different: It was detectable by the user. This particular flaw was in the floating-point unit (FPU) and caused a wrong answer when double-precision arithmetic, a very common operation, was performed.

A standard test was widely published to determine whether a user's microprocessor was flawed. Using spreadsheet software, the user was to take the number 4,195,835, multiply it by 3,145,727, and then divide that result by 3,145,727. As we all know from elementary math, when a number is multiplied and then divided by the same number, the result should be the original number. In this example, the result should be 4,195,835. However, with the flawed FPU, the result of this calculation was 4,195,579. [Infoworld, 1994] Depending on the application, this six-thousandths-of-a-percent error might be very significant.

At first, Intel's response to these reports was to deny that there was any problem with the chip. When it became clear that this assertion was not accurate, Intel switched its policy and stated that although there was indeed a defect in the chip, it was insignificant and the vast majority of users would never even notice it. The chip would be replaced for free only for users who could demonstrate that they needed an unflawed version of the chip. [Infoworld, 1994] There is some logic to this policy from Intel's point of view, since over two million computers had already been sold with the defective chip.

Of course, this approach didn't satisfy most Pentium owners. After all, how can you predict whether you might have a future application where this flaw might be significant? IBM, a major Pentium user, canceled the sales of all IBM computers containing the flawed chip. Finally, after much negative publicity in the popular personal computer literature and an outcry from Pentium users, Intel agreed to replace the flawed chip with an unflawed version for any customer who asked to have it replaced.

It should be noted that long before news of the flaw surfaced in the popular press, Intel was aware of the problem and had already corrected it on subsequent versions. It did, however, continue to sell the flawed version and, based on its early insistence that the flaw did not present a significant problem to users, seemingly planned to do so until the new version was available and the stocks of the flawed one were exhausted. Eventually, the damage caused by this case was fixed as the media reports of the problem died down and as customers were able to get unflawed chips into their computers. Ultimately, Intel had a write-off of 475 million dollars to solve this problem.

What did Intel learn from this experience? The early designs for new chips continue to have flaws, and sometimes these flaws are not detected until the product is already in use by consumers. However, Intel's approach to these problems has changed. It now seems to feel that problems need to be fixed immediately. In addition, the decision is now based on the consumer's perception of the significance of the flaw, rather than on Intel's opinion of its significance.

Indeed, similar flaws were found in 1997 in the early versions of the Pentium II and Pentium Pro processors. This time, Intel immediately confirmed that the flaw existed and offered customers software that would correct it. Other companies also seem to have benefited from Intel's experience. For example, Intuit, a leading manufacturer of tax preparation and financial software, called a news conference in March of 1995 to apologize for flaws in its TurboTax software that had become apparent earlier in that year. In addition to the apology, they offered consumers replacements for the defective software.

RUNWAY CONCRETE AT THE DENVER INTERNATIONAL AIRPORT

In the early 1990s, the city of Denver, Colorado embarked on one of the largest public works projects in history: the construction of a new airport to replace the aging Stapleton International Airport. The new Denver International Airport (DIA) would be the first new airport constructed in the US since the Dallas–Fort Worth Airport was completed in the early 1970s. Of course, the size and complexity of this type of project lends itself to many problems, including cost overruns, worker safety and health issues, and controversies over the need for the project. The construction of DIA was no exception.

Perhaps the most widely known problem with the airport was the malfunctioning of a new computer-controlled high-tech baggage handling system, which in preliminary tests consistently mangled and misrouted baggage and frequently jammed, leading to the shutdown of the entire system. Problems with the baggage handling system delayed the opening of the airport for over a year and cost the city millions of dollars in expenses for replacement of the system and lost revenues while the airport was unable to open. In addition, the baggage system made the airport the butt of many jokes, especially on late-night television. More interesting from the perspective of engineering ethics are problems during the construction of DIA involving the concrete used for the runways, taxiways, and aprons at the airport.

The story of concrete problems at DIA was first reported by the *Denver Post* in early August of 1993 as the airport neared completion. Two subcontractors filed lawsuits against the runway paving contractor, California construction company Ball, Ball, & Brosamer (known as 3Bs), claiming that 3Bs owed them money. Parts of these suits were allegations that 3Bs had altered the recipe for the concrete used in the runway and apron construction, deliberately diluting the concrete with more gravel, water, and sand (and thus less cement), thereby weakening it. 3Bs motivation for doing so would be to save money and thus to increase their profits. One of the subcontractors, CSI Trucking, whose job was to haul the sand and gravel used in the concrete, claimed that 3Bs hadn't paid them for materials that had been delivered. They claimed that these materials had been used to dilute the mixture, but hadn't been paid for, since the payment would leave a record of the improper recipe.

At first, Denver officials downplayed the reports of defective concrete, relying on the results of independent tests of the concrete. In addition, the city of Denver ordered core samples to be taken from the runways. Tests on these cores showed that the runway concrete had the correct strength. The subcontractors claimed that the improperly mixed concrete could have the proper test strength, but would lead to a severely shortened runway lifetime. The FBI also became involved in investigating this case, since federal transportation grants were used by Denver to help finance the construction of the runways.

The controversy seemed to settle down for a while, but a year later, in August of 1994, the Denver district attorney's office announced that it was investigating allegations that inspection reports on the runways were falsified during the construction. This announcement was followed on November 13, 1994 by a lengthy story in the *Denver Post* detailing a large number of allegations of illegal activities and unethical practices with regard to the runway construction.

The November 13 story revolved around an admission by a Fort Collins, Colorado company, Empire Laboratories, that test reports on the concrete had been falsified to hide results which showed that some of the concrete did not meet the specifications. Attorneys for Empire said that this falsification had happened five or six times in the course of this work, but four employees of Empire claimed that the altering of test data was standard operating procedure at Empire.

The nature of the test modifications and the rationale behind them illustrate many of the important problems we will discuss in this book, including the need for objectivity and honesty in reporting results of tests and experiments. One Empire employee said that if a test result was inconsistent with other tests, then the results would be changed to mask the difference. This practice was justified by Empire as being "based upon engineering judgment" [*Denver Post*, Nov. 13, 1994]. The concrete was tested by pouring test samples when the actual runways were poured. These samples were subjected to flexural tests, which consist of subjecting the concrete to an increasing force until it fails. The tests were performed at 7 days after pouring and also at 28 days. Many of the test results showed that the concrete was weaker at 28 days than at 7 days. However, the results should have been the opposite, since concrete normally increases in strength as it cures. Empire employees indicated that this apparent anomaly was because many of the 7-day tests had been altered to make the concrete seem stronger than it was.

Other problems with the concrete also surfaced. Some of the concrete used in the runways contained clay balls up to 10 inches in diameter. While not uncommon in concrete batching, the presence of this clay can lead to runways that are significantly weaker than planned.

Questions about the short cement content in 3Bs concrete mixture also resurfaced in the November *Denver Post* article. The main question was "given that the concrete batching operation was routinely monitored, how did 3Bs get away with shorting the cement content of the concrete?" One of the batch plant operators for 3Bs explained that they were tipped off about upcoming inspections. When an inspector was due, they used the correct recipe so that concrete would appear to be correctly formulated. The shorting of the concrete mixture could also be detected by looking at the records of materials delivered to the batch plants. However, DIA administrators found that this documentation was missing, and it was unclear whether it had ever existed.

A batch plant operator also gave a sworn statement that he had been directed to fool the computer that operated the batch plant. The computer was fooled by tampering with the scale used to weigh materials and by inputting false numbers for the moisture content of the sand. In some cases, the water content of the sand that was input into the computer was a negative number! This tampering forced the computer to alter the mixture to use less cement, but the records printed by the computer would show that the mix was properly constituted. In his statement, the batch plant operator also swore that this practice was known to some of the highest officials in 3Bs.

Despite the problems with the batching of the concrete used in the runways, DIA officials insisted that the runways built by 3Bs met the specifications. This assertion was based on the test results, which showed that although some parts of the runway were below standard, all of the runways met FAA specifications. 3Bs was paid for those areas that were below standard at a lower rate than for the stronger parts of the runway. Further investigations about misdeeds in the construction of DIA were performed by several groups, including a Denver grand jury, a federal grand jury, the FBI, and committees of Congress.

On October 19, 1995, the *Denver Post* reported the results of a lawsuit brought by 3Bs against the city of Denver. 3Bs contended that the city still owed them $2.3 million (in addition to the $193 million that 3Bs had already been paid) for the work they did. The city claimed that this money was not owed. The reduction was a penalty due to low test results on some of the concrete. 3Bs claimed that those tests were flawed and that the concrete was fine. A hearing officer sided with the city, deciding that Denver didn't owe 3Bs any more money. 3Bs said that they would take their suit to the next higher level.

As of the summer of 2003, DIA has been in operation for many years and no problems have surfaced regarding the strength of the runways. Unfortunately, problems with runway durability might not surface until after several more years of use. In the meantime, there is still plenty of litigation and investigation of this and other unethical acts surrounding the construction of this airport.

COMPETITIVE BIDDING AND THE PARADYNE CASE

Although competitive bidding is a well-established practice in purchasing, it can lead to many ethical problems associated with deception on the part of the vendor or with unfairness on the part of the buyer in choosing a vendor. The idea behind competitive bidding is that the buyer can get a product at the best price by setting up competition between the various suppliers. Especially with large contracts, the temptation to cheat on the bidding is great. Newspapers frequently report stories of deliberate underbidding to win contracts, followed by cost overruns that are unavoidable; theft of information on others' bids in order to be able to underbid them, etc. Problems also exist with buyers who make purchase decisions based on elements other than the advertised bid criteria, who leak information to a preferred bidder, or who give advance notice or detailed knowledge of evaluation procedures to preferred bidders. The Paradyne computer case is useful in illustrating some of the hazards associated with competitive bidding.

The Paradyne case began on June 10, 1980, when the Social Security Administration (SSA) published a request for proposals (RFP) for computer systems to replace the older equipment in its field offices. Its requirement was for computers that provide access to a central database. This database was used by field offices in the processing of benefit claims and in issuing new social security numbers. SSA intended to purchase an off-the-shelf system already in the vendor's product line, rather than a customized system. This requirement was intended to minimize the field testing and bugs associated with customized systems. In March of 1981, SSA let a contract for $115 million for 1,800 computer systems to Paradyne.

Problems occurred immediately upon award of the contract, when the Paradyne computers failed the acceptance testing. The requirements were finally relaxed so that the computers would pass. After delivery, many SSA field offices reported frequent malfunctions, sometimes multiple times per day, requiring manual rebooting of the system. One of the contract requirements was that the computers function 98% of the time. This requirement wasn't met until after 21 months of operation. After nearly two years of headaches and much wasted time and money, the system finally worked as planned. [Davis, 1988]

Subsequent investigation by SSA indicated that the product supplied by Paradyne was not an off-the-shelf system, but rather was a system that incorporated new technology that had yet to be built and was still under development. Paradyne had proposed selling SSA their P8400 model with the PIOS operating system. The bid was written as if this system currently existed. However, at the time that the bid was prepared, the 8400 system did not exist and had not been developed, prototyped, or manufactured. [Head, 1986]

There were other problems associated with Paradyne's performance during the bidding. The RFP stated that there was to be a preaward demonstration of the product, not a demonstration of a prototype. Paradyne demonstrated to SSA a different computer, a modified PDP 11/23 computer manufactured by Digital Equipment Corporation (DEC) placed in a cabinet that was labeled P8400. Apparently, many of the DEC labels on the equipment that was demonstrated to SSA had Paradyne labels pasted over them. Paradyne disingenuously claimed that since the DEC equipment was based on a 16-bit processor, as was the P8400 they proposed, it was irrelevant whether the machine demonstrated was the DEC or the actual P8400. Of course, computer users recognize that this statement is nonsense. Even modern "PC-compatible" computers with the same microprocessor chip and operating system can have widely different operating characteristics in terms of speed and the software that can be run.

There were also questions about the operating system. Apparently, at the time of Paradyne's bid, the PIOS system was under development as well and hadn't been tested on a prototype of the proposed system. Even a functioning hardware system will not operate correctly without the correct operating system. No software has ever worked correctly the first time, but rather requires extensive debugging to make it operate properly with a new system. Significantly, the DEC system with the P8400 label that was actually tested by SSA was not running with the proposed PIOS system.

Some of the blame for this fiasco can also be laid at the feet of the SSA. There were six bidders for this contract. Each of the bidders was to have an on-site visit from SSA inspectors to determine whether it was capable of doing the work that it included in its bid. Paradyne's capabilities were not assessed using an on-site visit. Moreover, Paradyne was judged based on its ability to manufacture modems, which was then its main business. Apparently, its ability to produce complete computer systems wasn't assessed. As part of its attempt to gain this contract, Paradyne hired a former

SSA official who, while still working for SSA, had participated in preparing the RFP and had helped with setting up the team that would evaluate the bids. Paradyne had notified SSA of the hiring of this person, and SSA decided that there were no ethical problems with this. However, when the Paradyne machine failed the initial acceptance test, this Paradyne official was directly involved in negotiating the relaxed standards with his former boss at SSA.

This situation was resolved when the Paradyne computers were finally brought to the point of functioning as required. However, as a result of these problems, there were many investigations by government agencies, including the Securities and Exchange Commission, the General Accounting Office, the House of Representative's Government Operations Committee, the Health and Human Services Department (of which SSA is part), and the Justice Department.

KEY TERMS

Code of ethics

Professions

Professional societies

REFERENCES

CHARLES E. HARRIS, JR., MICHAEL S. PRITCHARD, AND MICHAEL J. RABINS, *Engineering Ethics, Concepts and Cases,* Wadsworth Publishing Company, Belmont CA., 2000.

ROLAND SCHINZINGER AND MIKE W. MARTIN, *Introduction to Engineering Ethics,* McGraw-Hill, New York, 2000.

INTEL PENTIUM CHIP CASE

"When the Chips are Down," *Time*, Dec. 26–Jan. 2 1995, p. 126.

"The Fallout from Intel's Pentium Bug," *Fortune*, Jan. 16, 1995, p. 15.

"Pentium Woes Continue," *Infoworld*, Nov. 18, 1994, vol. 16, issue 48, p. 1.

"Flawed Chips Still Shipping," *Infoworld*, Dec. 5, 1994, vol. 16, issue 49, p. 1.

Numerous other accounts from late 1994 and early 1995 in The *Wall Street Journal, The New York Times*, etc.

DIA RUNAWAY CONCRETE

Lou Kilzer, Robert Kowalski, and Steven Wilmsen, "Concrete tests faked at airport," *Denver Post*, Nov. 13, 1994, Section A, p. 1.

PARADYNE COMPUTERS

J. STEVE DAVIS, "Ethical Problems in Competitive Bidding: The Paradyne Case," *Business and Professional Ethics Journal*, vol. 7, 1988, p. 3.

ROBERT V. HEAD, "Paradyne Dispute: A Matter of Using a Proper Tense," *Government Computer News*, February 14, 1986, p. 23.

Problems

1. What changes would have to be made for engineering to be a profession more like medicine or law?
2. In which ways do law, medicine, and engineering fit the social-contract and the business models of a profession?
3. The first part of the definition of a profession presented previously said that professions involve the use of sophisticated skills. Do you think that these skills are primarily physical or intellectual skills? Give examples from professions such as law, medicine, and engineering, as well as from nonprofessions.
4. Read about the space shuttle *Challenger* accident in 1986. (You can find information on this in magazines, newspapers, or on the internet). Apply one of the codes of ethics in Appendix A to this case. What guidance might one of the

engineering society codes of ethics have given the Thiokol engineers when faced with a decision to launch? Which specific parts of the code are applicable to this situation? Does a manager who is trained as an engineer still have to adhere to an engineering code of ethics?

5. Write a code of ethics for students in your college or department. Start by deciding what type of code you want: short, long, detailed, etc. Then, list the important ethical issues you think students face. Finally, organize these ideas into a coherent structure.
6. Imagine that you are the president of a small high-technology firm. Your company has grown over the last few years to the point where you feel that it is important that your employees have some guidelines regarding ethics. Define the type of company you are running; then develop an appropriate code of ethics. As in Question 2, start by deciding what type of code is appropriate for your company. Then, list specific points that are important—for example, relationships with vendors, treatment of fellow employees, etc. Finally, write a code that incorporates these features.

INTEL PENTIUM CHIP

7. Was this case simply a customer-relations and PR problem, or are there ethical issues to be considered as well?
8. Use one of the codes of ethics in Appendix A to analyze this case. Especially, pay attention to issues of accurate representation of engineered products and to safety issues.
9. When a product is sold, is there an implication that it will work as advertised?
10. Should you reveal defects in a product to a consumer? Is the answer to this question different if the defect is a safety issue rather than simply a flaw? (It might be useful to note in this discussion that although there is no apparent safety concern for someone using a computer with this flaw, PCs are often used to control a variety of instruments, such as medical equipment. For such equipment, a flaw might have a very real safety implication.) Is the answer to this question different if the customer is a bank that uses the computer to calculate interest paid, loan payments, etc., for customers?
11. Should you replace defective products even if customers won't recognize the defect?
12. How thorough should testing be? Is it ever possible to say that no defect exists in a product or structure?
13. Do flaws that Intel found previously in the 386 and 486 chips have any bearing on these questions? In other words, if Intel got away with selling flawed chips before without informing consumers, does that fact have any bearing on this case?
14. G. Richard Thoman, an IBM senior vice president was quoted as saying, "Nobody should have to worry about the integrity of data calculated on an IBM machine." How does this statement by a major Intel customer change the answers to the previous questions?
15. Just prior to when this problem surfaced, Intel had begun a major advertising campaign to make Intel a household name. They had gotten computer manufacturers to place "Intel Inside" labels on their computers and had spent money on television advertising seeking to increase the public demand for computers with Intel processors, with the unstated message that Intel chips were of significantly

higher quality than other manufacturers' chips. How might this campaign have affected what happened in this case?

16. What responsibilities did the engineers who were aware of the flaw have before the chip was sold? After the chips began to be sold? After the flaw became apparent?

DIA RUNAWAY CONCRETE

17. Using one of the codes of ethics in Appendix A, analyze the actions of the batch plant operators and Empire Laboratories.

18. Is altering data a proper use of "engineering judgment"? What alternative might have existed to altering the test data on the concrete?

19. Who is responsible for ensuring that the materials used in a project meet the specifications, the supplier or the purchaser?

PARADYNE COMPUTERS

20. Choose one of the codes of ethics from Appendix A and use it to analyze this case. Were the engineers and managers of Paradyne operating ethically?

21. In preparing their bid, Paradyne wrote in the present tense, as if the computer they proposed currently existed, rather than in the future tense, which would have indicated that the product was still under development. Paradyne claimed that the use of the present tense in its bid (which led SSA to believe that the P8400 actually existed) was acceptable, since it is common business practice to advertise products under development this way. Was this a new product announcement with a specified availability date? Is there a distinction between a response to a bid and company advertising? Is it acceptable to respond to a bid with a planned system if there is no indication when that system is expected to be available?

22. Paradyne also claimed that it was acting as a system integrator (which was allowed by the RFP), using components from other manufacturers to form the Paradyne system. These other components were mostly off-the-shelf, but they had never been integrated into a system before. Does this meet the SSA requirement for an existing system?

23. Once the Paradyne machine failed the initial test, should the requirements have been relaxed to help the machine qualify? If the requirements were going to be modified, should the bidding process have been reopened to the other bidders and others who might now be able to bid? Should bidding be reopened even if it causes a delay in delivery, increased work for the SSA, etc.?

24. Was it acceptable for Paradyne to submit another manufacturer's system for testing with a Paradyne label on it?

25. Was it acceptable to represent a proposed system as existing, if indeed that is what Paradyne did?

26. Is it ethical for a former SSA employee to take a job negotiating contracts with the SSA for a private company? Did this relationship give Paradyne an unfair advantage over its competition?

17

Understanding Ethical Problems

SECTIONS

3.1 Introduction
3.2 A Brief History of Ethical Thought
3.3 Ethical Theories

OBJECTIVES

After reading this chapter, you will be able to

- Discuss several ethical theories and
- See how these theories can be applied to engineering situations.

In late 1984, a pressure-relief valve on a tank used to store methyl isocyanate (MIC) at a Union Carbide plant in Bhopal, India accidentally opened. MIC is a poisonous compound used in the manufacture of pesticides. When the valve opened, MIC was released from the tank and a cloud of toxic gas formed over the area surrounding the plant. Unfortunately, this neighborhood was very densely populated. Some two thousand people were killed and thousands more injured as a result of the accident. Many of the injured have remained permanently disabled.

The causes of the accident are not completely clear, but there appear to have been many contributing factors. Pipes in the plant were misconnected, and essential safety systems were either broken or had been taken off-line for maintenance. The effects of the leak were intensified by the presence of so many people living in close proximity to the plant.

Among the many important issues this case brings up are questions of balancing risk to the local community with the economic benefits to the larger community of the state or nation. Undoubtedly, the presence of this chemical plant brought significant local economic benefit. However, the accident at the plant also brought disaster to the local community at an enormous cost in human lives and suffering. How can we decide if on balance the economic benefit brought by this plant out-weighed the potential safety hazards?

In order to answer this question and analyze other engineering ethics cases, we need a framework for analyzing ethical problems. In the previous chapter, we saw how codes of ethics can be used as an aid in analyzing ethical issues. In this chapter, we will examine moral theories and see how they can also be used as a means for analyzing ethical cases such as the Bhopal disaster.

3.1 INTRODUCTION

In this chapter, we will develop moral theories that can be applied to the ethical problems confronted by engineers. Unfortunately, a thorough and in-depth discussion of all possible ethical theories is beyond the scope of this text. Rather, some important theories will be developed in sufficient detail for use in analyzing cases.

Our approach to ethical problem solving will be similar to that taken in other engineering classes. To learn how to build a bridge, you must first learn the basics of physics and apply this physics to engineering statics and dynamics. Only when the basic theory and understanding of these topics has been acquired can problems in structures be solved and bridges built. Similarly, in ethical problem solving, we will need some knowledge of ethical theory to provide a framework for understanding and reaching solutions in ethical problems. In this chapter, we will develop this theoretical framework and apply it to an engineering case. We will begin by looking at the origins of Western ethical thinking.

3.2 A BRIEF HISTORY OF ETHICAL THOUGHT

It is impossible in this text to give a complete history of ethical thinking. Many books, some of them quite lengthy, have already been written on this subject. However, it is instructive to give a brief outline of the origins and development of the ethical principles that will be applied to engineering practice.

The moral and ethical theories that we will be applying in engineering ethics are derived from a Western cultural tradition. In other words, these ideas originated in the Middle East and Europe. Western moral thought has not come down to us from just a single source. Rather, it is derived both from the thinking of the ancient Greeks as well as from ancient religious thinking and writing, starting with Judaism and its foundations.

Although it is easy to think of these two sources as separate, there was a great deal of influence on ancient religious thought by the Greek philosophers. The written sources of the Jewish moral traditions are the Torah and the Old Testament of the Bible and their enumeration of moral laws, including the ten commandments. Greek ethical thought originated with the famous Greek philosophers that are commonly studied in freshman philosophy classes, principally Socrates and Aristotle, who discussed ethics at great length in his *Nichomachean Ethics*. Greek philosophic ideas were melded together with early Christian and Jewish thought and were spread throughout Europe and the Middle East during the height of the Roman Empire.

Ethical ideas were continually refined during the course of history. Many great thinkers have turned their attention to ethics and morals and have tried to provide insight into these issues through their writings. For example, philosophers such as Locke, Kant, and Mill wrote about moral and ethical issues. The thinking of these philosophers is especially important for our study of engineering ethics, since they did not rely on religion to underpin their moral thinking. Rather, they acknowledged that moral principles are universal, regardless of their origin, and are applicable even in secular settings.

Many of the moral principles that we will discuss have also been codified and handed down through the law. So, in discussing engineering ethics, there is a large body of thinking—philosophical, legal, and religious—to draw from. However, even though there are religious and legal origins of many of the moral principles that we will encounter in our study of engineering ethics, it is important to acknowledge that ethical conduct is fundamentally grounded in a concern for other people. It is not just about law or religion.

3.3 ETHICAL THEORIES

In order to develop workable ethical problem-solving techniques, we must first look at several theories of ethics in order to have a framework for decision making. Ethical problem solving is not as cut and dried as problem solving in most engineering classes. In most engineering classes, there is generally just one theory to consider when tackling a problem. In studying engineering ethics, there are several theories that will be considered. The relatively large number of theories doesn't indicate a weakness in theoretical understanding of ethics or a "fuzziness" of ethical thinking. Rather, it reflects the complexity of ethical problems and the diversity of approaches to ethical problem solving that have been developed over the centuries.

Having multiple theories to apply actually enriches the problem-solving process, allowing problems to be looked at from different angles, since each theory stresses different aspects of a problem. Even though we will use multiple theories to examine ethical problems, each theory applied to a problem will not necessarily lead to a different solution. Frequently, different theories yield the same solution. Our basic ethical problem-solving technique will utilize different theories and approaches to analyze the problem and then try to determine the best solution.

3.3.1 What Is a Moral Theory?

Before looking more closely at individual moral theories, we should start with a definition of what a moral theory is and how it functions. A moral theory defines terms in uniform ways and links ideas and problems together in consistent ways [Harris, Pritchard and Rabins, 1985]. This is exactly how the scientific theories used in other engineering classes function. Scientific theories also organize ideas, define terms, and facilitate problem solving. So, we will use moral theories in exactly the same way that engineering theories are used in other classes.

There are four ethical theories that will be considered here, each differing according to what is held to be the most important moral concept. *Utilitarianism* seeks to produce the most utility, defined as a balance between good and bad consequences of an action, taking into account the consequences for everyone affected. A different approach is provided by *duty ethics*. Duty ethics contends that there are duties that should be performed (for example, the duty to treat others fairly or the duty not to injure others) regardless of whether these acts lead to the most good. *Rights ethics* emphasizes that we all have moral rights, and any action that violates these rights is ethically unacceptable. Like duty ethics, the ultimate overall good of the actions is not taken into account. Finally, *virtue ethics* regards actions as right that manifest good character traits (virtues) and regards actions as bad that display bad character traits (vices); this ethical theory focuses on the type of person we should strive to be.

3.3.2 Utilitarianism

The first of the moral theories that will be considered is utilitarianism. Utilitarianism holds that those actions are good that serve to maximize human well-being. The emphasis in utilitarianism is not on maximizing the well-being of the individual, but rather on maximizing the well-being of society as a whole, and as such it is somewhat of a collectivist approach. An example of this theory that has been played out in this country many times over the past century is the building of dams. Dams often lead to great benefit to society by providing stable supplies of drinking water, flood control, and recreational opportunities. However, these benefits often come at the expense of people who live in areas that will be flooded by the dam and are required to find new homes. Utilitarianism tries to balance the needs of society with the needs of the individual, with an emphasis on what will provide the most benefit to the most people.

Utilitarianism is fundamental to many types of engineering analysis, including risk–benefit analysis and cost–benefit analysis, which we will discuss later. However, as good as the utilitarian principle sounds, there are some problems with it. First, as seen in the example of the building of a dam, sometimes what is best for everyone may be bad for a particular individual or group of individuals. An example of this problem is the proposed Waste Isolation Pilot Plant (WIPP) near Carlsbad, New Mexico. WIPP is designed to be a permanent repository for nuclear waste generated in the United States. It consists of a system of tunnels bored into underground salt formations, which are considered by geologists to be extremely stable, especially to incursion of water, which could lead to seepage of the nuclear wastes into groundwater. However, there are many who oppose the opening of this facility, principally on the grounds that transportation of the wastes across highways has the potential for accidents that might cause health problems for people living near these routes.

An analysis of WIPP using utilitarianism might indicate that the disposal of nuclear wastes is a major problem hindering the implementation of many useful technologies, including medicinal uses of radioisotopes and nuclear generation of electricity. Solution of this waste disposal problem will benefit society by providing improved health care and more plentiful electricity. The slight potential for adverse health effects for individuals living near the transportation routes is far outweighed by the overall benefits to society. So, WIPP should be allowed to open. As this example demonstrates, the utilitarian approach can seem to ignore the needs of individuals, especially if these needs seem relatively insignificant.

Another objection to utilitarianism is that its implementation depends greatly on knowing what will lead to the most good. Frequently, it is impossible to know exactly what the consequences of an action are. It is often impossible to do a complete set of experiments to determine all of the potential outcomes, especially when humans are involved as subjects of the experiments. So, maximizing the benefit to society involves guesswork and the risk that the best guess might be wrong. Despite these objections, utilitarianism is a very valuable tool for ethical problem solving, providing one way of looking at engineering ethics cases.

Before ending our discussion of utilitarianism, it should be noted that there are many flavors of the basic tenets of utilitarianism. Two of these are act utilitarianism and rule utilitarianism. Act utilitarianism focuses on individual actions rather than on rules. The best known proponent of act utilitarianism was John Stuart Mill (1806–1873), who felt that most of the common rules of morality (e.g., don't steal, be honest, don't harm others) are good guidelines derived from centuries of human experience. However, Mill felt that individual actions should be judged based on whether the most good was produced in a given situation, and rules should be broken if doing so will lead to the most good.

John Stuart Mill, a leading philosopher of utilitarianism. Courtesy of the Library of Congress.

Rule utilitarianism differs from act utilitarianism in holding that moral rules are most important. As mentioned previously, these rules include "do not harm others" and "do not steal." Rule utilitarians hold that although adhering to these rules might not always maximize good in a particular situation, overall, adhering to moral rules will ultimately lead to the most good. Although these two different types of utilitarianism can lead to slightly different results when applied in specific situations, in this text we will consider these ideas together and not worry about the distinctions between the two.

3.3.3 Cost–Benefit Analysis

One tool often used in engineering analysis, especially when trying to determine whether a project makes sense, is cost–benefit analysis. Fundamentally, this type of analysis is just an application of utilitarianism. In cost–benefit analysis, the costs of a project are assessed, as are the benefits. Only those projects with the highest ratio of benefits to costs will be implemented. This principle is similar to the utilitarian goal of maximizing the overall good.

As with utilitarianism, there are pitfalls in the use of cost–benefit analysis. While it is often easy to predict the costs for most projects, the benefits that are derived from them are often harder to predict and to assign a dollar value to. Once dollar amounts for the costs and benefits are determined, calculating a mathematical ratio may seem very objective and therefore may appear to be the best way to make a decision. However, this ratio can't take into account many of the more subjective aspects of a decision. For example, from a pure cost–benefit discussion, it might seem that the building of a dam is an excellent idea. But this analysis won't include other issues such as whether the benefits outweigh the loss of a scenic wilderness area or the loss of an endangered species with no current economic value. Finally, it is also important to determine whether those who stand to reap the benefits are also those who will pay the costs. It is unfair to place all of the costs on one group while another reaps the benefits.

3.3.4 Duty Ethics and Rights Ethics

Two other ethical theories—duty ethics and rights ethics—are similar to each other and will be considered together. These theories hold that those actions are good that respect the rights of the individual. Here, good consequences for society as a whole are not the only moral consideration.

A major proponent of duty ethics was Immanuel Kant (1724–1804), who held that moral duties are fundamental. Ethical actions are those actions that could be written down on a list of duties: be honest, don't cause suffering to other people, be fair to others, etc. These actions are our duties because they express respect for persons, express an unqualified regard for autonomous moral agents, and are universal principles [Martin and Schinzinger, 2000]. Once one's duties are recognized, the ethically correct moral actions are obvious. In this formulation, ethical acts are a result of proper performance of one's duties.

Rights ethics was largely formulated by John Locke (1632–1704), whose statement that humans have the right to life, liberty, and property was paraphrased in the Declaration of Independence of the soon-to-be United States of America in 1776. Rights ethics holds that people have fundamental rights that other people have a duty to respect.

Duty ethics and rights ethics are really just two different sides of the same coin. Both of these theories achieve the same end: Individual persons must be respected, and actions are ethical that maintain this respect for the individual. In duty ethics, people have duties, an important one of which is to protect the rights of others. And in rights ethics, people have fundamental rights that others have duties to protect.

As with utilitarianism, there are problems with the duty and rights ethics theories that must be considered. First the basic rights of one person (or group) may conflict with the basic rights of another group. How do we decide whose rights have priority? Using our previous example of the building of a dam, people have the right to use their property. If their land happens to be in the way of a proposed dam, then rights ethics would hold that this property right is paramount and is sufficient to stop the dam project. A single property holder's objection would require that the project be terminated. However, there is a need for others living in nearby communities to have a reliable water supply and to be safe from continual flooding. Who's rights are paramount here? Rights and duty ethics don't resolve this conflict very well; hence, the utilitarian approach of trying to determine the most good is more useful there.

Immanuel Kant, German philosopher whose work included early formulations of duty ethics.
Courtesy of the Library of Congress.

The second problem with duty and rights ethics is that these theories don't always account for the overall good of society very well. Since the emphasis is on the individual, the good of a single individual can be paramount compared to what is good for society as a whole. The WIPP case discussed before illustrates this problem. Certainly, people who live along the route where the radioactive wastes will be transported have the right to live without fear of harm due to accidental spills of hazardous waste. But the nation as a whole will benefit from the safe disposal of these wastes. Rights ethics would come down clearly on the side of the individuals living along the route despite the overall advantage to society.

Already it is clear why we will be considering more than one ethical theory in our discussion of engineering cases. The theories already presented clearly represent different ways of looking at ethical problems and can frequently arrive at different solutions.

Thus, any complete analysis of an ethical problem must incorporate multiple theories if valid conclusions are to be drawn.

3.3.5 Virtue Ethics

Another important ethical theory that we will consider is virtue ethics. Fundamentally, virtue ethics is interested in determining what kind of people we should be. Virtue is often defined as moral distinction and goodness. A virtuous person exhibits good and beneficial qualities. In virtue ethics, actions are considered right if they support good character traits (virtues) and wrong if they support bad character traits (vices) [Schinzinger and Martin, 2000]. Virtue ethics focuses on such words as responsibility, honesty, competence, and loyalty, which are virtues; and dishonesty, disloyalty, and irresponsibility, which are vices. As you can see, virtue ethics is closely tied to personal character. We do good things because we are virtuous people and seek to enhance these character traits in ourselves and in others.

In many ways, this theory may seem to be mostly personal ethics and not particularly applicable to engineering or business ethics. However, personal morality cannot, or at any rate should not, be separated from business morality. If a behavior is virtuous in the individual's personal life, the behavior is virtuous in his or her business life as well.

How can virtue ethics be applied to business and engineering situations? This type of ethical theory is somewhat trickier to apply to the types of problems that we will consider, perhaps because virtue ethics seems less concrete and less susceptible to rigorous analysis and because it is harder to describe nonhuman entities such as a corporation or government in terms of virtue. However, we can use virtue ethics in our engineering career by answering questions such as: Is this action honest? Will this action demonstrate loyalty to my community and/or my employer? Have I acted in a responsible fashion? Often, the answer to these questions makes the proper course of action obvious.

As with any ethical theory, it is important to be careful in applying virtue ethics. Problems can arise with words that on the face seem to be virtues, but can actually lead to vices. For example, the concept of "honor" has been around for centuries and is often viewed positively. One sense of the word "honor" is a code of dignity, integrity, and pride. Honor may seem like a very positive thing, especially the aspects related to integrity. But the aspects related to pride can often have negative consequences. There are numerous examples in history of wars that have been fought and atrocities committed in order to preserve the honor of an individual or a nation. Individuals have often committed crimes as a way of preserving their honor. In using virtue ethics, it is important to ensure that the traits you identify as virtues are indeed virtuous and will not lead to negative consequences.

3.3.6 Personal vs. Corporate Morality

This is an appropriate place to discuss a tricky issue in engineering ethics: Is there a distinction between the ethics practiced by an individual and the ethics practiced by a corporation? Put another way, can a corporation be a moral agent as an individual can? This is a question that is central to many discussions of business and engineering ethics. If a corporation has no moral agency, then it cannot be held accountable for its actions, although sometimes individuals within a company can be held accountable. The law is not always clear on the answer to this question and can't be relied upon to resolve the issue.

This dilemma comes most sharply into focus in a discussion of virtue ethics. Can a company truly be expected to display honesty or loyalty? These are strictly human traits and cannot be ascribed to a corporation. In the strictest definition of moral agency, a company cannot be a moral agent, and yet companies have many dealings with individuals or groups of people.

How, then, do we resolve this problem? In their capacity to deal with individuals, corporations should be considered pseudomoral agents and should be held accountable in the same way that individuals are, even if the ability to do this within the legal system is limited. In other words, with regard to an ethical problem, responsibility for corporate wrongdoing shouldn't be hidden behind a corporate mask. Just because it isn't really a moral agent like a person doesn't mean that a corporation can do whatever it pleases. Instead, in its interactions with individuals or communities, a corporation must respect the rights of individuals and should exhibit the same virtues that we expect of individuals.

3.3.7 Which Theory to Use?

Now that we have discussed four different ethical theories, the question arises: How do we decide which theory is applicable to a given problem? The good news is that in solving ethical problems, we don't have to choose from among these theories. Rather, we can use all of them to analyze a problem from different angles and see what result each of the theories gives us. This allows us to examine a problem from different perspectives to see what conclusion each one reaches. Frequently, the result will be the same even though the theories are very different.

Take, for example, a chemical plant near a small city that discharges a hazardous waste into the groundwater. If the city takes its water from wells, the water supply for the city will be compromised and significant health problems for the community may result. Rights ethics indicates that this pollution is unethical, since it causes harm to many of the residents. A utilitarian analysis would probably also come to the same conclusion, since the economic benefits of the plant would almost certainly be outweighed by the negative effects of the pollution and the costs required to ensure a safe municipal water supply. Virtue ethics would say that discharging wastes into groundwater is irresponsible and harmful to individuals and so shouldn't be done. In this case, all of the ethical theories lead to the same conclusion.

What happens when the different theories seem to give different answers? This scenario can be illustrated by the discussion of WIPP presented previously. Rights ethics indicated that transporting wastes through communities is not a good idea, whereas utilitarianism concluded that WIPP would be beneficial to society as a whole. This is a trickier situation, and the answers given by each of the theories must be examined in detail, compared with each other, and carefully weighed. Generally, rights and duty ethics should take precedence over utilitarian considerations. This is because the rights of individuals should receive relatively stronger weight than the needs of society as a whole. For example, an action that led to the death of even one person is generally viewed very negatively, regardless of the overall benefit to society. After thorough analysis using all of the theories, a balanced judgment can be formed.

3.3.8 Non-Western Ethical Thinking

It is tempting to think that the ethical theories that have been described here are applicable only in business relations within cultures that share our Western ethical traditions: Europe and the Americas. Since the rest of the world has different foundations for its ethical systems, it might seem that what we learn here won't be applicable in our business dealings in, for example, Japan, India, Africa, or Saudi Arabia. However, this thinking is incorrect. Ethics is not geographic or cultural. Indeed, ethical thinking has developed similarly around the world and is not dependent on a Western cultural or religious tradition. Ethical standards are similar worldwide.

For example, ethical principles in Arab countries are grounded in the traditions of their religion, Islam. Islam is one of the three major monotheistic religions, along with Christianity and Judaism. It is surprising to many Westerners that Islam developed in the Middle East, just as Judaism and Christianity did, and shares many prophets and religious concepts with the other two monotheistic religions. The foundations of ethical principles relating to engineering and business in Islamic countries are thus very similar to those in Western countries. Although cultural practices may vary when dealing with the many Islamic nations that stretch from Africa and the Middle East to Southeast Asia, the same ethical principles that apply in Western countries are applicable.

Similarly, ethical principles of Hindus, Buddhists, and practitioners of all the world's major religions are similar. Although the ethical principles in other cultures may be derived in different ways, the results are generally the same, regardless of culture.

Moreover, personal ethics are not determined by geography. Personal and business behavior should be the same regardless of where you happen to be on a given day.

For example, few would find the expression "When in Rome, do as the Romans do" applicable to personal morality. If you believe that being deceptive is wrong, certainly it is no less wrong when you are dealing with a (hypothetical) culture where this behavior is not considered to be bad. Thus, the ethics that we discuss in this book will be applicable regardless of where you are doing business.

APPLICATION: CASES

THE DISASTER AT BHOPAL

On the night of December 2, 1984, a leak developed in a storage tank at a Union Carbide chemical plant in Bhopal, India. The tank contained 10,000 gallons of methyl isocyanate (MIC), a highly toxic chemical used in the manufacture of pesticides, such as Sevin. The leak sent a toxic cloud of gas over the surrounding slums of Bhopal, resulting in the death of over 2,000 people, and injuries to over 200,000 more.

The leak was attributed to the accidental pouring of water into the tank. Water reacts very vigorously with MIC, causing heating of the liquid. In Bhopal, the mixing of water with MIC increased the temperature of the liquid in the tank to an estimated 400°F. The high temperature caused the MIC to vaporize, leading to a buildup of high pressure within the tank. When the internal pressure became high enough, a pressure-relief valve popped open, leaking MIC vapors into the air.

The water had probably been introduced into the tank accidentally. A utility station on the site contained two pipes side by side. One pipe carried nitrogen, which was used to pressurize the tank to allow the liquid MIC to be removed. The other pipe contained water. It appears that instead of connecting the nitrogen pipe, someone accidentally connected the water pipe to the MIC tank. The accident was precipitated when an estimated 240 gallons of water were injected into the MIC storage tank.

As with many of the disasters and accidents that we study in this book, there was not just one event that led to the disaster, but rather there were several factors that contributed to this accident. Any one of these factors alone probably wouldn't have led to the accident, but the combination of these factors made the accident almost inevitable and the consequences worse. A major factor in this accident was the curtailment of plant maintenance as part of a cost-cutting effort. The MIC storage tank had a refrigeration unit on it, which should have helped to keep the tank temperatures closer to normal, even with the water added, and might have prevented the vaporization of the liquid. However, this refrigeration unit had stopped working five months before the accident and hadn't yet been repaired.

The tank also was equipped with an alarm that should have alerted plant workers to the dangerous temperatures; this alarm was improperly set, so no warning was given. The plant was equipped with a flare tower. This is a device designed to burn vapors before they enter the atmosphere, and it would have been able to at least reduce, if not eliminate, the amount of MIC reaching the surrounding neighborhood. The flare tower was not functioning at the time of the accident. Finally, a scrubber that was used to neutralize toxic vapors was not activated until the vapor release was already in progress. Some investigators pointed out that the scrubber and flare systems were probably inadequate, even had they been functioning. However, had any of these systems been functioning at the time of the accident, the disaster could have at least been mitigated, if not completely averted. The fact that none of them were operating at the time ensured that once the water had been mistakenly added to the MIC tank, the ensuing reaction would proceed undetected until it was too late to prevent the accident.

It is unclear to whom the ultimate blame for this accident should be laid. The plant designers clearly did their job by anticipating problems that would occur and installing safety systems to prevent or mitigate potential accidents. The management of the plant seems obviously negligent. It is sometimes necessary for some safety features to be taken off-line for repair or maintenance. But to have all of the safety systems inoperative simultaneously is inexcusable. Union Carbide also seems negligent in not preparing a plan for notifying and evacuating the surrounding population in the event of an accident. Such plans are standard in the United States and are often required by local ordinance.

Union Carbide was unable to say that such an accident was unforeseeable. Leaky valves in the MIC system had been a problem at the Bhopal plant on at least six occasions before the accident. One of these gas leaks involved a fatality. Moreover, Union Carbide had a plant in Institute, West Virginia that also produced MIC. The experience in West Virginia was similar to that in Bhopal before the accident. There had been a total of 28 leaks of MIC over the previous five years, none leading to any serious problems. An internal Union Carbide memo from three months before the Bhopal accident warned of the potential for a runaway reaction in MIC storage tanks in West Virginia and called into question the adequacy of emergency plans at the plants. The memo concluded that "a real potential for a serious incident exists" [*US News and World Report*, Feb. 4, 1985, p. 12]. Apparently, these warnings had not been transmitted to the plant in India.

Ultimately, some share of the blame must be borne by the Indian government. Unlike in most Western nations, there was very little in the way of safety standards under which U.S. corporations must operate. In fact, third-world countries have often viewed pollution control and safety regulation as too expensive, and attempts by the industrialized nations to enforce Western-style safety and environmental regulations worldwide are regarded as attempts to keep the economies of developing countries backward [*Atlantic Monthly*, March 1987, p. 30]. In addition, the local government had no policy or zoning forbidding squatters and others from living so close to a plant where hazardous compounds are stored and used. The bulk of the blame goes to Union Carbide for failure to adequately train and supervise its Indian employees in the maintenance and safety procedures that are taken for granted in similar plants in the United States.

In the aftermath of the accident, lawsuits totaling over $250 billion were filed on behalf of the victims of the accident. Union Carbide committed itself to ensuring that the victims of the accident were compensated in a timely fashion. Union Carbide also helped set up job training and relocation programs for the victims of the accident.In addition, the chairman of Union Carbide, Warren Anderson, was charged in the Indian courts with culpable homicide, a criminal (rather than a civil) charge. As of early 2003, this case was still being decided in the courts. Ultimately, it has been estimated that approximately 10,000 of those injured in the accident will suffer some form of permanent damage [*Atlantic Monthly*, March 1987, p. 30].

THE ABERDEEN THREE

The Aberdeen Three is one of the classic cases often used in engineering ethics classes and texts to illustrate the importance of environmental protection and the safety of workers exposed to hazardous and toxic chemicals. The Aberdeen Proving Ground is a U.S. Army weapons development and test center located on a military base in Maryland with no access by civilian non-employees. Since World War II, Aberdeen has been used to develop and test chemical weapons. Aberdeen has also been used for the storage and disposal of some of these chemicals.

This case involves three civilian managers at the Pilot Plant at the Proving Grounds: Carl Gepp, manager of the Pilot plant; William Dee, who headed the chemical weapons development team; and Robert Lentz, who was in charge of developing manufacturing processes for the chemical weapons. [Weisskopf, 1989] Between 1983 and 1986, inspections at the Pilot Plant indicated that there were serious safety hazards. These hazards included carcinogenic and flammable substances left in open containers, chemicals that can become lethal when mixed together being stored in the same room, barrels of toxic chemicals that were leaking, and unlabeled containers of chemicals. There was also an external tank used to store sulfuric acid that had leaked 200 gallons of acid into a local river. This incident triggered state and federal safety investigations that revealed inadequate chemical retaining dikes and a system for containing and treating chemical hazards that was corroded and leaking.

In June of 1988, the three engineer/managers were indicted for violation of RCRA, the Resource Conservation and Recovery Act. RCRA had been passed by Congress in 1976 and was intended to provide incentives for the recovery of important resources from wastes, the conservation of resources, and the control of the disposal of hazardous wastes. RCRA banned the dumping of solid hazardous wastes and included criminal penalties for violations of hazardous-waste disposal guidelines. The three managers claimed that they were not aware that the plant's storage practices were illegal and that they did things according to accepted practices at the Pilot Plant. Interestingly, since this was a criminal prosecution, the Army could not help defray the costs of the manager's defense, and each of them incurred great costs defending themselves.

In 1989, the three engineer/managers were tried and convicted of illegally storing, treating, and disposing of hazardous wastes. There was no indication that these three were the ones who actually handled chemicals in an unsafe manner, but as managers of the plant, the three were ultimately responsible for how the chemicals were stored and for the maintenance of the safety equipment. The potential penalty for these crimes was up to 15 years in prison and a fine of up to $750,000. Gepp, Dee, and Lentz were each found guilty and sentenced to three years probation and 1,000 hours of community service. The relative leniency of the sentences was based partly on the large court costs each had already incurred.

PROFESSIONAL SUCCESS: TEAMWORK

Ethical issues can arise when working on projects in groups or teams. Many of your engineering classes are designed so that labs or projects are performed in groups. Successful performance in a group setting is a skill that is best learned early in your academic career since most projects in industry involve working as part of a team.

In order for a project to be completed successfully, cooperation among team members is essential. Problems can arise when a team member doesn't do a good job on his part of the project, doesn't make a contribution at all, or doesn't complete his assignments on time. There can also be a problem when one team member tries to do everything. This shuts out teammates who want to contribute and learn. An analogy can be made here to team sports: clearly one individual on the team who is not performing his role can lead to a loss for the entire team. Equally true, individuals who try to do it all—"ballhogs"—can harm the team. Ethical teamwork includes performing the part of the work that you are assigned, keeping to schedules, sharing information with other team members, and helping to foster a cooperative and supportive team atmosphere so everyone can contribute.

KEY TERMS

Cost–benefit analysis
Duty ethics
Rights ethics
Utilitarianism
Virtue ethics

REFERENCES

Charles E. Harris, Jr., Michael S. Pritchard, and Michael J. Rabins, *Engineering Ethics, Concepts and Cases*, Wadsworth Publishing Company, Belmont, CA, 2000.

Roland Schinzinger and Mike W. Martin, *Introduction to Engineering Ethics,* McGraw-Hill, New York, 2000.

BHOPAL

Philip Elmer-DeWitt, "What Happened at Bhopal?" *Time Magazine*, April 1, 1985, p. 71.

"Bhopal Disaster—New Clues Emerge," *US News and World Report*, Feb. 4, 1985, p. 12.
Peter Stoler, "Frightening Findings at Bhopal," *Time Magazine*, Feb. 18, 1985, p. 78.
Fergus M. Bordewich, "The Lessons of Bhopal," *Atlantic Monthly*, March 1987, p. 30.

ABERDEEN THREE

Steven Weisskopf, "The Aberdeen Mess," *The Washington Post Magazine*, Jan. 15, 1989, p. 55.

Problems

1. Analyze the space shuttle *Challenger* accident in 1986 using the ethical theories developed in this chapter. What does utilitarianism tell us about this case? In your analysis, be sure to include issues regarding benefits to the United States and mankind that might result from the space shuttle program. You might also include benefits to Morton Thiokol and the communities where it operates if the program is successful.
2. What do duty and rights ethics tell us about the *Challenger* case? How do your answers to this question and to the previous question influence your ideas on whether the *Challenger* should have been launched?
3. Use contemporary newspaper accounts to find information on problems with Intel's Pentium computer chip (1995) and with runway concrete at the Denver International Airport (1994). Analyze these cases, using virtue ethics. Start by deciding what virtues are important for people in these businesses (e.g., honesty, fairness, etc.). Then see if these virtues were exhibited by the engineers working for these companies.

BHOPAL

4. Use the ethical theories discussed in this chapter to analyze the Bhopal case. Topics to be considered should include the placing of a hazardous plant in a populated area, decisions to defer maintenance on essential safety systems, etc. Especially important theories will be rights and duty ethics and utilitarianism.
5. Use the code of ethics of the American Institute of Chemical Engineers, reprinted in Appendix A, to analyze what a process engineer working at this plant should have done. What does the code say about the responsibilities of the engineers who designed the plant and the engineers responsible for making maintenance decisions?
6. What responsibility does Union Carbide have for the actions of its subsidiaries? Union Carbide India was 50.9% owned by the parent company.
7. What duty did Union Carbide have to inform local officials in India of the potential dangers of manufacturing and storing MIC in India?
8. Some of Union Carbide's reports hinted strongly that part of the fault lay with the inadequate workforce available in a third-world country such as India. How valid is this statement? What are the ethical implications for Union Carbide if this statement is true?
9. What responsibility should the national and local government in Bhopal have for ensuring that the plant is operated safely?
10. What relative importance should be placed on keeping safety systems operating as compared to maintaining other operations? (Note: From the reports on this accident, there is no indication that Union Carbide skimped on safety to keep production going. Rather, this is a hypothetical question.)

11. In the absence of environmental or safety laws in the locality where it operates, what responsibility does a U.S. corporation have when operating overseas? Does the answer change if the locality does have laws, but they are less strict than ours? What about the ethics of a U.S. corporation selling products overseas that are banned in the United States, such as DDT?

THE ABERDEEN THREE

12. What does utilitarianism tell us about the behavior of the Aberdeen Three? What do duty and rights ethics tell us? In analyzing this, start by determining who is harmed or potentially harmed by these activities and who benefits or potentially benefits from them.
13. Can the actions of these engineer/managers be classified as engineering decisions, management decisions, or both? Ethically, does it matter whether these decisions were engineering or management decisions?
14. Do you think that the Aberdeen Three knew about RCRA? If not, should they have? Does it really matter if they knew about RCRA or not?
15. Do you think that the Aberdeen Three were knowledgeable about the effects of these chemicals and proper storage methods? Should they have been?
16. Were the actions of the Aberdeen Three malicious?
17. In the course of this case, it came out that cleaning up the chemical storage at Aberdeen would have been paid for out of separate Army funds and would not have come from the budgets of the three managers. What bearing does this information have on the case?
18. What should the Aberdeen Three have done differently? Should the lower level workers at the plant have done anything to solve this problem?
19. The bosses of the Aberdeen Three claimed to have no idea about the conditions at the Pilot Plant. Should they have done anything differently? Should they have been prosecuted as well?
20. Apply the code of ethics of one of the professional societies given in Appendix A to this situation. Were the managers guilty of ethical violations according to the code?

18

Ethical Problem-Solving Techniques

In the early 1990s, newspapers began to report on studies indicating that living near electrical-power distribution systems leads to an increased risk of cancer, especially in children. The risk was attributed to the effects of the weak, low-frequency magnetic fields present near such systems. Further reports indicated that there might also be some risk associated with the use of common household items such as electric blankets and clock radios. Predictably, there was much concern among the public about this problem, and many studies were performed to verify these results. Power companies began to look into methods for reducing the fields, and many engineers sought ways to design products that emitted reduced amounts of this radiation.

In designing products and processes, engineers frequently encounter scenarios like the one just described. Nearly everything an engineer designs has some health or safety risk associated with it. Often, as with the case of the weak magnetic fields, the exact nature of the hazard is only poorly understood. How then does an engineer decide whether it is ethical to work on a particular product or process? What tools are there for an engineer who needs to decide which is the ethically correct path to take?

SECTIONS

4.1 Introduction
4.2 Analysis of Issues in Ethical Problems
4.3 Line Drawing
4.4 Flow Charting
4.5 Conflict Problems
4.6 An Application of Problem-Solving Methods: Bribery/Acceptance of Gifts

OBJECTIVES

After reading this chapter, you will be able to

- Apply these methods to hypothetical and real cases
- See how flow charting can be used to solve ethical problems, and
- Learn what bribery is and how to avoid it.

In this chapter, we will develop analysis and problem-solving strategies to help answer these questions. These techniques will allow us to put ethical problems in the proper perspective and will point us in the direction of the correct solution.

4.1 INTRODUCTION

Now that we have discussed codes of ethics and moral theories, we are ready to tackle the problem of how to analyze and resolve ethical dilemmas when they occur. In solving engineering problems, it is always tempting to look for an appropriate formula, plug in the numbers, and calculate an answer. This type of problem-solving approach, while sometimes useful for engineering analysis problems, is useless for ethical problem solving. As we have seen in the previous chapters, there are theories that help us to frame our understanding of the problem, but there are no formulas and no easy "plug-and-chug" methods for reaching a solution.

In this chapter, we will examine methods for analyzing ethical problems and see how to apply them. Obviously, some problems are easily solved. If you are tempted to embezzle money from your employer, it is clear that this action is stealing and is not morally acceptable. However, as mentioned previously, many of the situations encountered by practicing engineers are ambiguous or unclear, involving conflicting moral principles. This is the type of problem for which we will most need analysis and problem-solving methods.

4.2 ANALYSIS OF ISSUES IN ETHICAL PROBLEMS

A first step in solving any ethical problem is to completely understand all of the issues involved. Once these issues are determined, frequently a solution to the problem becomes apparent. The issues involved in understanding ethical problems can be split into three categories: factual, conceptual, and moral [Harris, Pritchard, and Rabins, 2000]. Understanding these issues helps to put an ethical problem in the proper framework and often helps point the way to a solution.

4.2.1 Types of Issues in Ethical Problem Solving

Let's begin by examining in depth each of the types of issues involved in ethical problems. Factual issues involve what is actually known about a case—i.e., what the facts are. Although this concept seems straightforward, the facts of a particular case are not always clear and may be controversial. An example of facts that are not necessarily clear can be found in the controversy in contemporary society regarding abortion rights. There is great disagreement over the point at which life begins and at which point a fetus can be legally protected. *Roe v. Wade*, the original decision legalizing abortion in the United States, was decided by the Supreme Court by a split decision. Even the justices of the Supreme Court were unable to agree on this "fact."

In engineering, there are controversies over facts as well. For example, global warming is of great concern to society as we continue to emit greenhouse gases into the atmosphere. Greenhouse gases, such as carbon dioxide, trap heat in the atmosphere. This is thought to lead to a generalized warming of the atmosphere as emissions from automobiles and industrial plants increase the carbon dioxide concentration in the atmosphere. This issue is of great importance to engineers, since they might be required to design new products or redesign old ones to comply with stricter environmental standards if this warming effect indeed proves to be a problem. However, the global warming process is only barely understood, and the need to curtail emission of these gases is a controversial topic. If it were known exactly what the effects of emitting greenhouse gases into the atmosphere would be, the engineer's role in reducing this problem would be clearer.

Conceptual issues have to do with the meaning or applicability of an idea. In engineering ethics, this might mean defining what constitutes a bribe as opposed to an acceptable gift, or determining whether certain business information is proprietary. In the case of the bribe, the value of the gift is probably a well-known fact. What isn't known is whether accepting it will lead to unfair influence on a business decision. For example, conceptually it must be determined if the gift of tickets to a sporting event by a potential supplier of parts for your project is meant to influence your decision or is just a nice gesture between friends. Of course, like factual issues, conceptual issues are not always clear-cut and will often result in controversy as well.

Once the factual and conceptual issues have been resolved, at least to the extent possible, all that remains is to determine which moral principle is applicable to the situation. Resolution of moral issues is often more obvious. Once the problem is defined, it is usually clear which moral concept applies, and the correct decision becomes obvious. In our example of a "gift" offered by a sales representative, once it is determined whether it is simply a gift or is really a bribe, then the appropriate action is obvious. If we determine that it is indeed a bribe, then it cannot ethically be accepted.

Given that the issues surrounding an ethical problem can be controversial, how can these controversies be resolved? Factual issues can often be resolved through research to establish the truth. It is not always possible to achieve a final determination of the "truth" that everyone can agree on, but generally, further research helps clarify the situation, can increase the areas of agreement, and can sometimes achieve consensus on the facts. Conceptual issues are resolved by agreeing on the meaning of terms and concepts. Sometimes agreement isn't possible, but as with factual issues, further analysis of the concepts at least clarifies some of the issues and helps to facilitate agreement. Finally, moral issues are resolved by agreement as to which moral principles are pertinent and how they should be applied.

Often, all that is required to solve a particular ethical problem is a deeper analysis of the issues involved according to the appropriate principles. Once the issues are analyzed and agreement is reached on the applicable moral principles, it is clear what the resolution should be.

4.2.2 Application to a Case Study: Paradyne Computers

To illustrate the use of this problem-solving method, let's analyze a case study. In 1980, Paradyne, a computer company, bid to supply the Social Security Administration with new computer systems. Details of this case can be found in a previous chapter and in the references listed at the end of the chapter. We'll look at the factual issues first. The request for proposals clearly specified that only existing systems would be considered. Paradyne did not have any such system running and had never tested the operating system on the product they actually proposed to sell to the Social Security Administration (SSA). The employment of a former SSA worker by Paradyne to help lobby SSA for the contract is also clear. In this case, the factual issues do not appear particularly controversial.

The conceptual issues involve whether bidding to provide an off-the-shelf product when the actual product is only in the planning stages is lying or is an acceptable business practice. Is placing a Paradyne label over the real manufacturer's label deceptive? Does lobbying your former employer on behalf of your current employer constitute a conflict of interest? These questions will certainly generate discussion. Indeed, Paradyne asserted that it had done nothing wrong and was simply engaging in common business practices. The issue of the conflict of interest is so hard to decide that laws have been enacted making it illegal for workers who have left government employ to lobby their former employers for specified periods of time.

The moral issues then include the following: Is lying an acceptable business practice? Is it alright to be deceptive if doing so allows your company to get a contract? The answers to these questions are obvious: Lying and deceit are no more acceptable in your business life than in your personal life. So, if conceptually we decide that Paradyne's practices were deceptive, then our analysis indicates that their actions were unethical.

4.3 LINE DRAWING

The line-drawing technique that will be described in this section is especially useful for situations in which the applicable moral principles are clear, but there seems to be a great deal of "gray area" about which ethical principle applies. Line drawing is performed by drawing a line along which various examples and hypothetical situations are placed. At one end is placed the "positive paradigm," an example of something that is unambiguously morally acceptable. At the other end, the "negative paradigm," an example of something that is unambiguously not morally acceptable, is placed. In between is placed the problem under consideration, along with other similar examples. Those examples that more closely conform to the positive paradigm are placed near it, and examples closer to the negative paradigm are placed near that paradigm. By carefully examining this continuum and placing the moral problem under consideration in the appropriate place along the line, it is possible to determine whether the problem is more like the positive or negative paradigm and therefore whether it is acceptable or unacceptable.

Let's illustrate this technique using a hypothetical situation. Our company would like to dispose of a slightly toxic waste by dumping it into a local lake from which a nearby town gets its drinking water. How can we determine if this practice is acceptable? Let's start by defining the problem and the positive and negative paradigms.

Problem: It is proposed that our company dispose of a slightly hazardous waste by dumping it into a lake. A nearby town takes its drinking water supply from this lake. Our research shows that with the amount of waste we plan to put into the lake, the average concentration of the waste in the lake will be 5 parts per million (ppm). The EPA limit for this material has been set at 10 ppm. At the 5-ppm level, we expect no health problems, and consumers would not be able to detect the compound in their drinking water.

Positive paradigm: The water supply for the town should be clean and safe.

Negative paradigm: Toxic levels of waste are put into the lake.

Let's start by drawing a line and placing the positive and negative paradigms on it:

Figure 4.1. Example of line drawing showing the placement of the negative and positive paradigms.

Now let's establish some other hypothetical examples for consideration:

1. The company dumps the chemical into the lake. At 5 ppm, the chemical will be harmless, but the town's water will have an unusual taste.
2. The chemical can be effectively removed by the town's existing water-treatment system.
3. The chemical can be removed by the town with new equipment that will be purchased by the company.
4. The chemical can be removed by the town with new equipment for which the taxpayer will pay.
5. Occasionally, exposure to the chemical can make people feel ill, but this only lasts for an hour and is rare.
6. At 5 ppm, some people can get fairly sick, but the sickness only lasts a week, and there is no long-term harm.
7. Equipment can be installed at the plant to further reduce the waste level to 1 ppm.

Obviously, we could go on for a long time creating more and more test examples. Generally, where your problem fits along the line is obvious with only a few examples, but the exercise should be continued with more examples until it is clear what the proper resolution is. Now let's redraw our line with the examples inserted appropriately:

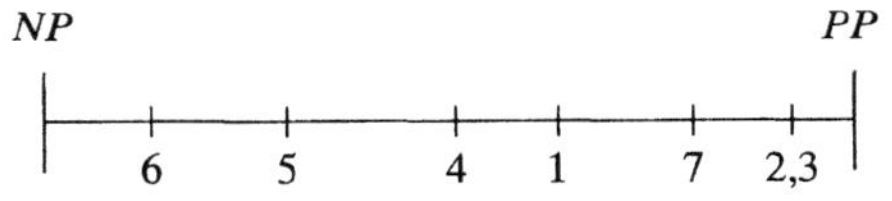

Figure 4.2. Same as Figure 4.1, with the addition of the examples to the line.

After setting up the examples, it may be clear that there is a gap in the knowledge. For example, in our case, we might need more information on seasonal variations in waste concentration and water usage of the town. We also could use information on potential interactions of the chemical with other pollutants, such as runoff of pesticides from local farms. Note that there is some subjectivity in determining exactly where along the line each of the examples fits.

Now let's complete the exercise by denoting our problem by a "P" and inserting it at the appropriate place along the line. As with the previous examples, placement of the problem along the line is somewhat subjective.

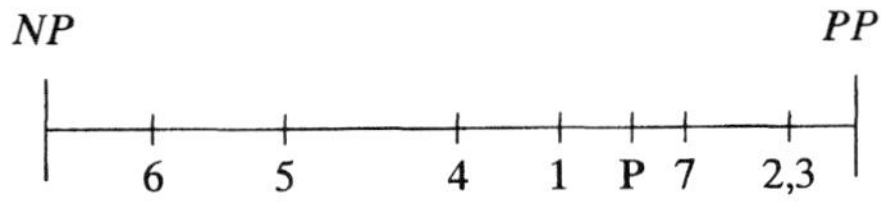

Figure 4.3. Final version of the line-drawing example, with the problem under consideration added.

As drawn here, it is clear that dumping the toxic waste is probably a morally acceptable choice, since no humans will be harmed and the waste levels will be well below those that could cause any harm. However, since it is somewhat far from the positive paradigm, there are probably better choices that can be made, and the company should investigate these alternatives.

It should be noted that although this action seems ethically acceptable, there are many other considerations that might be factored into the final decision. For example, there are political aspects that should also be considered. Many people in the community are likely to regard the dumping of a toxin at any level as unacceptable. Good community relations might dictate that another solution should be pursued instead. The company also might want to avoid the lengthy amount of time required to obtain a permit for the dumping and the oversight by various government units. This example illustrates that line drawing can help solve the ethical aspects of a problem, but a choice that appears morally acceptable still might not be the best choice when politics and community relations are considered as well. Of course, the immoral choice is never the correct choice.

Although this problem-solving method seems to help with problem analysis and can lead to solutions, there are many pitfalls in its use. If not used properly, line drawing can lead to incorrect results. For example, line drawing can easily be used to prove that something is right when it is actually wrong. Line drawing is only effective if it is used objectively and honestly. The choice of where to put the examples and how to define the paradigms is up to you. You can reach false conclusions by using incorrect paradigms, by dishonest placement of the examples along the line, and by dishonest placement of the problem within the examples. In our example, we might have decided that the problem is somewhat like example 2 and thus placed our problem closer to the positive paradigm, making this solution seem more acceptable. Line drawing can be a very powerful analytic tool in ethical problems, but only if used conscientiously.

There is a long history of the improper use of this technique. In its early days, this method was known as "casuistry," a term that eventually came to be pejorative. In the Middle Ages, casuistry was often used in religious debates to reach false conclusions. Indeed, one of the definitions of casuistry from the American Heritage Dictionary implies the use of false and subtle reasoning to achieve incorrect solutions. Because of this negative connotation, the term "casuistry" is rarely used any more. This emphasizes the hazards of using line drawing: It is useful only if properly applied.

4.3.1 Application of Line Drawing to the Pentium Chip Case

In 1994–95, it was discovered and widely reported that the latest version of the Intel Pentium chip had flaws. At first, Intel sought to hide this information, but later came around to a policy of offering consumers chips in which the flaw had been corrected. We can use line drawing to get some insight into this problem.

For our positive paradigm, we will use the statement that "products should perform as advertised." The negative paradigm will be: "Knowingly sell products that are defective and that will negatively affect customer's applications." A few examples that we can add to the line are as follows:

1. There is a flaw in the chip, but it truly is undetectable and won't affect any customer's applications.
2. There are flaws in the chip, the customer is informed of them, but no help is offered.
3. A warning label says that the chip should not be used for certain applications.
4. Recall notices are sent out, and all flawed chips are replaced.
5. Replacement chips are offered only if the customer notices the problem.

Of course, there are many other possible examples. One view of the line, then, is as follows:

Figure 4.4. Application of line drawing to the Pentium case. Negative and positive paradigms are provided along with the examples.

Where does our situation—there is a flaw, customers aren't informed, and the magnitude of the problem is minimized—fit on this line? One possible analysis is the following:

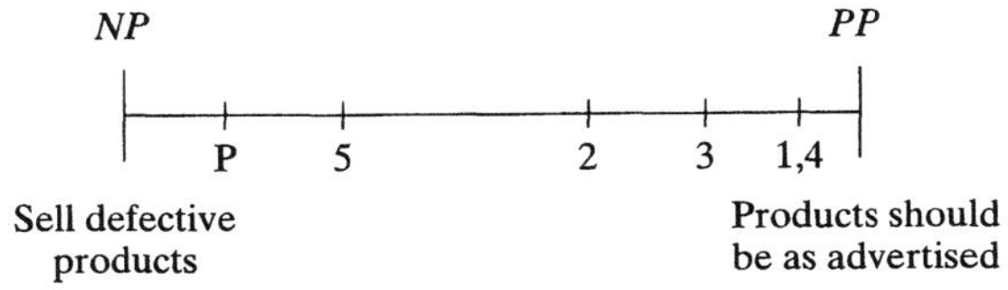

Figure 4.5. Final version of the Pentium chip line-drawing example, with the problem added to the line.

According to this line-drawing analysis, the approach taken by Intel in this case wasn't the best ethical choice.

4.4 FLOW CHARTING

Flow charts are very familiar to engineering students. They are most often used in developing computer programs, and also find application in other engineering disciplines. In engineering ethics, flow charting will be helpful for analyzing a variety of cases, especially those in which there is a sequence of events to be considered or a series of consequences that flows from each decision. An advantage of using a flow chart to analyze ethical problems is that it gives a visual picture of a situation and allows you to readily see the consequences that flow from each decision.

As with the line-drawing technique described in the previous section, there is no unique flow chart that is applicable to a given problem. In fact, different flow charts can be used to emphasize different aspects of the same problem. As with line drawing, it will be essential to be as objective as possible and to approach flow charting honestly. Otherwise, it will be possible to draw any conclusion you want, even one that is clearly wrong.

We can illustrate this technique by applying a simple flow chart to the disaster at Bhopal. One possible flow chart, illustrated in Figure 4.6, deals with the decision-making process that might have gone on at Union Carbide as they decided whether or not to build a plant at Bhopal. This chart emphasizes safety issues for the surrounding community. As indicated on the chart, there were several paths that might have been taken and multiple decisions that had to be made. The flow chart helps to visualize the consequences of each decision and indicates both the ethical and unethical choices.

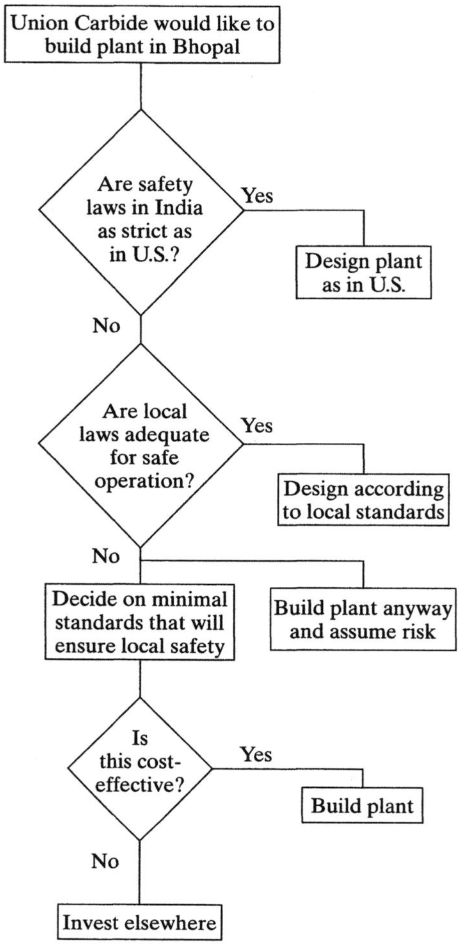

Figure 4.6. Application of a simple flow chart to the Bhopal case, emphasizing potential decisions made during consideration of locating a plant in India.

Of course, this flow chart should be much larger and more complex to thoroughly cover the entire problem.

Another possible flow chart is shown in Figure 4.7. This chart deals with the decisions required during the maintenance of the flare tower, an essential safety system. It considers issues of whether the MIC tank was filled at the time that the flare tower was taken off-line for maintenance, whether other safety systems were operating when the flare tower was taken out of operation, and whether the remaining safety systems were sufficient to eliminate potential problems. Using such a flow chart, it is possible to decide whether the flare tower can be taken off-line for maintenance or whether it should remain operating.

The key to effective use of flow charts for solving ethical problems is to be creative in determining possible outcomes and scenarios and also to not be shy about getting a negative answer and deciding to stop the project.

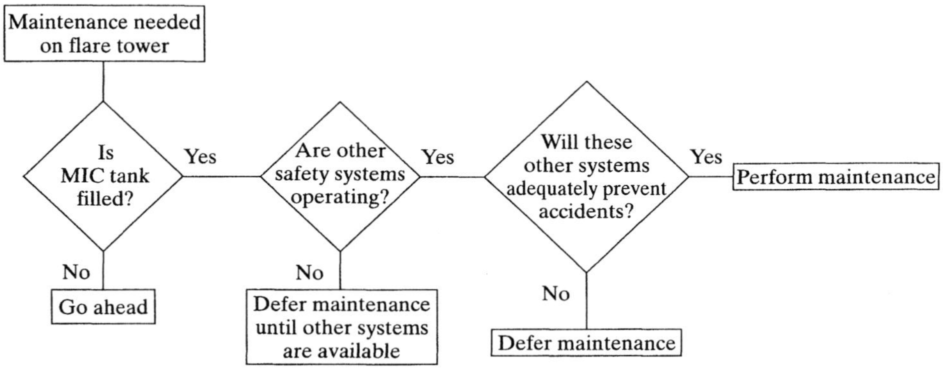

Figure 4.7. An alternative flow chart for the Bhopal case, emphasizing decisions made when considering deactivating the flare tower for maintenance.

4.5 CONFLICT PROBLEMS

An area of ethical problem solving that we will frequently encounter in this book relates to problems that present us with a choice between two conflicting moral values, each of which seems to be correct. How do we make the correct choice in this situation?

Conflict problems can be solved in three ways [Harris, Pritchard, and Rabins, 2000]. Often, there are conflicting moral choices, but one is obviously more significant than the other. For example, protecting the health and safety of the public is more important than your duty to your employer. In this type of case, the resolution of the conflict involves an easy choice.

A second solution is sometimes called the "creative middle way" [Harris, Pritchard, and Rabins, 2000]. This solution is an attempt at some kind of a compromise that will work for everyone. The emphasis here should be on the word "creative," because it takes a great deal of creativity to find a middle ground that is acceptable to everyone and a great deal of diplomacy to sell it to everyone. The sales job is especially difficult because of the nature of compromise, which is often jokingly defined as "the solution where nobody gets what they want." An example of a creative middle ground would be that rather than dumping a toxic waste into a local lake, one finds ways to redesign the production process to minimize the waste product, finds ways to pretreat the waste to minimize the toxicity, or offers to pay for and install the equipment at the municipal water system necessary to treat the water to remove this chemical before it is sent to homes. Obviously, no one will be completely satisfied with these alternatives, since redesigns and pretreatment cost money and take time. Some people will not be satisfied with even a minimized dumping of toxics.

Finally, when there is no easy choice and attempts to find a middle ground are not successful, all that is left is to make the hard choice. Sometimes, you have to bite the bullet and make the best choice possible with the information available at the time. Frequently, you must rely on "gut feelings" for which path is the correct one.

Let's illustrate the resolution-of-conflict problems by examining the *Challenger* explosion, focusing on the dilemma faced by the engineering manager, Bob Lund. The conflict was clear: There was an unknown probability that the shuttle would explode, perhaps killing all aboard. On the other hand, Lund had a responsibility to his company and the people who worked for him. There were consequences of postponing the launch, potentially leading to the loss of future contracts from NASA, the loss of jobs to many Thiokol workers, and perhaps even bankruptcy of the company. For many, the easy choice here is simply to not launch. The risk to the lives of the astronauts is too great and far outweighs any other considerations. It is impossible to balance jobs against lives. After all, most people who lose their jobs will be able to find other employment. However, not everyone will find this to be such an easy choice; clearly, Lund didn't find it to be so.

The creative middle ground might involve delaying the launch until later in the day, when the temperature will have warmed up. Of course, this option might not be possible for many reasons associated with the timing of rocket launches and the successful completion of the planned missions. Instead, perhaps, the astronauts could be informed of the engineer's concerns and be allowed to make the choice whether to launch or not. If a risk is informed and a choice is made by those taking the risk, it somewhat relieves the company of the responsibility if an accident occurs.

The hard choice is what Lund made. He chose to risk the launch, perhaps because the data were ambiguous. He might also have wanted to help ensure the future health of the shuttle program and to save the jobs of the Thiokol workers. As we know, his gamble didn't pay off. The shuttle did explode, causing the deaths of the astronauts and leading to lengthy delays in the shuttle program, political problems for NASA, and business difficulties for Thiokol.

4.6 AN APPLICATION OF PROBLEM-SOLVING METHODS: BRIBERY/ACCEPTANCE OF GIFTS

One of the many gray areas of engineering ethics is the acceptance of gifts from vendors or the offering of gifts to customers to secure business. The difficulty here comes because of the potential for gifts to become bribes or to be perceived of as bribes. Frequently, engineers find themselves in the position of either dealing with vendors who wish to sell them products for incorporation into the engineer's work or acting as vendors themselves and working on sales to other engineers or companies. In this section, we will look at what bribery is and see how some of the problem-solving techniques developed in this chapter can be used to decide when a gift is really a bribe.

Bribery is illegal in the United States and, contrary to popular opinion, is also illegal everywhere in the world. There are some places where bribery may be overlooked, or even expected, but it always takes place "under the table" and is never a legitimate business practice. Moreover, United States federal law forbids American businesses from engaging in bribery overseas, regardless of the local customs or expectations. In many cases, there is a fine line between bribery and a simple gift. Sometimes, the distinction has to do with the value of the gift. Always, it has to do with the intent of the gift. It is important to ensure that no matter how innocent the gift may be, the appearance of impropriety is avoided.

By definition, a bribe is something, such as money or a favor, offered or given to someone in a position of trust in order to induce him to act dishonestly. It is something offered or serving to influence or persuade. What are the ethical reasons for not tolerating bribery? First, bribery corrupts our free-market economic system and is anticompetitive. Unlike the practice of buying the best product at the best price, bribery does not reward the most efficient producer. One can argue the virtues or vices of the free-market economy, but it is the system under which our economy operates, and anything that subverts this system is unfair and unethical. Second, bribery is a sellout to the rich. Bribery corrupts justice and public policy by allowing rich people to make all the rules. In business, it guarantees that only large, powerful corporations will survive, since they are more capable of providing bribes. A small start-up company doesn't have the resources to compete in an environment where expensive favors are required to secure business. Finally, bribery treats people as commodities that can be bought and sold. This practice is degrading to us as human beings and corrupts both the buyer and the seller [Harris, Pritchard, and Rabins, 2000].

4.6.1 When Is a Gift a Bribe?

Frequently, the boundary between a legitimate gift and a bribe is very subtle. Gifts of nominal value, such as coffee mugs or calendars with a vendor's logo and phone number on it, are really just an advertising tool. Generally, there is no problem with accepting these types of items. Dining with a customer or a supplier is also an acceptable practice, especially if everyone pays his or her own way. It is important from the point of view of both suppliers and customers that good relations be maintained so that good service can be provided. Social interaction, such as eating together, often facilitates the type of close and successful interactions required by both sides. However, when meals or gifts are no longer of low cost and the expense of these items are not shared equally, the possibility for abuse becomes large.

4.6.2 Examples of Gifts vs. Bribes

To help illustrate the difference between bribes and legitimate gifts, let's look at a few potential scenarios to see how fuzzy this boundary can be. No answer will be given to the questions posed, but rather the solution of these questions will be left to the reader.

- During a sales visit, a sales representative offers you a coffee mug with his company's name and logo on it. The value of the mug is five dollars. Can you accept this item? Does the answer to this question change if this item is a $350 crystal bowl with the name of the company engraved on it? How about if there is no engraving on it?
- Your meeting with a sales representative is running into the lunch hour. She invites you to go out for lunch. You go to a fast-food restaurant and pay for your own lunch. Is this practice acceptable? Does the answer to this question change if you go to an expensive French restaurant? If she pays for lunch?
- A sales representative from whom you often purchase asks if you would like to play tennis with him this weekend at one of the local municipal courts. Should you go? Is the answer to this question different if the match is at an exclusive local club to which he belongs? What if he pays the club's guest fee for you?
- A company sales representative would like for you to attend a one-day sales seminar in Cleveland. Your company will pay for your trip. Should you go? How about if the meeting is in Maui? What if the sales representative's company is going to pay for you to go? What if your family is invited as well?

Do the answers to any of these questions change if the gift is offered before you purchase anything from the company, as opposed to after you are already a steady customer? (A more detailed version of these types of scenarios can be found in [Harris, Pritchard, and Rabins, 2000].)

Keep in mind that gifts accepted even after the purchase of something from a company might be a bribe directed at securing future sales from you or might be aimed at engineers at other companies. Although nothing was said about a gift up front, now that you have received one, the expectation of gifts might affect your future purchase decisions. Similarly, an employee of a company like yours might become aware of the gift that you received. He now realizes that if he orders parts from the same supplier that you did, he will receive a gift similar to yours. He will be tempted to order from this supplier even if there is a better supplier of that product on the market. These types of gifts tend to shut out smaller companies that can't necessarily afford gifts and might also cause an increase in everyone's costs, since if everyone now expects to receive gifts, the product cost must go up. Clearly, bribery is pernicious, and even the appearance of bribery should be avoided.

4.6.3 Problem Solving

How can the analysis methods described in this chapter be applied to these examples concerning bribery and the acceptance of gifts? We won't go into the answer to this question in depth here, but will rather save it for the questions at the end of the chapter. However, some general ideas can be presented now. Bribery can easily be analyzed by looking at the factual, conceptual, and moral issues described previously. Frequently, the facts will be obvious: who offered a gift, what its value was, and what its purpose was. Conceptual issues will be somewhat more difficult, since it must be determined whether the gift is of sufficient value to influence a decision or whether that influence is the intent of the gift. Once the conceptual issues have been worked out and it is clear whether or not the gift is a bribe, the moral issue is often very clear.

Line drawing can be very effectively applied to the examples given previously. The subtle differences between the value of the gift, the timing of the gift, etc., are easily visualized using line drawing, and often it will be very clear what the ethical choice will be based on a well-drawn line. Likewise, flow charting can be used to examine the consequences that will result from the acceptance or offer of a gift.

4.6.4 Avoiding Bribery Problems

How does one ensure that accepting a gift doesn't cross the line into bribery? The first and most important method for determining this is to look at company policy. All large corporations and many smaller companies have very clear rules about what is acceptable. Some companies have very strict policies. For example, some companies say that employees are not allowed to accept anything from a vendor and that any social interaction with vendors or customers must be paid for by your company. Any deviation from this rule requires approval from appropriate supervisors. This philosophy is rooted in a sense of trying to avoid any conflict of interest and any appearance of impropriety.

Other companies realize the importance of social interactions in business transactions and allow their employees more discretion in determining what is proper. In the absence of strict corporate guidelines, a preapproval from one's management is the best guide to what is acceptable.

In the absence of any corporate guidelines, another method for determining the acceptability of an action is sometimes referred to as the "New York Times Test": Could your actions withstand the scrutiny of a newspaper reporter? Could you stand to see your name in the newspaper in an article about the gift you received? If you couldn't easily defend your action without resorting to self-serving rationalizations, then you probably shouldn't do it.

APPLICATION: CASES

CELLULAR PHONES AND CANCER

This case will seem different from many of the other cases we will study, since there is no disaster or wrongdoing that has to be analyzed after the fact. Rather, this is a case about the experimental nature of engineering and deals with issues of what engineers should do early in the design cycle for a new product or system in order to avoid possible harm to customers or the public in general. It also deals with what engineers should do after a product has been released when possible dangers are brought up.

Concerns about potential adverse health effects of cell phones began in 1992 with a lawsuit filed in Florida. In this suit, David Reynard claimed that his wife's fatal brain cancer had been caused by her use of a cell phone. Although the suit was dismissed in 1995 due to a lack of scientific evidence to support Reynard's claim, this and other similar suits received a great deal of media attention and caused some concern among frequent cell phone users.

The possible problems with cell phones are clear. In using a cell phone, you are placing a source of electromagnetic radiation in close proximity to your brain. It doesn't take much imagination to see the potential for problems: Microwave ovens use electromagnetic radiation to cook food. Of course, cell phones operate at a different frequency and at much lower power levels than do microwave ovens, but the analogy is clear. The human body evolved in an environment that did not contain significant levels of radio-frequency (rf) radiation, so it is plausible that the ubiquity of rf fields in our modern industrial world might cause some adverse health effects.

The biological effects of rf energy have been studied for many years. Some of the early studies go back to the 1940s. What types of studies were performed? Typically, these were epidemiological studies and were retrospective looks at people who have used cell phones. The goal of these studies was to try to determine the levels of exposure to rf radiation from cell phones of every person in the study and to try to correlate the levels with subsequent health effects, especially cancers. While the studies all generally indicated that there is no harm in cell phone use, problems remain. Many of the problems are due to the fact that the studies relied on self-reporting of cell phone use. They asked people to report how much time they spent talking on their phones. Of course, many people reported their phone use accurately, but many others either didn't really know how much they used their phones or misestimated their use. Epidemiological studies are also difficult to analyze, since it is hard to know the power levels each individual has been exposed to. The power emitted by the phone depends on what model of phone you use and how far you are from the base station while talking. Also, brain cancers generally take a long time to develop. There may not have been enough time since the widespread use of cell phones for a significant number of cancers to have developed. Solid links between cell phone use and brain cancers might not show up for another 10 to 20 years.

Studies have also been performed on laboratory animals. Typically, these are done by placing the animals in an environment containing rf fields designed to mimic those of cell phones. Like the epidemiological studies, the research studies on laboratory animals have not indicated any significant increase in health problems for the animals. Of course, since laboratory animals are not humans, the results may not be directly applicable to humans.

There have been some studies of the effects of rf radiation on laboratory tissue and cell cultures. The results of these studies and their applicability to human health are controversial. Some theoretical studies have examined how rf energy might be deposited into a human brain during cell phone use. These studies are very difficult to benchmark because it is difficult to make measurements of energy deposition directly into a human brain.

What is an engineer working for a cell phone company or some other company making products that emit rf radiation to do when confronted with the ongoing concerns about the health effects of rf fields? Cell phones can certainly be redesigned to reduce or eliminate this problem, but, of course, any design that will lead to reduced emission will probably cost more. We won't know for many years what the final answer is regarding cell phone health effects. For now, it seems that cell phones are probably safe to use. What is the prudent and ethical thing to do in designing such products in an atmosphere where some doubt about safety exists? This case illustrates the problems that engineers have in dealing with and managing the unknown.

Many of the designs that engineers produce are experimental in nature or deal with effects that aren't fully understood. It is incumbent on the designer to be informed about the potential risks to users of her designs and to seek to minimize these risks to the extent possible.

VICE PRESIDENT SPIRO AGNEW AND CONSTRUCTION KICKBACKS IN MARYLAND

In January of 1973, architects and consulting engineers all over Baltimore, Maryland were seeking out any available attorneys with experience in criminal law. This activity was brought on by subpoenas issued by the U.S. Attorney for Maryland, George Beall, who was looking into charges of bribes and kickbacks given to elected officials by engineers working in the construction industry. The subpoenas required these engineers to submit the records of their firms to the U.S. attorney. One of these engineers was Lester Matz, a partner in Matz, Childs and Associates, a Baltimore engineering firm. The subsequent events described by Richard Cohen and Jules Witcover in their book *A Heartbeat Away* eventually led to the disgrace and resignation of Spiro Agnew, then the Vice President of the United States.

Matz was an engineer trained at Johns Hopkins University in Baltimore. Although his firm was doing well, it always seemed to lose out to other firms on big public-works contracts. In Maryland, engineering and architectural services for government projects were not put out for bid, but rather were awarded to individual firms using various criteria, including the firm's ability to do the work, its performance on past contracts, etc. Interestingly, unlike the situation for engineering services, the contractor for government projects was chosen through a competitive bidding process. It became clear to Matz that in acquiring government contracts, his talents and those of his firm were unimportant. What was required to get the contracts for public works was contacts in government and the requisite bribes and kickbacks.

In 1961, Matz began courting Spiro T. Agnew, an ambitious and rising politician. In 1962, Matz donated $500 to Agnew's campaign for Baltimore county executive, a post that is roughly equivalent to mayor for the areas of the county outside the city limits of Baltimore. The county executive wielded great power in determining who received contracts for the engineering services required for the numerous public-works projects undertaken by the county. The campaign contribution was given by Matz and his partner in the hopes of receiving some of the county engineering contracts that they had been locked out of. After Agnew won the election, the contribution made by Matz's engineering firm was rewarded with contracts for county engineering work. In return, the firm paid Agnew 5% of their fees from the county work, which apparently was the kickback paid by other engineering firms at the time.

With this arrangement, Matz, Childs and Associates prospered and Matz became relatively wealthy. At its peak, the firm employed nearly 350 people. Matz was able to rent an apartment in Aspen for his winter ski vacations and also had a beach condo at St. Croix in the Virgin Islands. Matz's St. Croix condo was near a condo owned by his friend, Spiro Agnew. The "business" arrangement between Agnew and Matz continued when Agnew was elected governor of Maryland, only now Matz, Childs and Associates received contracts for state work. The financial arrangement remained the same: Agnew received a payment for every contract awarded.

These payments continued even after Agnew was elected vice president on the Republican ticket with Richard Nixon in 1968. Matz testified that he met with Agnew in his office in the White House and had given him an envelope containing $10,000 in cash. Indeed, Matz also indicated that he had given $2,500 dollars to Agnew for a federal contract that a subsidiary of Matz, Child and Associates had received. All told, Matz described payments that he had given Agnew over the years totaling over $100,000.

As a brief aside, it is interesting to describe how the money paid to Agnew was generated. Clearly, these payments had to be made in cash in order to avoid leaving records of the transactions. However, engineering firms are not paid in cash for their services and thus don't typically have large amounts of cash on hand. One method of generating cash was to give cash "bonuses" to key employees. After retaining a sufficient amount to pay the income taxes on the bonus, the employee returned the cash to the firm, where it was placed in a safe until needed. Of course, this practice is a violation of the tax code: The company books record the transaction as a bonus, yet much of the money is retained by the firm. This practice subjected Matz, Child and Associates to prosecution under the federal tax code. This method didn't always generate the required amount of cash, so other means were also used. For example, large "loans" were made to colleagues, who cashed the money and returned it to the firm. These loans were then "repaid" slowly over a long period of time to make the books appear right.

With federal prosecutors threatening to indict Matz and Childs for income-tax evasion and other charges, they decided to provide evidence to the government of the wrongdoing of Agnew and his successor as county executive. Agnew's lawyers and the prosecutors reached an agreement whereby Agnew would resign as vice president and plead *nolo contendere* (no contest) to a single count of income-tax evasion, a felony, for payments received in 1967. This plea is the legal equivalent of a plea of guilty; the defendant doesn't admit to the crime, but does acknowledge that there is enough evidence to convict him. On October 10, 1973, Agnew resigned as vice president, the first vice president to have resigned in disgrace. Later that day, in a dramatic appearance in a Maryland courtroom, he entered his plea. The judge fined him $10,000 and honored the plea agreement whereby Agnew received no jail term, but only three years unsupervised probation. For agreeing to cooperate with the prosecution, Matz and Childs were not prosecuted.

These events took place against the backdrop of one of the most intense government crises in U.S. history. Although Nixon and Agnew had been reelected in a landslide in the 1972 election, the Watergate scandal hung over the administration. Shortly after the events of this case, the Watergate scandal intensified, culminating in the resignation of Richard Nixon from the presidency.

PROFESSIONAL SUCCESS: LOOKING FOR A JOB

Many ethical issues arise in the course of looking for a job. Even though as you approach graduation you are still an "amateur," ethical and professional behavior is expected during your job search. There are many ways to be unethical in searching for a job, such as exaggerating or falsifying your resume, or overstating expenses when getting reimbursed for an interview trip.

Other, less obvious, ethical concerns can occur during interview trips. For example, suppose you have had an on-campus interview with a large corporation. After the interview you have decided that you aren't really interested in this company. The company calls you later and asks you to come to company headquarters in Cleveland for a plant visit. You have a friend in Cleveland who you would like to visit. Is it acceptable to go on the plant trip? Why? Does the situation change if the plant trip is to Hawaii? Does it change if your interest level in the company is low, but you honestly feel that you could be persuaded?

How do you decide what is acceptable during your job search? The easiest thing to do is to honestly discuss your plans with the recruiter. If she feels that what you want to do isn't acceptable, then you shouldn't do it. If, however, your plans are acceptable to the company then you can proceed. In addition, the ethical analysis and problem-solving methods that we developed in Chapter 4 and have applied to cases thus for are equally applicable to job searches.

PROFESSIONAL SUCCESS: CHEATING ON ASSIGNMENTS

The intense pressure to get good grades in college often leads to temptations to cheat on exams or assignments. Cheating is an issue that is likely to have arisen in educational settings even before you began your study of engineering. Of course the stakes become higher in a college or university setting, so the temptation to cheat might seem larger now than in high school. Cheating can take many forms, including copying someone else's work or using "cheat sheets" during an exam.

Although it can be analyzed using utilitarianism or rights and duty ethics, it is perhaps easiest to examine cheating using virtue ethics. Honesty is a virtue. Honesty facilitates trust between individuals whereas dishonesty causes friction. People rarely want to associate with others who they feel don't behave fairly and can't be trusted. Cheating or falsifying work is a form of dishonesty. We should seek to enhance virtues such as honesty within ourselves and others, so virtue ethics clearly tells us that cheating is unethical.

As with many of the problems discussed in this book, you may be unsure whether something you are doing really is cheating. When faced with these types of problems, the same analysis and problem-solving methods that are discussed in this book can be used to determine the correct course of action.

KEY TERMS

Bribery
Flow charting
Line drawing

REFERENCES

CHARLES E. HARRIS, JR., MICHAEL S. PRITCHARD, AND MICHAEL J. RABINS, *Engineering Ethics, Concepts and Cases*, Wadsworth Publishing Company, Belmont, CA, 2000.

ROLAND SCHINZINGER AND MIKE W. MARTIN, *Introduction to Engineering Ethics,* McGraw-Hill, New York, 2000.

CELLULAR PHONES AND CANCER

KENNETH R. FOSTER AND JOHN E. MOULDER, "Are mobile phones safe?" IEEE Spectrum, August 2000, pp.23–28.

SPIRO AGNEW

RICHARD M. COHEN AND JULES WITCOVER, *A Heartbeat Away: The Investigation and Resignation of Vice President Spiro T. Agnew,* Viking, New York, 1974.

New York Times, October 11, 1973. Numerous articles, starting with the front-page article about Agnew's resignation and his appearance in court. Articles leading up to this event can also be found in copies of the *New York Times* up to several weeks before this date.

Problems

1. Analyze the factual, conceptual, and moral issues in the following cases: a) the Denver International Airport case, b) the Pentium chip case, and c) the space shuttle *Challenger* case. What conclusion do you come to about the ethically correct choice that should have been made in each case?
2. Use the line-drawing technique to analyze the actions of the following corporations: a) the 3Bs construction company in the Denver International Airport case and b) Paradyne in its dealings with the Social Security Administration. What conclusions do you draw about the actions of these companies?
3. Use flow charting to decide whether the space shuttle *Challenger* should have been launched. You can start at the beginning of the shuttle program and include some of the problems encountered in the development of the booster rockets, include consequences for NASA and Thiokol of delays in the launch, etc.
4. Use flow charting to analyze the actions of the Aberdeen Three. What conclusions can be drawn from this chart regarding their actions?
5. Use line drawing to assess whether the scenarios of bribery/gift giving under Examples in Section 4.6 are acceptable. What other examples can you think of to add to these scenarios?
6. Use flow charting to analyze whether the examples given in Section 4.6 are legitimate gifts or bribes. Be sure to indicate what consequences will flow from each decision.

CELLULAR PHONES AND CANCER

7. What does utilitarianism tell us about this case? What do rights and duty ethics tell us? Consider these questions from the point of view of a design engineer who must work on a product that might emit hazardous radiation. Which ethical theory applies best in this case? What does the code of ethics of the IEEE tell us about this case?
8. Analyze this case by determining the factual issues, determining the conceptual issues, and deciding which moral issues apply. Hint: This case is a perfect instance

of what we discussed previously in this chapter when we said that the factual issues can be controversial.

9. If there are potential, but not well-understood, hazards in building a product, what are the future consequences of doing nothing—i.e., of making no changes in the design? Will warnings to the consumer suffice to get the designer off the hook? Must the product be engineered to be totally safe at all costs?
10. How can one best balance safety with economics in this case?
11. In their book *Ethics in Engineering*, Martin and Schinzinger state that "[e]ngineering, more than any other profession, involves social experimentation." How applicable is this statement to this case? Do you think that this statement is true in general?
12. In light of the results of various panels that indicate that there is no hazard associated with cell phone use, what should an engineer do today when designing products that will emit this rf radiation?
13. Many of the studies researching cell phone safety have been funded by the cell phone industry. What are the ethical implications of this?
14. Similar concerns about the safety of powerlines and low-frequency magnetic fields were voiced in the early 1990s. Research this case and compare it to the case study on cell phones and cancer. A good starting point is the article "Today's View of Magnetic Fields" in the December 1994 issue of IEEE Spectrum.

SPIRO AGNEW

15. Does the fact that paying government officials for receiving contracts seemed to be a common-place business practice in Maryland at the time make this practice ethically acceptable?
16. What should an engineer do in the face of competition from others who are willing to resort to bribery?
17. What issues does this case raise regarding competitive bidding for engineering services? Would competitive bidding for the engineering contracts in Baltimore County have solved this problem?
18. What is the ethical status of a campaign contribution given to a politician to secure future business? Is this a bribe? Is it the same as a kickback? Perhaps line drawing would help answer this question.

Appendix A
Codes of Ethics of Professional Engineering Societies

THE INSTITUTE OF ELECTRICAL AND ELECTRONICS ENGINEERS, INC.* (IEEE)

We, the members of the IEEE, in recognition of the importance of our technologies affecting the quality of life throughout the world, and in accepting a personal obligation to our profession, its members and the communities we serve, do hereby commit ourselves to the highest ethical and professional conduct and agree:

1. to accept responsibility in making engineering decisions consistent with the safety, health and welfare of the public, and to disclose promptly factors that might endanger the public or the environment;
2. to avoid real or perceived conflicts of interest whenever possible, and to disclose them to affected parties when they do exist;
3. to be honest and realistic in stating claims or estimates based on available data;
4. to reject bribery in all its forms;
5. to improve the understanding of technology, its appropriate application, and potential consequences;
6. to maintain and improve our technical competence and to undertake technological tasks for others only if qualified by training or experience, or after full disclosure of pertinent limitations;
7. to seek, accept, and offer honest criticism of technical work, to acknowledge and correct errors, and to credit properly the contributions of others;
8. to treat fairly all persons regardless of such factors as race, religion, gender, disability, age, or national origin;
9. to avoid injuring others, their property, reputation, or employment by false or malicious action;
10. to assist colleagues and co-workers in their professional development and to support them in following this code of ethics.

Approved by the IEEE Board of Directors, August 1990

NATIONAL SOCIETY OF PROFESSIONAL ENGINEERS (NSPE) CODE OF ETHICS FOR ENGINEERS

Preamble

Engineering is an important and learned profession. As members of this profession, engineers are expected to exhibit the highest standards of honesty and integrity. Engineering has a direct and vital impact on the quality of life for all people. Accordingly, the services provided by engineers require honesty, impartiality, fairness and equity, and must be dedicated to the protection of the public health, safety, and welfare. Engineers must perform under a standard of professional behavior that requires adherence to the highest principles of ethical conduct.

I. Fundamental Canons

Engineers, in the fulfillment of their professional duties, shall:

1. Hold paramount the safety, health and welfare of the public.
2. Perform services only in areas of their competence.
3. Issue public statements only in an objective and truthful manner.
4. Act for each employer or client as faithful agents or trustees.
5. Avoid deceptive acts.
6. Conduct themselves honorably, responsibly, ethically, and lawfully so as to enhance the honor, reputation, and usefulness of the profession.

II. Rules of Practice

1. Engineers shall hold paramount the safety, health, and welfare of the public.
   a. If engineers' judgment is overruled under circumstances that endanger life or property, they shall notify their employer or client and such other authority as may be appropriate.
   b. Engineers shall approve only those engineering documents that are in conformity with applicable standards.
   c. Engineers shall not reveal facts, data or information without the prior consent of the client or employer except as authorized or required by law or this Code.
   d. Engineers shall not permit the use of their name or associate in business ventures with any person or firm that they believe are engaged in fraudulent or dishonest enterprise.
   e. Engineers having knowledge of any alleged violation of this Code shall report thereon to appropriate professional bodies and, when relevant, also to public authorities, and cooperate with the proper authorities in furnishing such information or assistance as may be required.
2. Engineers shall perform services only in the areas of their competence.
   a. Engineers shall undertake assignments only when qualified by education or experience in the specific technical fields involved.
   b. Engineers shall not affix their signatures to any plans or documents dealing with subject matter in which they lack competence, nor to any plan or document not prepared under their direction and control.
   c. Engineers may accept assignments and assume responsibility for coordination of an entire project and sign and seal the engineering

documents for the entire project, provided that each technical segment is signed and sealed only by the qualified engineers who prepared the segment.

3. Engineers shall issue public statements only in an objective and truthful manner.
    a. Engineers shall be objective and truthful in professional reports, statements, or testimony. They shall include all relevant and pertinent information in such reports, statements, or testimony, which should bear the date indicating when it was current.
    b. Engineers may express publicly technical opinions that are founded upon knowledge of the facts and competence in the subject matter.
    c. Engineers shall issue no statements, criticisms, or arguments on technical matters that are inspired or paid for by interested parties, unless they have prefaced their comments by explicitly identifying the interested parties on whose behalf they are speaking, and by revealing the existence of any interest the engineers may have in the matters.
4. Engineers shall act for each employer or client as faithful agents or trustees.
    a. Engineers shall disclose all known or potential conflicts of interest that could influence or appear to influence their judgment or the quality of their services.
    b. Engineers shall not accept compensation, financial or otherwise, from more than one party for services on the same project, or for services pertaining to the same project, unless the circumstances are fully disclosed and agreed to by all interested parties.
    c. Engineers shall not solicit or accept financial or other valuable consideration, directly or indirectly, from outside agents in connection with the work for which they are responsible.
    d. Engineers in public service as members, advisors, or employees of a governmental or quasi-governmental body or department shall not participate in decisions with respect to services solicited or provided by them or their organizations in private or public engineering practice.
    e. Engineers shall not solicit or accept a contract from a governmental body on which a principal or officer of their organization serves as a member.
5. Engineers shall avoid deceptive acts.
    a. Engineers shall not falsify their qualifications or permit misrepresentation of their or their associates' qualifications. They shall not misrepresent or exaggerate their responsibility in or for the subject matter of prior assignments. Brochures or other presentations incident to the solicitation of employment shall not misrepresent pertinent facts concerning employers, employees, associates, joint venturers, or past accomplishments.
    b. Engineers shall not offer, give, solicit or receive, either directly or indirectly, any contribution to influence the award of a contract by public authority, or which may be reasonably construed by the public as having the effect of intent to influencing the awarding of a contract. They shall not offer any gift or other valuable consideration in order to secure work. They shall not pay a commission, percentage,

or brokerage fee in order to secure work, except to a bona fide employee or bona fide established commercial or marketing agencies retained by them.

III. Professional Obligations

1. Engineers shall be guided in all their relations by the highest standards of honesty and integrity.
    a. Engineers shall acknowledge their errors and shall not distort or alter the facts.
    b. Engineers shall advise their clients or employers when they believe a project will not be successful.
    c. Engineers shall not accept outside employment to the detriment of their regular work or interest. Before accepting any outside engineering employment they will notify their employers.
    d. Engineers shall not attempt to attract an engineer from another employer by false or misleading pretenses.
    e. Engineers shall not actively participate in strikes, picket lines, or other collective coercive action.
    f. Engineers shall not promote their own interest at the expense of the dignity and integrity of the profession.
2. Engineers shall at all times strive to serve the public interest.
    a. Engineers shall seek opportunities to participate in civic affairs; career guidance for youths; and work for the advancement of the safety, health and well-being of their community.
    b. Engineers shall not complete, sign, or seal plans and/or specifications that are not in conformity with applicable engineering standards. If the client or employer insists on such unprofessional conduct, they shall notify the proper authorities and withdraw from further service on the project.
    c. Engineers shall endeavor to extend public knowledge and appreciation of engineering and its achievements.
3. Engineers shall avoid all conduct or practice that deceives the public.
    a. Engineers shall avoid the use of statements containing a material misrepresentation of fact or omitting a material fact.
    b. Consistent with the foregoing, Engineers may advertise for recruitment of personnel.
    c. Consistent with the foregoing, Engineers may prepare articles for the lay or technical press, but such articles shall not imply credit to the author for work performed by others.
4. Engineers shall not disclose, without consent, confidential information concerning the business affairs or technical processes of any present or former client or employer, or public body on which they serve.
    a. Engineers shall not, without the consent of all interested parties, promote or arrange for new employment or practice in connection with a specific project for which the Engineer has gained particular and specialized knowledge.

b. Engineers shall not, without the consent of all interested parties, participate in or represent an adversary interest in connection with a specific project or proceeding in which the Engineer has gained particular specialized knowledge on behalf of a former client or employer.

5. Engineers shall not be influenced in their professional duties by conflicting interests.

   a. Engineers shall not accept financial or other considerations, including free engineering designs, from material or equipment suppliers for specifying their product.

   b. Engineers shall not accept commissions or allowances, directly or indirectly, from contractors or other parties dealing with clients or employers of the Engineer in connection with work for which the Engineer is responsible.

6. Engineers shall not attempt to obtain employment or advancement or professional engagements by untruthfully criticizing other engineers, or by other improper or questionable methods.

   a. Engineers shall not request, propose, or accept a commission on a contingent basis under circumstances in which their judgment may be compromised.

   b. Engineers in salaried positions shall accept part-time engineering work only to the extent consistent with policies of the employer and in accordance with ethical considerations.

   c. Engineers shall not, without consent, use equipment, supplies, laboratory, or office facilities of an employer to carry on outside private practice.

7. Engineers shall not attempt to injure, maliciously or falsely, directly or indirectly, the professional reputation, prospects, practice, or employment of other engineers. Engineers who believe others are guilty of unethical or illegal practice shall present such information to the proper authority for action.

   a. Engineers in private practice shall not review the work of another engineer for the same client, except with the knowledge of such engineer, or unless the connection of such engineer with the work has been terminated.

   b. Engineers in governmental, industrial, or educational employ are entitled to review and evaluate the work of other engineers when so required by their employment duties.

   c. Engineers in sales or industrial employ are entitled to make engineering comparisons of represented products with products of other suppliers.

8. Engineers shall accept personal responsibility for their professional activities provided, however, that Engineers may seek indemnification for services arising out of their practice for other than gross negligence, where the Engineer's interests cannot otherwise be protected.

   a. Engineers shall conform with state registration laws in the practice of engineering.

b. Engineers shall not use association with a nonengineer, a corporation, or partnership as a "cloak" for unethical acts.

9. Engineers shall give credit for engineering work to those to whom credit is due, and will recognize the proprietary interests of others.

a. Engineers shall, whenever possible, name the person or persons who may be individually responsible for designs, inventions, writings, or other accomplishments.

b. Engineers using designs supplied by a client recognize that the designs remain the property of the client and may not be duplicated by the Engineer for others without express permission.

c. Engineers, before undertaking work for others in connection with which the Engineer may make improvements, plans, designs, inventions, or other records that may justify copyrights or patents, should enter into a positive agreement regarding ownership.

d. Engineers' designs, data, records, and notes referring exclusively to an employer's work are the employer's property. Employer should indemnify the Engineer for use of the information for any purpose other than the original purpose.

As Revised July 1996

"By order of the United States District Court for the District of Columbia, former Section 11(c) of the NSPE Code of Ethics prohibiting competitive bidding, and all policy statements, opinions, rulings or other guidelines interpreting its scope, have been rescinded as unlawfully interfering with the legal right of engineers, protected under the antitrust laws, to provide price information to prospective clients; accordingly, nothing contained in the NSPE Code of Ethics, policy statements, opinions, rulings or other guidelines prohibits the submission of price quotations or competitive bids for engineering services at any time or in any amount."

Statement by NSPE Executive Committee

In order to correct misunderstandings which have been indicated in some instances since the issuance of the Supreme Court decision and the entry of the Final Judgment, it is noted that in its decision of April 25, 1978, the Supreme Court of the United States declared: "The Sherman Act does not require competitive bidding."

It is further noted that as made clear in the Supreme Court decision:

1. Engineers and firms may individually refuse to bid for engineering services.
2. Clients are not required to seek bids for engineering services.
3. Federal, state, and local laws governing procedures to procure engineering services are not affected, and remain in full force and effect.
4. State societies and local chapters are free to actively and aggressively seek legislation for professional selection and negotiation procedures by public agencies.
5. State registration board rules of professional conduct, including rules prohibiting competitive bidding for engineering services, are not affected and remain in full force and effect. State registration boards with authority to

adopt rules of professional conduct may adopt rules governing procedures to obtain engineering services.

6. As noted by the Supreme Court, "nothing in the judgment prevents NSPE and its members from attempting to influence governmental action . . ."

NOTE: In regard to the question of application of the Code to corporations vis-à-vis real persons, business form or type should not negate nor influence conformance of individuals to the Code. The Code deals with professional services, which services must be performed by real persons. Real persons in turn establish and implement policies within business structures. The Code is clearly written to apply to the Engineer and items incumbent on members of NSPE to endeavor to live up to its provisions. This applies to all pertinent sections of the Code.

NOTE: There is also the NSPE Ethics Reference Guide, which fleshes out some of this stuff.

AMERICAN SOCIETY OF MECHANICAL ENGINEERS (ASME)

Ethics

ASME requires ethical practice by each of its members and has adopted the following Code of Ethics of Engineers as referenced in the ASME Constitution, Article C2.1.1.

Code of Ethics of Engineers

The Fundamental Principles

Engineers uphold and advance the integrity, honor and dignity of the engineering profession by:

I. Using their knowledge and skill for the enhancement of human welfare;
II. Being honest and impartial, and serving with fidelity the public, their employers and clients; and
III. Striving to increase the competence and prestige of the engineering profession.

1. Engineers shall hold paramount the safety, health and welfare of the public in the performance of their professional duties.
2. Engineers shall perform services only in the areas of their competence.
3. Engineers shall continue their professional development throughout their careers and shall provide opportunities for the professional and ethical development of those engineers under their supervision.
4. Engineers shall act in professional matters for each employer or client as faithful agents or trustees, and shall avoid conflicts of interest or the appearance of conflicts of interest.
5. Engineers shall build their professional reputation on the merit of their services and shall not compete unfairly with others.
6. Engineers shall associate only with reputable persons or organizations.
7. Engineers shall issue public statements only in an objective and truthful manner.

The ASME Criteria for Interpretation of the Canons

The ASME criteria for interpretation of the Canons are guidelines and represent the objectives toward which members of the engineering profession should strive. They are principles which an engineer can reference in specific situations. In addition, they provide interpretive guidance to the ASME Board on Professional Practice and Ethics on the Code of Ethics of Engineers.

1. Engineers shall hold paramount the safety, health and welfare of the public in the performance of their professional duties.

    a. Engineers shall recognize that the lives, safety, health and welfare of the general public are dependent upon engineering judgments, decisions and practices incorporated into structures, machines, products, processes and devices.

    b. Engineers shall not approve or seal plans and/or specifications that are not of a design safe to the public health and welfare and in conformity with accepted engineering standards.

    c. Whenever the Engineers' professional judgments are overruled under circumstances where the safety, health, and welfare of the public are endangered, the Engineers shall inform their clients and/or employers of the possible consequences.

        (1) Engineers shall endeavor to provide data such as published standards, test codes, and quality control procedures that will enable the users to understand safe use during life expectancy associated with the designs, products, or systems for which they are responsible.

        (2) Engineers shall conduct reviews of the safety and reliability of the designs, products, or systems for which they are responsible before giving their approval to the plans for the design.

        (3) Whenever Engineers observe conditions, directly related to their employment, which they believe will endanger public safety or health, they shall inform the proper authority of the situation.

    d. If engineers have knowledge of or reason to believe that another person or firm may be in violation of any of the provisions of these Canons, they shall present such information to the proper authority in writing and shall cooperate with the proper authority in furnishing such further information or assistance as may be required.

2. Engineers shall perform services only in areas of their competence.

    a. Engineers shall undertake to perform engineering assignments only when qualified by education and/or experience in the specific technical field of engineering involved.

    b. Engineers may accept an assignment requiring education and/or experience outside of their own fields of competence, but their services shall be restricted to other phases of the project in which they are qualified. All other phases of such project shall be performed by qualified associates, consultants, or employees.

3. Engineers shall continue their professional development throughout their careers, and should provide opportunities for the professional and ethical development of those engineers under their supervision.

4. Engineers shall act in professional matters for each employer or client as faithful agents or trustees, and shall avoid conflicts of interest or the appearance of conflicts of interest.

   a. Engineers shall avoid all known conflicts of interest with their employers or clients and shall promptly inform their employers or clients of any business association, interests, or circumstances which could influence their judgment or the quality of their services.

   b. Engineers shall not undertake any assignments which would knowingly create a potential conflict of interest between themselves and their clients or their employers.

   c. Engineers shall not accept compensation, financial or otherwise, from more than one party for services on the same project, or for services pertaining to the same project, unless the circumstances are fully disclosed to, and agreed to, by all interested parties.

   d. Engineers shall not solicit or accept financial or other valuable considerations, for specifying products or material or equipment suppliers, without disclosure to their clients or employers.

   e. Engineers shall not solicit or accept gratuities, directly or indirectly, from contractors, their agents, or other parties dealing with their clients or employers in connection with work for which they are responsible. Where official public policy or employers' policies tolerate acceptance of modest gratuities or gifts, engineers shall avoid a conflict of interest by complying with appropriate policies and shall avoid the appearance of a conflict of interest.

   f. When in public service as members, advisors, or employees of a governmental body or department, Engineers shall not participate in considerations or actions with respect to services provided by them or their organization(s) in private or product engineering practice.

   g. Engineers shall not solicit an engineering contract from a governmental body or other entity on which a principal, officer, or employee of their organization serves as a member without disclosing their relationship and removing themselves from any activity of the body which concerns their organization.

   h. Engineers working on codes, standards or governmental sanctioned rules and specifications shall exercise careful judgment in their determinations to ensure a balanced viewpoint, and avoid a conflict of interest.

   i. When, as a result of their studies, Engineers believe a project(s) will not be successful, they shall so advise their employer or client.

   j. Engineers shall treat information coming to them in the course of their assignments as confidential, and shall not use such information as a means of making personal profit if such action is adverse to the interests of their clients, their employers or the public.

      (1) They will not disclose confidential information concerning the business affairs or technical processes of any present or former employer or client or bidder under evaluation, without his consent, unless required by law or court order.

(2) They shall not reveal confidential information or finding of any commission or board of which they are members unless required by law or court order.

(3) Designs supplied to Engineers by clients shall not be duplicated by the Engineers for others without the express permission of the client(s).

k. Engineers shall act with fairness and justice to all parties when administering a construction (or other) contract.

l. Before undertaking work for others in which Engineers may make improvements, plans, designs, inventions, or other records which may justify seeking copyrights, patents, or proprietary rights, Engineers shall enter into positive agreements regarding the rights of respective parties.

m. Engineers shall admit their own errors when proven wrong and refrain from distorting or altering the facts to justify their mistakes or decisions.

n. Engineers shall not accept professional employment or assignments outside of their regular work without the knowledge of their employers.

o. Engineers shall not attempt to attract an employee from other employers or from the marketplace by false or misleading representations.

5. Engineers shall build their professional reputation on the merit of their services and shall not compete unfairly with others.

a. Engineers shall negotiate contracts for professional services on the basis of demonstrated competence and qualifications for the type of professional service required.

b. Engineers shall not request, propose, or accept professional commissions on a contingent basis if, under the circumstances, their professional judgments may be compromised.

c. Engineers shall not falsify or permit misrepresentation of their, or their associates, academic or professional qualification. They shall not misrepresent or exaggerate their degrees of responsibility in or for the subject matter of prior assignments. Brochures or other presentations used to solicit personal employment shall not misrepresent pertinent facts concerning employers, employees, associates, joint venturers, or their accomplishments.

d. Engineers shall prepare articles for the lay or technical press which are only factual. Technical Communications for publication (theses, articles, papers, reports, etc.) which are based on research involving more than one individual (including students and supervising faculty, industrial supervisor/researcher or other co-workers) must recognize all significant contributors. Plagiarism, the act of substantially using another's ideas or written materials without due credit, is unethical. (See Appendix.)

e. Engineers shall not maliciously or falsely, directly or indirectly, injure the professional reputation, prospects, practice or employment of another engineer, nor shall they indiscriminately criticize another's work.

f. Engineers shall not use equipment, supplies, laboratory or office facilities of their employers to carry on outside private practice without consent.

6. Engineers shall associate only with reputable persons or organizations.

   a. Engineers shall not knowingly associate with or permit the use of their names or firm names in business ventures by any person or firm which they know, or have reason to believe, are engaging in business or professional practices of a fraudulent or dishonest nature.

   b. Engineers shall not use association with non-engineers, corporations, or partnerships to disguise unethical acts.

7. Engineers shall issue public statements only in an objective and truthful manner.

   a. Engineers shall endeavor to extend public knowledge, and to prevent misunderstandings of the achievements of engineering.

   b. Engineers shall be completely objective and truthful in all professional reports, statements or testimony. They shall include all relevant and pertinent information in such reports, statements or testimony.

   c. Engineers, when serving as expert or technical witnesses before any court, commission, or other tribunal, shall express an engineering opinion only when it is founded on their adequate knowledge of the facts in issue, their background of technical competence in the subject matter, and their belief in the accuracy and propriety of their testimony.

   d. Engineers shall issue no statements, criticisms, or arguments on engineering matters which are inspired or paid for by an interested party, or parties, unless they preface their comments by identifying themselves, by disclosing the identities of the party or parties on whose behalf they are speaking, and by revealing the existence of any financial interest they may have in matters under discussion.

   e. Engineers shall be truthful in explaining their work and merit, and shall avoid any act tending to promote their own interest at the expense of the integrity and honor of the profession or another individual.

8. Engineers accepting membership in The American Society of Mechanical Engineers by this action agree to abide by this Society Policy on Ethics and procedures for its implementation.

Responsibility: Council on Member Affairs/Board on Professional Practice and Ethics

Adopted: March 7, 1976

Revised : December 9, 1976
December 7, 1979
November 19, 1982
June 15, 1984
(editorial changes 7/84)
June 16, 1988
September 12, 1991
September 11, 1994

AMERICAN SOCIETY OF CIVIL ENGINEERS* (ASCE)

The Engineering Code of Ethics

Fundamental Principles

Engineers uphold and advance the integrity, honor and dignity of the engineering profession by:

using their knowledge and skill for the enhancement of human welfare;

being honest and impartial and serving with fidelity the public, their employers and clients;

striving to increase the competence and prestige of the engineering profession; and supporting the professional and technical societies of their disciplines.

Fundamental Canons

Engineers shall hold paramount the safety, health and welfare of the public in the performance of their professional duties.

Engineers shall perform services only in areas of their competence.

Engineers shall issue public statements only in an objective and truthful manner.

Engineers shall act in professional matters for each employer or client as faithful agents or trustees, and shall avoid conflicts of interest.

Engineers shall build their professional reputation on the merit of their services and shall not compete unfairly with others.

Engineers shall act in such a manner as to uphold and enhance the honor, integrity, and dignity of the engineering profession.

Engineers shall continue their professional development throughout their careers, and shall provide opportunities for the professional development of those engineers under their supervision.

AMERICAN INSTITUTE OF CHEMICAL ENGINEERS (AICHE)

AIChE Code of Ethics

American Institute of Chemical Engineers

The Council of the American Institute of Chemical Engineers adopted this Code of Ethics to which it expects that the professional conduct of its members shall conform, and to which every applicant attests by signing his or her membership application.

Members of the American Institute of Chemical Engineers shall uphold and advance the integrity, honor, and dignity of the engineering profession by: being honest and impartial and serving with fidelity their employers, their clients, and the public; striving to increase the competence and prestige of the engineering profession; and

* Courtesy of ASCE

using their knowledge and skill for the enhancement of human welfare. To achieve these goals, members shall:

1. Hold paramount the safety, health, and welfare of the public in performance of their professional duties.
2. Formally advise their employers or clients (and consider further disclosure, if warranted) if they perceive that a consequence of their duties will adversely affect the present or future health or safety of their colleagues or the public.
3. Accept responsibility for their actions and recognize the contributions of others; seek critical review of their work and offer objective criticism of the work of others.
4. Issue statements or present information only in an objective and truthful manner.
5. Act in professional matters for each employer or client as faithful agents or trustees, and avoid conflicts of interest.
6. Treat fairly all colleagues and co-workers, recognizing their unique contributions and capabilities.
7. Perform professional services only in areas of their competence.
8. Build their professional reputations on the merits of their services.
9. Continue their professional development throughout their careers, and provide opportunities for the professional development of those under their supervision.

Appendix B
Bibliography

GENERAL BOOKS ON ENGINEERING ETHICS

CHARLES E. HARRIS, JR., MICHAEL S. PRITCHARD, AND MICHAEL J. RABINS, *Engineering Ethics, Concepts and Cases*, Wadsworth Publishing Company, Belmont, CA, 2000.

DEBORAH G. JOHNSON, *Ethical Issues In Engineering*, Prentice Hall, Englewood Cliffs, NJ, 1991.

ROLAND SCHINZINGER AND MIKE W. MARTIN, *Introduction to Ethics*, 2d. ed., McGraw-Hill, New York, 2000.

JOURNALS WITH ARTICLES ON ENGINEERING ETHICS AND CASES

Business and Professional Ethics Journal, published by the Center for Applied Philosophy, University of Florida, Gainesville, FL.

Journal of Business Ethics, Kluwer Academic Publishers.

Science and Engineering Ethics, Opragen Publications

WEBSITES

There are many websites that deal with issues of engineering ethics. Some of the better ones are listed here. Of course, the addresses of websites tend to be very unstable, so these addresses may no longer be valid at the time you read this text. Many of these websites offer links to other Web-based ethics resources as well.

Ethics Center for Engineering and Science, Casewestern University:
http://onlineethics.org

The IEEE Society on Social Implications of Technology (SSIT):
http://radburn.rutgers.edu/andrews/projects/ssit/default.htm

Index

A

Aberdeen Proving Ground, 317
Aberdeen Three, 317–318, 320, 336
ABET, 30
"About" page, Web sites, 81
Abscissa, 128, 130
Academic engineers, compared to workplace engineers, 66
Academic lab reports, 71–72
acceptance of gifts, 330, 332
act utilitarianism, 310
Active voice, 97–99, 111
Adobe® Acrobat® software, 184
Advisors, 65–66
  and decision makers, 66
aeronautical engineers, 6
aerospace engineers, 5
Agnew, Spiro, 334–336
AIChE Code of Ethics, 349
American Bar Association (ABA), 293–295
American Institute of Chemical Engineers, 319
American Medical Association (AMA), 293–295
American Society of Civil Engineers, 349
American Society of Mechanical Engineers (ASME), 294, 344
  Code of Ethics of Engineers, 344
  Criteria for Interpretation of the Canons, 344
analog, 21, 49
Analysis, 42, 57
Analysis report, 72–73, 76
analytical problems, 43
Anderson, Paul, 65
Apollo moon landing, 241
Appendices, 235–237
  including tables in, 129
Applied Science and Technology Abstracts, 79
apprenticeships, 292–293
Aristotle, 308
ASCE (American Society of Civil Engineering), 221
  website, 234
ASCE Engineering Code of Ethics, 349
Associative thinking, 88
athlete, 293
Atlantic Monthly, 317
Audience:
  as client, 62
  changes in, 64
  differences in, 66
  making eye contact with, 170

B

background ideas, 278
Backmatter:
  appendices, 235–237
  List of References section, 233–234
Backups, 261
baggage handling system, 301
ballhogs, 318
Bar graphs, 129
Bay Area Rapid Transit (BART), 299
Beall, George, 334
Beyond Engineering: Essays and Other Attempts to Figure without Equations (Petroski), 62, 68
Bhopal case, 307, 316–317, 319, 327–329
Bible, The, 308
Biomedical, 8, 13, 22
biomedical engineering, 8, 22
block design, 43, 45
blue–ribbon commission, 286
Body language, 169–170
Body of a report:
  central sections, 214–229
  conclusions, 229–232
  executive summary, 208–210
  introduction, 211–213
  recommendation, 232
Boisjoly, Roger, 285–287, 289
Bookmarks: A Guide to Research and Writing 2d. ed. (Ruszkiewicz/Walker/Pemberton), 87
brainstorming, 55, 58
bribery, 330, 332, 336
  problems, avoiding, 332
bribery problems, avoiding, 332
Buddhists, 316
Bullet points, 165–166
Bush, George, 285
business models, 295
business vs. personal ethics, 280

C

C++, 46
Cancer, 333
Capital letters, 181
Captions, punctuation of, 134
carbon dioxide concentration, 322
carpenters, 292–293
cell phones, 333
Chabris, Christopher, 141
Challenger, 281–282, 288, 319, 329, 336
Charts, 124–126
cheating, 335
Checklists, 250
chemical engineering, 13, 19
chemical engineers, 7, 19, 22
chemical engineers, manufacturing engineers, 19
chemical plant, 307
Chicago Manual of Style 15th ed. (Univ. of Chicago Press), 86, 115, 117, 134, 144, 155, 238
Child and Associates, 334
Christian thought, 308
Christianity, 316
circuit simulation, 11, 46
circuit simulation software, 54
circuits, 19
civil engineering, 12
civil engineers, 5, 10, 30
Clarity, 91–93, 97–98
clinical training, 296
clock radios, 321
code of ethics, 10, 32, 50–51
codes of ethics, 291, 296, 300, 304–305, 322
  other types of, 299–300
  professional engineering societies, 338–350
codes of the engineering societies, 297–298
codes, internal conflicts in, 298
codes, objections to, 297
codes, protection of employees, 299
Cognition and Complexity (Reeves), 69
Cognitive Style of PowerPoint, The (Tufte), 178
Cohen, Richard, 334
Collaboration, 92
Colleagues, as audience, 66
Color, in tables, 128–129

Columbia Guide to Online Style, The (Walker/Taylor)), 87, 238
.com, 81
Combining sentences, 101
Communication:, 39, 54
and engineering work, 59–60
defined, 59
company code, 299
Comparison section, analysis report, 77
competitive bidding, 302, 304
computer programming, 5
computer science, physics, chemistry, 42
computer software, 17, 46, 56
computer–aided design (CAD), 294
Computers, and revision, 152
Conclusions and Recommendations section, 77, 231
Conclusions section, 77, 229–232
lab reports, 72
concurrent engineering, 49–50
conflict problems, 329–330
Congress, 284, 288, 302, 318
Content, revising for, 145–147
checklist, 146
continuing education, 9, 30
control systems, 19
Controlling nervousness, 172–173
deep breathing, 172
mental imaging, 172
mind games, 172–173
Conversations, recording, 79
Copyediting, 139
copyrights, 42, 56
corporate codes, 299
corporate guidelines, 332
correct solutions, 281
cost–benefit analysis, 310–312, 318
Cover page, 200–201
Cray, Seymour, 18
creative middle way, 329
creativity, 43, 54–55, 57
Criteria:
evaluation, 76, 82
section of report, 76
Critical path method (CPM), 251
Critical pathways, 253
CSI Trucking, 301
Customers, as audience, 66

D

Dallas–Fort Worth Airport, 301
Data, 63, 67
citing in graphics, 136
collecting, 117–118, 134
comparisons of, 128
complex, 117
deviations of, 119
display of, 116, 118–122, 128, 130, 132, 134, 138
in tables, 120
points, 119–124
relevancy of, 128
representing graphically, 116, 118–119
scatter in, 119
sets, 131
summarizing, 129
understanding, 118
debugging, 303
Decision makers, 65–66
and advisors, 66
research and recommendation reports for, 72–77
Dee, William, 317–318
Deep breathing, 172
Delivery:
body language, 169–170
planning, 168–171
rehearsing, 168–169
transitions, 171–172
visuals, using, 171
voice quality, 169–170
Denver International Airport (DIA), 301–302
Denver Post, 301–302
Description of Alternatives section, 77
Description of Methods section, analysis report, 77
Design:, 44–45, 48
and writing, 87
as a process, 258
of documents, 179–184
of slides, 166
report and proposal, 88
starting with, 88
text as, 180–183
Design examples
motorized wheelchair safety flat, 240–241
design patents, 56
design problems, 26, 42–43, 281
design process, 46–47, 49, 51
design report, 53
Development team, 247
DIA Runaway Concrete, 306, 336
Digital Equipment Corporation (DEC), 303
digital logic, 15
discretion, 292, 294
Displaying data in written documents, 116–138
figures:
charts, 124–126
combinations, 127
drawings and schematics, 126–127
graphs, 121–124
integrating text with, 134–136
graphs, guidelines for designing, 128
tables, 120–121
guidelines for designing, 128–129
integrating text with, 134–136
taking control of the design, 129–132
Documentation, 255–262
paper vs. electronic, 256
schematics and drawings, 261
technical reports and memoranda, 261
trail, 256, 262
Documenting information, 81–83
Documents, designing, 179–184
Documents, disseminating digitally, 183–184
Doing Science: Design, Analysis, and Communication of Scientific Research (Valiela), 138
Domains, 81
defined, 78
Drafts, 61, 77, 90, 139
of introduction, 77
Drawings, 126–127, 261
duty ethics, 309, 312, 314, 318
Dynamics, 169

E

easy choice, 329
economic benefits, 307
Editing, 139
for logic and flow, 147–149
checklist, 149
for sense and meaning, 149–152
Editing Technical Writing (Samson), 155
.edu, 81
Effective teams, building, 242

EI Compendex Plus, 79
Elbow, Peter, 90–91, 93
electric blankets, 321
electrical–power distribution systems, 321
electromagnetic radiation, 333
electromagnetics, 13, 20
Electronic database, searches via, 78
electronics, 13, 16, 19–22, 47, 56
Elements of Graphing Data, The (Cleveland), 138
Elements of Style, The (Strunk and White), 110, 114–115, 117
Empire Laboratories, 302, 306
Empty pronouns, 101–102, 115
Engineer's logbook (notebook), 257–258
  defined, 257
  example, 259–260
  format for, 258
  good logbook habits, developing, 265
  importance of (case study), 263–265
  recordkeeping, 262
  software documentation and the role of, 262
  techniques for, 258
  using, 258–259
engineering, 1–2, 53
  as a profession, 291, 296
  defined, 304
  differences between it and other professions, 294–295
  ethics, 278
  ethics, study of, 278–279
  societies, 296
engineering communications, 52
Engineering conventions, 130
engineering design, 42–44, 57
engineering design process, 42–43, 45, 57
engineering design report, 53
Engineering documents, producing, 179–238
engineering drawing, 53
Engineering Fluid Mechanics 7th ed. (Crowe/Elger/Robertson), 68
Engineering Index (EI Compendex Plus), 79
Engineering information, best sources for, 79
engineering sciences, 2
Engineers:
  and writing competence, 60–61
  as technical communicators, 59
  intellectual property, 266
  legal issues, 266–272
  patents, 266–272
entrepreneur, 11
entrepreneurship, 2, 10
environment, 51
Envisioning Information (Tufte), 138
Error bars, 119
esthetics, 51
Estrin, Jodith, 11
ethical actions, 312
ethical dilemmas, 322
ethical problem
  types of issues in, 322–323
ethical problems, 281, 308
  analysis of issues in, 322–324
  understanding, 307–320
ethical problem–solving techniques, 321–337
ethical thought
  a brief history of, 309–316
  origins of, 280
European Space Agency, 284
"Evaluation of Wall Designs" section, 222–225
Executive Summary, 77, 208
  benchmark for length of, 210
  sample in report, 208–210
Experts, 79
Eye contact, with audience, 170

F

"Factual Summary and Rating Procedure" section, 226–229
FE exam, 30
feasibility study, 53
Federal Aviation Association (FAA), 302
Federal Bureau of Investigations (FBI), 301–302
Feedback, seeking, 91
Feynman, Richard, 286
Figures:
  charts, 124–126
  combinations, 127
  drawings and schematics, 126–127
  graphs, 121–124
  in correspondence, 186
  integrating text with, 134–136
Final copy, proofreading, 152–153
Financial institutions, as audience, 66
flare tower, 316
floating–point unit (FPU), 300
flow charting, 336
Focus, maintaining, 107–109
Fonts, 180
  for engineering reports, 183
Ford Motor Company, 277–279, 282
Fort Collins, Colorado, 302
Four–step process of managing/documenting information, 82–84
framework, 296
Free–answer questions, in surveys, 80
Freewriting, 90, 93
Front matter:
  cover page, 200–201
  letter of transmittal, 201–202
  list of illustrations, 206–207
  Table of Contents (TOC), 205–207
  title, 200
  title page, 204–206
Fundamentals of Engineering (FE)
  exam, 30
Future tense, 113
fuzzy logic, 20

G

Gantt chart, 250–251
General Accounting Office (GAO), 304
Gepp, Carl, 317–318
gifts vs. bribes, 331–332
Gladwell, Malcolm, 141
global warming, 322
Goals, and teams, 242
.gov, 81
government agencies, 299
government regulations, 299
graduate school, 29
Grammar and sentence structure, 110–112
Grammar checkers, 143–145
Graphical and verbal presentation of data, 63
Graphics:
  creating, 90
  designed by others, using, 134–136
  display of data, 116, 118–122, 128, 130, 132, 134, 138
  labeling, 127–128
  process, 135
  technical, guidelines for, 135
  titles, 132–134
Graphs, 121–124
  guidelines for designing, 128
Greek philosophers, 308

Greeks, 280
greenhouse gases, 322–323

H

Halley's comet, 284
Handbook for Preparing Engineering Documents: From Concept to Completion (Nagle), 115
hard choice, 329
hardware, 4, 17, 43, 50
Headings, 182–183
Health and Human Services Department, 304
Heartbeat Away, 334
Hindus, 316
HMOs, 294–295
House of Representative's Government Operations Committee, 304
How to Write and Publish Engineering Papers and Reports 3d. ed. (Michaelson), 95
Hubble space telescope, 241

I

IBM, 305
Ideas/facts, organizing, 88–95
IEEE, 1, 10, 20, 26, 32, 39, 50, 298
IEEE code of ethics, 10, 32, 51
Implementers, 65–66
Incremental Method (Michaelson), 94
Indentation, 180
Industry lab reports, 72
Informal writing, 61
Information, 67
  evaluating sources of, 81–82
  how people read/understand, 97
  managing, 82–84
    evaluating the sources and the information itself, 82–83
    informally documenting all sources of information, 83
    organizing the information, 84
    summarizing the information from each source, 83–84
Information collection/organization, 70
Information sources, evaluating, 81–82
Institute of Electrical and Electronics Engineers (IEEE), 294, 298–299
Institute of Electrical and Electronics Engineers (IEEE), 9, 15
integrated, 2
integrated circuits, 2, 7, 13, 16, 18–19, 23, 45
Integrity of writing, 81
Intel, 290–291, 300–301
Intel Pentium chip case, 300–301, 305
intellectual, 55
intellectual property, 42, 55–57, 266
International Space Station, 241
Internet, 1, 11, 18, 31, 38, 46–48, 299
Internet robots, 81
Interpersonal relationships, and teams, 242
interview trips, 335
Interviewing, 79–80
Introduction, 77
  draft of, 77
  reports, 211–213
  surveys, 80
Islam, 316
"It", using at the beginning of a sentence, 102

J

Jewish moral traditions, 308
Jewish thought, 308
Job description, 244–246
job searches, 291, 293, 335
Johns Hopkins University, 334
Johnson, Arnie, 285
joint rotation, 283
Judaism, 316
Judeo–Christian, 280
judgment, 292
Justice Department, 304
Justified text, 180

K

Kant, Immanuel, 308, 312–313
Kennedy Space Center, 283, 285
Keywords, in searches, 78
kickbacks, 334, 337
Kilminster, Joe, 285

L

Lab reports, 71–72
  academic, 71–72
  formats for, 72
  industry, 72
  sections of, 71
  steps for writing, 71
Labels, graphics, 127–128
Lanham, Richard, 110
Laser pointer, 171
Last date of modification, 82
law, 304
Leadership roles, and teams, 242
Learning and Cognition 4th ed. (Leahey and Harris), 86
Lentz, Robert, 317–318
Letter of transmittal, 201–202
Levels of edit, 140
license, 10, 30
licensure, 10–12, 26, 30, 32
Lifelong Learning, 26, 30, 32
line drawing, 324, 327, 336
Line graph, tables compared to, 129
Line spacing, 180
liquid–fuel booster, 286
List of illustrations, 206–207
List of References section, 233–234
Listener, as client, 62
Literal patent claims, 268
Locke, John, 308, 312
logbooks, 52
logic circuits, 15
Logical linkage, maintaining, 103–107
Logic–and–flow level of edit, 147–149
  checklist, 149
Long noun phrases, 102–103
Lund,Bob, 285, 329

M

magnetic fields, 321, 336
Managing information, 82–84
  evaluating the sources and the information itself, 82–83
  informally documenting all sources of information, 83
  organize the information, 84
  summarizing the information from each source, 83–84
managing the unknown, 279–280
manufacturability, 51, 57
manufacturing, 2, 47, 49–50
manufacturing engineers, 5, 7, 16, 46, 50
Manufacturing team, 247
Marketing team, 247
Mars Exploration Rover, 241
Marshall Space Flight Center, 285
Mason, Jerald, 285–286
mathematical model, 2
mathematics, 42

Matz, Childs and Associates, 334
Matz, Lester, 334
McAuliffe, Christa, 284
McDonald, Alan, 285–286
mechanical, 5, 11, 13–14
mechanical engineering, 5, 8, 49
mechanical engineers, 5, 23, 48
medicine, 304
Meeting "on demand", 246
Meetings:
  teams, 246–247
Memo:
  notes on, 186
  sample, 185
Mental imaging, 172
Method claims, 268
Methodology section, 77
methyl isocyanate (MIC), 307, 316, 319, 328
Michaelson, Herbert, 94
Micro-electromechanical systems (MEMS), 268
microelectronics, 13, 19, 23
Microsoft Excel, 118, 136
Microsoft PowerPoint, 162, 164–166
Microsoft Word, 184
  creating an outline view using, 207
  Style command, 181
Mill, John Stuart, 310–311
Mills, 308
Mind games, 172–173
Mind mapping/mind dumping, 90
Missing words, 152
Misspelled/misused words, 113
Misspellings, 152
modeling, 2, 20
moral theories, 308–309, 322
Morton Thiokol, 283, 285, 287, 289, 305, 319, 330, 336
Motorized wheelchair safety flat, 240–241
Mulloy,Larry, 285
multidisciplinary teams, 5, 50

N

NASA, 282–283, 285–289, 329, 336
National Aeronautics and Space Administration (NASA), organization chart, 244–245
National Society of Professional Engineers (NSPE), 294, 298
negative paradigm, 324, 326
Nervousness:
  before you talk, 172–173
  controlling, 172–173
    deep breathing, 172
    mental imaging, 172
    mind games, 172–173
  just before/during talk, 173
networking, 32, 34, 39–40
New York Times Test, The, 332
New York Times, The, 290
News media, as audience, 66
Newspaper article, example of, 65
Nichomachean Ethics, 308
Nixon, Richard, 334–335
Nobel prize, 286
Node, 252
nolo contendere, 335
Non–Designer's Design Book, The (Williams), 184
Non–Designer's Web Book, The (Williams), 184
non–Western cultures, 280
non–Western ethical thinking, 315–316
Note taking, 79
Noun compression, 102–103
nuclear engineering, 7
nuclear engineers, 7, 14
Numbering mistakes, 152

O

occupations, 291, 293
of cell phones, 333
off–the–shelf product, 323
off–the–shelf system, 303
Old Testament, 308
On Writing Well: An Informal Guide to Writing Nonfiction (Zinsser), 117
on–campus interview, 335
open–ended problems, 43
optoelectronics, 13, 22–23
Oral Presentations for Technical Communication (Gurak), 178
oral reports, 49, 54
Ordinate, 128
.org, 81
Organizational chart, 244
O–rings, 281–283, 285–286, 289
Outline, 90
  checking relevance of information against, 84
Outlining, 157–160
  audience, 157
  Body, 159
  Conclusions, 159
  highlighting, 157–158
  Introduction, 158–159
  sample outline, 160
Outtakes file, 84
oxide, 286

P

P8400 model, 303
Pages, designing, 181
Paper documentation, electronic documentation vs., 256
paradigm, 324
Paradyne computers case, 303–304, 323–324, 337
Paragraph construction, 103–110, 115
  focus, maintaining, 107–109
  logical linkage,
    maintaining, 103–107
  parallelism, maintaining, 109–110
  topic sentences, 108
  transitions, 105–106
Parallel processing of information, 117
Parallelism, maintaining, 109–110
Paramedic Method, 110
Passive voice, 111
Past tense, 113
Patent disclosure, 268
patents, 15, 40, 42, 56–57, 266–272
  abstract, 272
  anatomy of, 269–270
  assignee, 272
  background of the invention, 272
  claim limitation, 272
  claims, 272
  defined, 266
  dependent claims, 272
  description of the invention, 272
  disclosure, 268
  illustrations, 272
  inventor of, 272
  jargon, 266–267
  patent number, 270
  patent prosecution, 268
  patent title, 270
  primary claim, 272
  prior art, concept of, 266–268
  process of obtaining, 268–270
  prosecution history, 268
patents, copyrights, 42, 56

PC–compatible computers, 303
PDF (portable document format), 184
Peer review:
organizing, 141–143
sample peer–review questions for a research proposal, 143
Pentium chip case, 291, 304, 326–327, 336
Pentium microprocessors, 290, 300
personal ethics
vs. business ethics, 280
vs. corporate morality, 314–315
personal vs. business ethics, 280
PERT chart, 251–255
compared to Gantt chart, 252
Petroski, Henry, 62
physician–patient privilege, 293
physics, 2, 13
Pilot Plant, 317
Pinto, Ford, 277, 279, 282
PIOS operating system, 303
Pitch (voice), 169
positive paradigm, 324, 326
post mortem examinations, 282
power, 13–14
power companies, 321
Powerful writing, keys to, 98
preengineering degree, 296
Prentice Hall Reference Guide to Grammar and Usage (Harris), 115
Prepositional phrases, determining the number of, 112
Prepositions, and compressed noun phrases, 102–103
Present tense, 113
Presentation software, 164–165
designing with, 165
Prior art, concept of, 266–268
problems, 57
problem–solving
approach, 322
techniques, 309
Procedure definitions, and teams, 242
profession, 8–10
profession, defined, 292–293, 304
professional, 8
professional athletes, 292–293
Professional documents, designing, 179–184
professional engineer, 292
professional societies, 9–10, 32, 295, 304
professional societies, protection of employees, 299
Professions, 304
Progress report:
organization of, 197
sample, 193–196
progress reports, 52
Project Description section, Progress Report, 197
Project management, 240–276
continuous interaction between the components of an organizational chart, 247
job description, 244–246
team contact list, 246
teams, 241–249
project plan, 43–44
Pronouns:
and compressed noun phrases, 102–103
empty, 101–102, 115
Proofreading, 139, 141
final copy, 152–153
suggestions for, 153
property, 42, 55–57
Proposals, 53, 60
and writing competence, 60–61
sample, 187–192
prototype, 46, 51
Pruning, 100–101, 103
pseudomoral agents, 315
Published information, 78–79
Punctuation, 114–115, 152
of titles/captions, 134
rules, 113

R

radiation, 321
Readable style, 96–103
active voice, 97–99
empty pronouns, 101–102
noun compression, 102–103
passive voice, 97–99
wordiness, 99–101
Reader, as client, 62
Reagan, Ronald, 284, 286, 288
recall notices, 326
Recommendation, 232
"Recommendations" section, 77
Referencing style, 83
Regulatory agencies, as audience, 66
Rehearsal reviewer, questions for, 168
replacement chips, 326
Reporting Technical Information 10th ed. (Houp/Pearsall/Tebeaux/Dragga), 178
request for proposals (RFP), 303
Research, 70–87
interviews, 79–80
methods, 74–75
proposal, sample, 186–192
published information, 78–79
strategies for, 78
summaries, 90
unpublished information, 79–80
Research and recommendation reports for decision makers, 72–73
Resonance (voice), 169
Resource Conservation and Recovery Act (RCRA), 318, 320
"Results and Discussion" section, 77
Revising, 139–155
and computers, 152
editing:
for logic and flow, 147–149
for sense and meaning, 149–152
for structure/content, 145–147
grammar checkers, 143–145
proofreading the final copy, 152–153
Revising Business Prose 4th Ed. (Lanham), 117
"Revising Functional Documents: The Scenario Principle" (Flower/Hayes/Swarts), 68
Rewriting, 139–141
guidelines for, 140–141
rights ethics, 309, 312, 314, 318
risk, 307
risk–benefit analysis, 310
robotics, 13, 23
robots, 5, 7, 23
Roe v. Wade, 322
Rogers commission, 286
Role definition, and teams, 242
Roman Empire, 308
rule utilitarianism, 310–311
runway concrete, 301–302, 304
Russian probe, 284

S

safety, 51
Sales and marketing staff, as audience, 66
Sample correspondence, 184–185

Sample feasibility report, 197–237
- backmatter:
  - appendices, 235–237
  - "List of References" section, 233–234
- body of:
  - central sections, 214–229
  - conclusions, 229–232
  - executive summary, 208–210
  - introduction, 211–213
  - recommendation, 232
- central sections:
  - "Evaluation of Wall Designs" section, 222–225
  - "Factual Summary and Rating Procedure" section, 226–229
  - "Solution Criteria" section, 219–221
  - "Wall Design Alternatives" section, 214–218
- front matter:
  - cover page, 200–201
  - letter of transmittal, 201–202
  - list of illustrations, 206–207
  - Table of Contents (TOC), 205–207
  - title, 200
  - title page, 204–206
- organization of, 197

Sample memo, 185
- notes on, 186

Sample progress report, 193–196
Sample proposal (research), 186–192
San-serif fonts, 180
schematic drawing, 54
Schematics, 126–127, 261
Scope of work, defining, 75, 78
scrubber, 317
Searches, 78
Securities and Exchange Commission, 304
"Selection Criteria" section, 77
Self-criticism, 91–92
semiconductor, 56
Sense-and-meaning level of edit, 149–152
- checklist, 151

Sentences, combining, 101
Serif fonts, 180–181
Set of instructions, example of, 65
short courses, 30–31
SigmaGraph, 128
Simons, Daniel, 141
Simple-to-answer questions, in surveys, 80
Sketches:
- treated as part of the text, 218

skills, 292
Slack time, 253
Slides, designing, 166
Social Security Administration (SSA), 303, 323, 336
social-contract model, 295
societies, 9–10, 32
Socrates, 308
software, 4, 11, 16, 43, 45–46, 49–50, 54, 56–57
solid booster, 286
solid rocket boosters, 283–284
Solid Rocket Motor Project, 285
solid rocket-propellant combustion, 286
solid-propellant boosters, 281
"Solution Criteria" section, 219–221
Spacing problems, 152
Speaking, 156–178
- delivery:
  - body language, 169–170
  - planning, 168–171
  - rehearsing, 168–169
  - transitions, 171–172
  - visuals, using, 171
  - voice quality, 169–170
- handling questions, 174–175
- nervousness, controlling, 172–173
- outlining, 157–160
- team presentation,
  - preparing, 175–176
- visuals, planning, 160–167

specifications, 50–51
Spelling and Grammar function, Microsoft Word, 144
Spiders, 81
Spiro Agnew, 337
Spreadsheet programs, 136
standards for admission, 292
Stapleton International Airport, 301
state-of-the-union address, 284
Structure:
- revising for, 145–147
  - checklist, 146

study skills, 34–35
Style guides, 114–115, 183
Style sheet checklist, 183
Subheadings:
- report introduction, 213

supercomputers, 18
Support staff, as audience, 66
Supreme Court, 322
Surveys, 80

T

Table of Contents (TOC), 205–207
Tables, 120–121
- color in, 128–129
- guidelines for designing, 128–129
- in correspondence, 186
- integrating text with, 134–136

tang and clevis joint, 284
Task management:
- checklist, 250
- Gantt chart, 250–251
- PERT chart, 251–253
- time line, 250

Teaching the Research Paper: From Theory to Practice, from Research to Writing (Ford), 86
team, 3, 27, 38, 48–50, 55
Team contact list, 246
Team leader, 243
Team presentation, preparing, 175–176
Teams, 241–249
- development team, 247
- effective, building, 242
- manufacturing team, 247
- marketing team, 247
- meetings, 246–247
- working with other teams in the organization, 247–248

Teams and organizational chart, 244
teamwork, 42, 48, 57, 241–242, 318
technical, 3, 11, 35, 53
Technical Communication 2d. ed. (Wohlstrom/Rude/Selzer/Selfe), 86
technical manual, 53
Technical processes/concepts/vocabulary, explaining, 62
Technical report, example of, 64
Technical reports, 261
Technical writing, 61
- active voice, 111
- clarity in, 91–93, 97–98
- passive voice, 111
- readable style, 96–103

Tesla, Nikola, 15
testing, 11, 47, 51
Text:
- integrating with figures, 134–136
- using as a design
  - element, 180–183

Thank-you statement, in surveys, 80
theoretical framework, 308
"This", using at the beginning of a sentence, 102

Thoman, G.Richard, 305
3Bs construction company, 302, 336
Tilde (~), in URLs, 81
Time estimation, laws of, 255
Time line, 250
Time management, 249–250
Titan missile, 283
Title, 182, 200
Title page, 204–206
Titles, punctuation of, 132–134
To-do list, 250
Tone, 169
Topic, defining, 78
Topic sentences, 108
Torah, 308
toxic gas, 307
toxic vapors, 317
trade secrets, 56
trademarks, 42, 56–57
training, 292
Transcutaneous energy transfer device (TET), 264
transistors, 16, 18–19, 45
Transitions, 105–106, 115
  and organization, 106
  in speech, 171–172
  position in sentences, 107
  words/phrases, 106
Trendline, 119
"Turf wars", 248

U

U.S. News and World Report, 317
under the table, 330
Union Carbide, 307, 317, 319, 327
United States Patent and Trademark Office (USPTO), 266
Unpublished information:
  interviews, 79–80
  surveys, 80
utilitarianism, 309–310, 318
utility patents, 56

V

Vendors, as audience, 66
Verb tenses, 112–113
Verbal presentation of data, 63
virtue ethics, 309, 314, 318
Visual Display of Quantitative Information, The (Tufte), 138
Visual elements of a page, 181
Visuals:
  bullet points, 165–166
  creating, 161
  designing, 161
    guidelines for, 166
  laser pointer, 171
  mapping visuals, including, 162–164
  modes of delivery, 164
  number to use, 171
  planning, 160–167
  simplifying hard–copy graphics, 161–162
  thinking big, 162
Voice quality, 169–170
voltage, 15, 21
Volume, voice, 169

W

"Wall Design Alternatives" section, 214–218
warning label, 326
Waste Isolation Pilot Plant (WIPP), 310, 315
Web sites, 299
Web–based search engine, searches via, 78
Western ethical thinking, 280, 308
"What Survey Research Tells Us about Writing at Work" (Anderson), 68
White space, using to set off items, 180
wind shear, 286
Witcover, Jules, 334
word processing, 46
Word processor, 181
Wordiness, 99–101
  pruning, 100–101, 103
Words, verb tenses, 112–113
"Work Accomplished" section, Progress Report, 197
"Work Remaining" section, Progress Report, 197
Workplace engineers, compared to academic engineers, 66
"Works Cited" section, 234
Writing:
  and engineering competence, 60–61
  and planning/developing/revising your engineering solutions, 61
  as a process, 88–89
    handling, 91
  as act of translation, 88
  collaboration, 92
  compared to a project–management procedure, 89
  define objectives stage, 89
  discuss subject matter with others, 91
  draft stage, 89–92
  for control, 96–117
  freewriting, 90, 93
  grammar and sentence structure, 110–112
  graphics, creating, 90
  hardest part of, 139
  outline, 90
  plan stage, 89
  powerful paragraphs, constructing, 103–110
  powerful writing, keys to, 98
  punctuation, 114–115
  readable style, 96–103
  research summaries, 90
  starting to write, 88–89
  using to solve problems, 93–94
  using to think clearly, 92–93
  words, 112–113
Writing in the Workplace: New Research Perspectives (Spilka), 69
Writing projects:
  framing, 71–78
  lab reports, 71–72
  research and recommendation reports for decision makers, 72–77
Writing to Learn (Zinnser), 69
Writing with Power: Techniques for Mastering the Writing Process (Elbow), 95, 155
Writing without Teachers (Elbow), 95

X

x–y graph, 63

Z

Zadeh, Lofti, 20